Fischer / Achten / Launhardt

Plastics for Additive Manufacturing

Andreas Fischer
Dirk Achten
Martin Launhardt

Plastics for Additive Manufacturing

Properties, Processing, and Applications of Thermoplastics

HANSER

Print-ISBN: 978-1-56990-229-5
E-Book-ISBN: 978-1-56990-284-4

All information, procedures, and illustrations contained in this work have been compiled to the best of our knowledge and is believed to be true and accurate at the time of going to press. Nevertheless, errors and omissions are possible. Neither the authors, editors, nor publisher assume any responsibility for possible consequences of such errors or omissions. The information contained in this work is not associated with any obligation or guarantee of any kind. The authors, editors, and publisher accept no responsibility and do not assume any liability, consequential or otherwise, arising in any way from the use of this information – or any part thereof. Neither do the authors, editors, and publisher guarantee that the described processes, etc., are free of third party intellectual property rights. The reproduction of common names, trade names, product names, etc., in this work, even without special identification, does not justify the assumption that such names are to be considered free in the sense of trademark and brand protection legislation and may therefore be used by anyone.

The final determination of the suitability of any information for the use contemplated for a given application remains the sole responsibility of the user.

Bibliographic information of the German National Library:
The German National Library lists this publication in the German National Bibliography; detailed bibliographic data are available on the Internet at http://dnb.d-nb.de.

Vilshofener Straße 10 | 81679 Munich | info@hanser.de
www.hanserpublications.com
www.hanser-fachbuch.de
Editor: Dr. Mark Smith
Production Management: Eberl & Koesel Studio, Kempten
Cover concept: Marc Müller-Bremer, www.rebranding.de, München
Cover design: Max Kostopoulos
Cover picture: © Leonie Fensterle
Typesetting: le-tex publishing services GmbH, Leipzig

Contents

1 Introduction to Thermoplastic-based Additive Manufacturing

The term additive manufacturing (AM) refers to manufacturing processes that produce three-dimensional components layer by layer by means of material application and accumulation. At times, the term rapid prototyping (RP) is also used for this type of component production, while the colloquial term "3D printing" is used.

When comparing the terms rapid prototyping and additive manufacturing, it becomes clear how the processes have developed over time, particularly with regard to their range of applications. In the initial phase of digital AM, the first systems used data sets to enable the direct – and, compared to conventional manufacturing, rapid – production of prototypes or sample parts. By using AM, product developments can be improved or shortened. Through the improvement and new development of processes and systems, the proportion of directly produced components and products using AM increased. This development also initiated the necessary development or adaptation of new materials for the various AM processes. In the early 1990s, the first commercial thermoplastic-based additive systems in the extrusion and sintering process cluster came onto the market.

In this chapter, we describe the basic strategy on which AM processes are based and on which process clusters are founded. We will then describe the basic principles of fused deposition modeling (FDM), fused layer modeling (FLM), and selective laser sintering (SLS) will be explained, before explaining the concept of hybrid additive manufacturing with thermoplastics.

1.1 Additive Manufacturing Building Strategy

The building strategy of components by means of additive processes can be explained by the analog manufacturing strategy of contour relief maps patented in

1892 by J. E. Blanther. The terrain models built up with this manufacturing strategy consisted of wax plates, which were cut to size based on the respective contour line of the terrain and then placed on top of each other and positioned in a further step. The thickness of the wax plates resulted in the so-called layer thickness of the additively built model. In order to obtain a coherent wax component from the individual contoured wax plates, they were joined together using targeted heat. With a relatively large connection between the individual plates or layers, it was also possible to ensure layer adhesion corresponding to the material and the resulting stability. Accordingly, the basic idea for component building in AM is the layering and bonding of filled contours. This can be done in a manual process that is analogous to that for contour relief maps, or digitally and automatically. In the latter case, the component to be manufactured is available in a digital form, from which the layers with the respective contours are then derived. These layers have a specific constant thickness and follow a geometric axis through the component. After the virtual layers are generated, they are converted into control commands for the specific AM system. The additive system, in turn, generates the contoured layers in a specific order and their connection using specific materials.

Physically, especially for components with overhangs of more than 45°, there is a need to protect them against bending or breaking with a support structure. These support structures are generated digitally in advance in some additive processes (as are the virtual layers), depending on the geometry and the process. Depending on the process, these contours are in turn converted into control commands. Depending on the additive process used, these support structures lead to higher material consumption and longer manufacturing times, as well as to the need for subsequent removal of the support structure from the component.

Due to the layered structure of the additive processes, perceptible steps form on the surface in the build direction. The shape and visibility of these steps depends on the layer thickness. The thinner the respective layers can be built, the lower their perceptibility. At the same time, however, a thinner layer thickness also leads to longer manufacturing times and thus higher manufacturing costs per component. The thinnest layer thicknesses are achieved with additive processes whose materials are in a liquid or gaseous state. However, special processes in which the material is in powder form, such as laser micro-sintering, can also achieve layer thicknesses of 15 µm down to 1 µm.

The surfaces of additively produced components that are not characterized by the step effect are characterized by process-specific principles. In fused deposition modeling (FDM) and fused layer modeling (FLM) processes, there are melting nozzles for thermoplastic material with a corresponding nozzle diameter. In contrast, in selective laser sintering (SLS), the surfaces are imprinted by a pulsed laser, which locally melts thermoplastic powder. Laminated object manufacturing (LOM), selective adhesive and hot press process (SAHP) and layer milling process (LMP) processes are an exception with regard to both step formation and process-specific surface embossing.

In principle, the step formation in the processes can be eliminated by 5-axis machining. Furthermore, the surfaces that are not affected by layering are characterized by the surface of the film or sheet material used and not by the process. Additive processes, which primarily process thermoplastic materials, are anchored in the solid initial state (Figure 1.1). The thermoplastic to be processed can be present here as wire, round cord (filament), powder, or in film or sheet form.

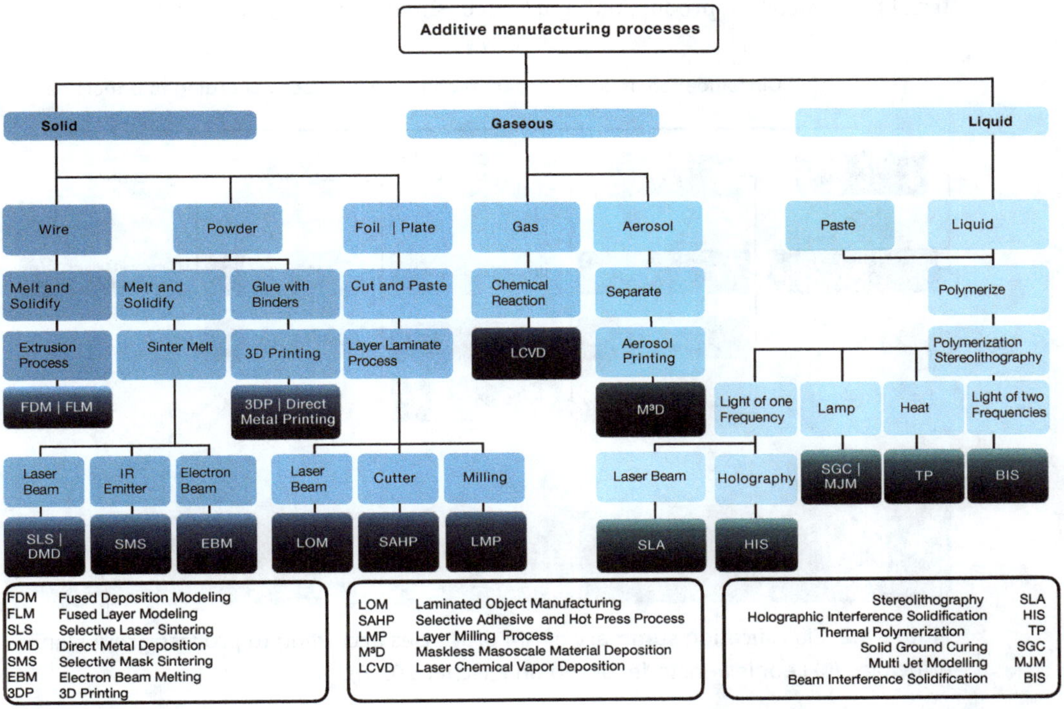

Figure 1.1 Subdivision of additive processes according to the initial state of the material (© Leonie Fensterle, adapted from [Sehrt 2016])

1.2 Process Groups of Additive Manufacturing

For historical reasons, various terms have become established for layer-building processes. As a result of technological development and thus the expanded application possibilities of additive systems, the following terms have evolved over time and are still used specifically today:

- Rapid prototyping (RP): Manufacturing of sample or prototype parts
- Rapid tooling (RT): Manufacturing of tools for primary forming and forming production processes
- Additive manufacturing (AM): Production of components and small batches as marketable products.

In addition to the classification according to DIN 8580, additive systems can also be subdivided on the basis of the condition of the starting material, the material of the component produced, the use in the product development process, or on the basis of the process principles. The differentiation according to process principles is summarized in Figure 1.2. The reduction was made according to the criteria "plastic as a material for component production" and "establishment of the systems on the market".

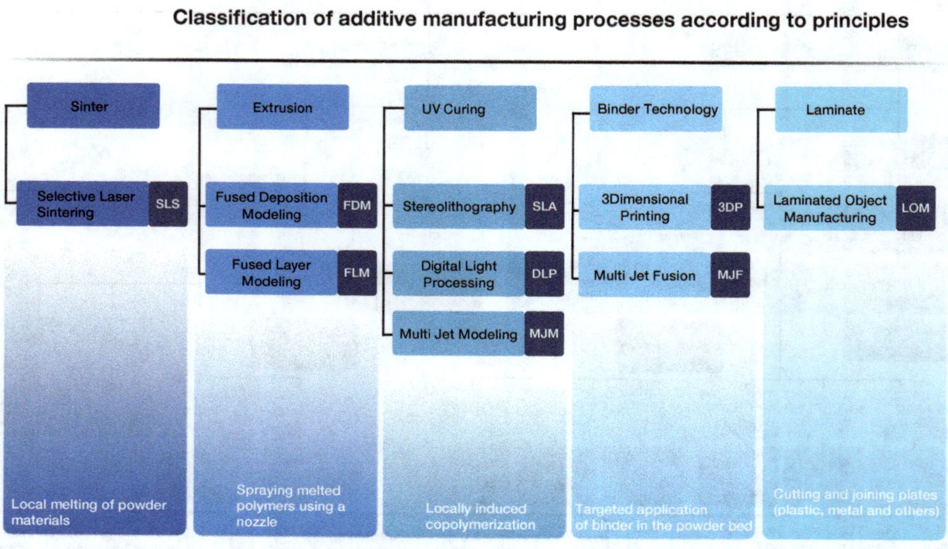

Figure 1.2 Classification summary of AM processes according to process groups and principles (© Leonie Fensterle, based on [Fischer 2020])

1.2.1 Sintering and Extrusion

The sintering and extrusion process groups are based on thermoplastic materials (Figure 1.2). SLS was commercially launched by DTM in 1992. At that time, the systems were called Mod A and Mod B. In 2001, DTM was acquired by 3D Systems. At that time, 3D Systems specialized in stereolithography (SLA) of the UV curing process group and was the market leader in AM systems.

In the extrusion process group, the following two processes are included in Figure 1.2: FDM and FLM. Both processes use the principle of spraying molten polymers by means of a nozzle. The differences between the systems of the two processes are minor in places. FDM systems are often priced higher than fused layer modeling systems, some of which are also designed for private users. FDM systems also often have a more robust manufacturing process and a higher degree of automation. The term FDM was introduced by Stratasys, which launched its first additive system, the 3D

Modeler, in 1992. Focusing on low-end systems in AM, Stratasys became the market leader in 2003, selling nearly half of all additive systems at that time and representing the largest contingent of additive systems installed. In principle, additive systems manufactured by Stratasys and belonging to the Extrusion group are listed under FDM. This also applies to systems from manufacturers such as MakerBot, which belong to Stratasys. FLM includes all systems that belong to the extrusion process group and are not manufactured by Stratasys or associated companies. This also includes many additive systems that, for example, are not commercially available and are still in the research or development stage.

1.2.2 Binding Technology and Laminating

Two further thermoplastic-based additive processes are included in the process group: binder technology (multi-jet fusion, MJF) and lamination (laminated object manufacturing, LOM) (Figure 1.2).

Multi-jet fusion (MJF)

The parts that can be produced by MJF are similar to those produced by SLS and can currently be produced using polyamide (PA) 12 and thermoplastic polyurethane (TPU) 90A. Due to the nature of the process, layer building is faster than in Selective Laser Sintering, which, combined with lower or equal equipment costs, results in lower part costs. In the MJF process launched by HP, heat-conducting liquid, called fusing agent, is sprayed onto the active powder-based layer in the part area via a print head. Another liquid, the detailing agent, is applied in parallel to the contours of the component in the layer and is responsible for creating sharp edges. After the two liquids are sprayed on, energy is added to the respective layer via IR light. The fusing agent increases the energy by absorbing the thermoplastic powder, causing the material to fuse in the wetted area. The detailing agent counteracts this process as an insulator and limits the melting of the material (Figure 1.3).

In the current systems, 300 million drops of liquid are sprayed onto the powder layer every second with an accuracy of 21 µm. Thus, the system has about half the accuracy of multi-jet modeling (MJM) systems in the "UV curing" process group, which are among the most accurate additive systems. Both processes use print heads similar to those used in inkjet printers. As a result, the XY resolution of the MJF systems is specified as 1200 dpi. The use of liquid and its optimum introduction into the powder lead to uniform fusion and a relative dense component, which also largely eliminates the problem of layer adhesion in additively produced components. The high speed at which the layers are created is generated by a high number of print heads and the incorporation of energy input in the wetting process. According to HP, multi jet modeling can achieve 3 cm of part height per hour, which is three times the build rate of SLS.

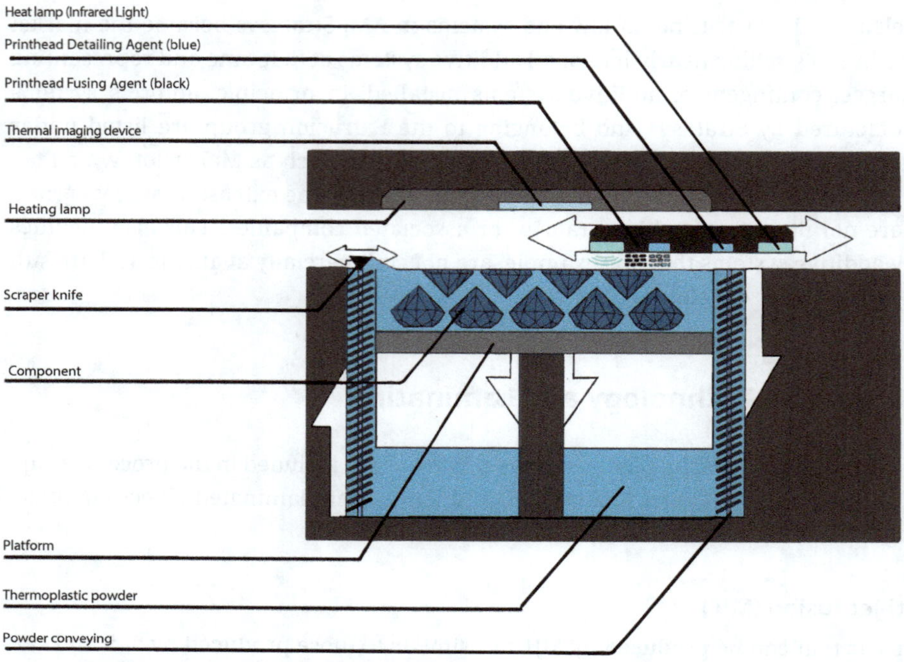

Figure 1.3 Functional principle of the MJF process from HP (© Leonie Fensterle, from [3Druck 2016])

A special feature of the process is the effect of sinking, which leads to a burr in the order of 0.1 mm on the end surface of the component in the Z-direction. This effect is caused by the introduction of the fluids. The installation space of the current HP systems is 380 × 380 × 28 mm. The post-processing of the components is similar to that of SLS, in that the components must be cleaned of residual powder after removal from the system. The surface can be smoothed, for example, via vibratory grinding. As with SLS, the components can also be colored. Compared to SLS and FDM, MJF is a young additive process. As of 2016, the first commercial systems were available under the name HP Multi Jet Fusion 4200. HP cooperated with the companies Nike, BMW, and Johnson & Johnson [3Faktur 2020] from 2014.

High-speed sintering (HSS)

Another inkjet-based additive process with thermoplastic materials is HSS from the company Voxeljet. HSS differs only to a limited extent from HP's MJF process (Table 1.1). As with the HP system, an energy-absorbing liquid is applied to a powder bed via a print head and the energy is supplied via IR light. The printed powder bed areas are thereby fused by the energy input, with the remaining powder area remaining loose (this can be reintroduced into the manufacturing process after preparation). The main difference to MJF is that no second liquid (detailing agent) is required, as the temperature of both powder areas can be controlled independently of each other using two different IR light emitters with different wavelengths. In addition, the system from Voxeljet allows full access to the pressure parameters in order to optimize the production for the respective material and application. Thus, it is a so-called open system, which can also be used for AM with customer-specific materials or new material developments. Differences exist between the print heads used. HP's MJF system uses bubble-jet print heads with a resolution of with a resolution of around 1200 dpi. In contrast, HSS uses piezo print heads with a resolution of around 360 dpi (Table 1.1). However, the resolution of the components is significantly influenced by the grain size of the material powder. This is around 55 µm for both systems. The piezo print heads of high-speed sintering can process oil-based, water-based, and solvent-based fluids. This option further increases flexibility in terms of material selection [Voxeljet 2022].

Table 1.1 MJF and HSS in Comparison [Voxeljet 2022]

	High-speed sintering (HSS)	Multi-jet fusion (MJF)
Applications	Material development and research, prototyping, production applications, service centers, manufacturing, and injection molding companies	Prototyping, service centers, and production applications
Installation space	VX200 HSS: 290 × 140 × 180 mm VX1000 HSS: 1000 × 540 × 400 mm	MJF 540/580: 322 × 190 × 248 mm MJF 4200: 380 × 285 × 380 mm MJF 5200: 380 × 294 × 380 mm
Speed	VX200 HSS: 580 cm^3/h VX1000 HSS: 6500 cm^3/h	MJF 540/580: 1817 cm^3/h MJF 4200: 4115 cm^3/h MJF 5200: 505 cm^3/h
Resolution	360 dpi	1200 dpi
Materials	PA12, TPU, PP, PEBA, and EVA Open system	PA12, PA11, TPA, and TPU
Minimum layer thickness	80 µm	80 µm
Cooling times	14 h (depending on the headroom)	48 h without "fast cooling" (depending on material and grain size)

Laminated object manufacturing (LOM)

LOM, from the laminating process group, is similar to the process principle of J. E. Blanther's contour relief maps (see Section 1.1). The process group is based on layer generation, in which semi-finished products like sheet or film materials are contoured by means of knives, milling cutters, hot wires, or lasers, and are joined and bonded simultaneously or downstream specifically depending on the material. Since this group of processes also includes separating and ablative processing steps, it was also referred to as a subtractive-additive process by Bernard and Taillander in 1998.

Another term used for the process group is layer laminate manufacturing (LLM). This term is due to the fact that the term "laminated object manufacturing" was created by the company Helisys, similar to "fused deposition modeling" by the company Stratasys. In contrast to Stratasys, Helisys was dissolved in 2000, at which point their production of LOM systems ceased. Companies such as Mcor Technologies Ltd, founded in 2005, reintroduced A4 paper-based systems to the market. The first digital LOM systems were launched by Helisys in 1991, making them commercially available a year before SLS and Fused Deposition Modeling systems. The Helisys system was based on special paper which was laminated on one side with a thermally activated adhesive. This coated paper was in roll form and contoured using a CO_2 laser. The special paper was pulled over the build platform. A roller heated to 330 °C activates the adhesive and uses additional pressure to bond the current layer to the partially finished component. After this step, the CO_2 laser contours the current and joined layer, which corresponds to the specific cross-sectional layer of the underlying CAD model. This procedure increases the positioning accuracy of the respective layers relative to each other, but also increases the waste volume. Furthermore, the laser cuts a constant frame from the paper, which also corresponds to the maximum XY dimensions of the build space. Through this frame, the current layer is detached from the roll material. Paper that is in this frame but does not belong to the component is cut into squares. The elements resulting from this can be more easily mechanically removed from the component after completion. The elements are comparable to FDM support structures. The build speed in the Z-direction can be increased by laser-processing a maximum of four layers at a time. However, this reduces the resolution of the component in the Z-direction, and the machined contours become more discolored due to the energy input of the laser.

LOM has been further developed in various directions. The materials that can be used are generally unlimited, as long as they are in sheet or film form and can be joined. Plastics, metals, ceramics, and wood can all be processed using LOM. It is also possible to produce multi-material components made of several materials. The material change can be done layer by layer or locally in the layer. Various classic separating processes can be combined with different joining processes such as gluing, welding, clamping, and ultrasonic welding. In places, LOM systems can also be categorized according to the principle of joining.

The Cirtes company offers LOM solutions under the name Stratoconception. For contouring, micro-CNC milling, laser machining, hot wire, or an oscillating knife are used. The system behind Stratoconception was developed by Claude Barlier, and a patent was filed in 1991. The special feature of Stratoconception is the subsequent positioning of the layers with pins or recesses. Subsequent positioning allows maximum utilization of the sheet material for the creation of the components. Another special feature is the ability to choose between 2-axis, 2½-axis, or 5-axis machining of the plates. In comparison, the layers are clearly visible in 2-axis machining, depending on the thickness of the plate. With 2½-axis machining, this visibility is reduced to the maximum. With the Stratoconception system, both small and large components can be produced. In the xy-direction, dimensions of 2000 × 3000 mm are possible, and in the z-direction, there are only physical or static limitations.

One of the main advantages of LOM is the higher process speed compared to other additive processes for large components with high volume and low complexity of the geometry. Furthermore, the technology of the equipment is easy to master and partially independent of the material used. By bonding the layers, stresses are reduced and the components are thus produced primarily without distortion. A problem is the different load capacity of the components in and across the layer direction, which occurs in particular when joining the layers with adhesives. Ultrasonic- or diffusion-welded metal components are an exception here and exhibit constant load-bearing capacity in all directions. Furthermore, geometries with cavities or cutouts that prevent the subsequent removal of segments that do not belong to the component pose a problem. One solution to this is to remove the respective segment part in the layer. However, this is not possible with every LOM system, and it slows down the process speed. Furthermore, the waste material produced, which can be a factor of 1:10, can offset the same or lower material price. If LOM is used with a consistent thermoplastic material, the resulting waste could be reformed back into sheet material, ameliorating the environmental and economic disadvantages.

1.3 Functional Principle of Fused Deposition Modeling

As described in Section 1.2.1, the name fused deposition modeling (FDM)in the extrusion process group is closely associated with the company Stratasys. This is a US company that was originally founded in Minnesota and has since moved its headquarters to Rehovot in Israel; the headquarters in Eden Prairie was nevertheless retained. In addition to MakerBot in 2013, Stratasys acquired equipment manufacturer Objet Geometries in 2012. MakerBot also covers the consumer thermoplastic extrusion market with its low-cost systems. The MJM systems from Objet Geometries belong to the UV curing process group and are marketed by Stratasys under the name PolyJet. Also, the service area, which was covered by RedEye at Stratasys, has been expanded through the integration of Solid Concepts and Harvest Technologies. Stratasys' various FDM systems differ in a number of aspects, including build volume, material bandwidth, support material, and additional requirements (such as washers to remove support structures or set-up location). Both the Stratasys and MakerBot systems specialize in processing thermoplastics, which come in specific-diameter round cord. This round cord is referred to as "filament" in the additive field and is produced via extrusion from thermoplastic pellets. Thermoplastic material is used in FDM both for the creation of the components and for the necessary support structures.

1.3.1 Additive Extrusion with Filament

The systems in the extrusion process group are distinguished from the other groups primarily by their accessibility. No other additive systems can be introduced into companies in such a timely and cost-effective manner. Due to the comparatively inexpensive and easy-to-produce thermoplastic consumables, product concept tests in particular can be carried out without major financial and time outlay.

The thermoplastic filaments for the production of the components and necessary support structures are available with a round cord diameter of 1,75 mm. The filament is produced by extruding thermoplastic granules. In this continuous process, the thermoplastic materials are pressed through a die by means of a screw press. The granules are melted and homogenized by heat input and internal friction. The pressure built up at the same time causes the thermoplastic to be ejected through the forming die. The thermoplastic exiting through the die then solidifies in a water-cooled calibration. Additional application of vacuum presses the thermoplastic profile against the walls of the calibration, thus completing the molding process. The manufacturing tolerances during extrusion are in the range of ±0,05 mm. The extruded thermoplastic filament is then wound onto spools with a view to unobstructed and smooth delivery. These filament spools are fed to some FDM systems in encapsulated form or via adapters. This method of feeding filament, in addition to recording consumption, also provides protection from unwanted particles and moisture. Furthermore, such filament storage supports the automatic loading and unloading process widely used in FDM systems. In this process, the filament is conveyed by conveyor motors through a hose system to the actual melting head and automatically threaded into or out of it. After the thermoplastic filament has been automatically threaded in, a certain length of it is extruded through the melting nozzle into a container. This procedure also serves to convey particles or residual material out of the nozzle. After this short extrusion, the nozzle is moved over a rubber lip and metal brush, which removes or reduces adhering material and contaminants. This cleaning process is also carried out during the build process when switching between active nozzles, thereby increasing the quality of the components.

With the F123 systems from Stratasys, the so-called purge tower is additionally is generated. This represents an automatically generated and separate component that is used in addition to the nozzle cleaning. This system was adapted from FLM, where stationary nozzle cleaning areas are sometimes missing. The idea behind the purge tower is that, on the one hand, the nozzle is cleaned of particles and residual material and, at the same time, is filled with sufficient material before it applies material to the actual component again. The disadvantage of this additionally generated component is the higher consumption of thermoplastic material and the increase in manufacturing time for components.

In principle, color changes in components can only be generated layer by layer in FDM systems. For this purpose, the build process is specifically stopped after the completion of a certain layer and the active thermoplastic filament is automatically threaded out of the melting head. Depending on the system, this filament can then be removed together with the coil and the adapter. The stop can be scheduled in advance by setting a pause or can be performed manually on the system. After the new filament has been inserted, it is automatically conveyed to the melting head and threaded in. The build process then starts again with the new different colored filament in the next layer of the part to be produced. Residual particles in the nozzle create a smooth color transition in the component. The same system is also used when further filament is needed to complete a component. In this case, however, stopping the build process and unthreading is initiated automatically by the system.

With a few exceptions in MakerBot systems, FDM systems are equipped with two melting heads. In each case, one melting head is intended for the component thermoplastic. The second head is designed for the thermoplastic support material. On the F-Series FDM systems introduced at Stratasys in 2017, the two melt heads are easily interchangeable. This is also due to the polylactic acid (polylactide, PLA) material, which is new to Stratasys systems. The thermoplastic material is very common in the additive private sector as a consumable material and is produced from renewable raw materials. The use of PLA as a material in FDM can be traced back to systems from MakerBot, which primarily use this material for the building process. The cost of a kilogram of PLA is approximately 50 % less than the thermoplastic acrylonitrile-butadiene-styrene (ABS), which is the most commonly used material in FDM systems. Accordingly, many of the Stratasys systems are optimized to process ABS or similar thermoplastics. For fast and cost-effective concept and verification models, PLAs are provided on the F-Series. To process these on the F-Series, the melt head is changed to create the part because PLAs have lower processing temperatures than the other system-compatible thermoplastics. If this change of melt heads was not made, residual particles of the thermoplastics with higher processing temperatures could lead to faulty extrusion or clogging. In the case of PLAs, the support structure is provided with the same material and additional targeted cooling via a special second attachment, which replaces the exchangeable second melt head for the support material. The PLA support structure must be mechanically removed in a downstream step.

FDM systems are basically designed as a 3-axis system in a gantry arrangement (Figure 1.4). The resolution in the xy-direction is determined by the nozzle diameter of the melt units. In Stratasys systems, this diameter averages around 0,45 mm and is not variable. The resolution in the z-direction is generated by an axis and can be changed in some FDM systems via the control software. On average, this is around 0,25 mm. The thermoplastic filament is mechanically conveyed by pinions into the melting nozzles and extruded by permanent mechanical push conveying. The resulting thin extrusion beads are then are then applied in a targeted manner and in layers via axis movements to produce the component or necessary support structures. In the process, the warmer applied material bonds with the material already present. The layer is usually created by tracing the respective outer contour and filling this contour with a 45° hatch. If the extrusion beads are placed directly next to each other in this process, a fully filled component is produced. By deliberately offsetting the extrusion beads in the 45° hatching, hollow structures can be created in the components and the support structure. In the z-direction, the small cavities are closed by placing several fully filled layers on top of each other.

In the majority of FDM systems, thermoplastic-based building plates or films are used as the substrate base for the component to be produced. These are designed to ensure good adhesion of the component material and the support material, and are sometimes provided with an adhesion-promoting structure. The adhesion between substrate, support material, and component material prevent both the component from detaching during the build process and the deformation of the layers in the z-direction, either of which can lead to the abortion of the build process.

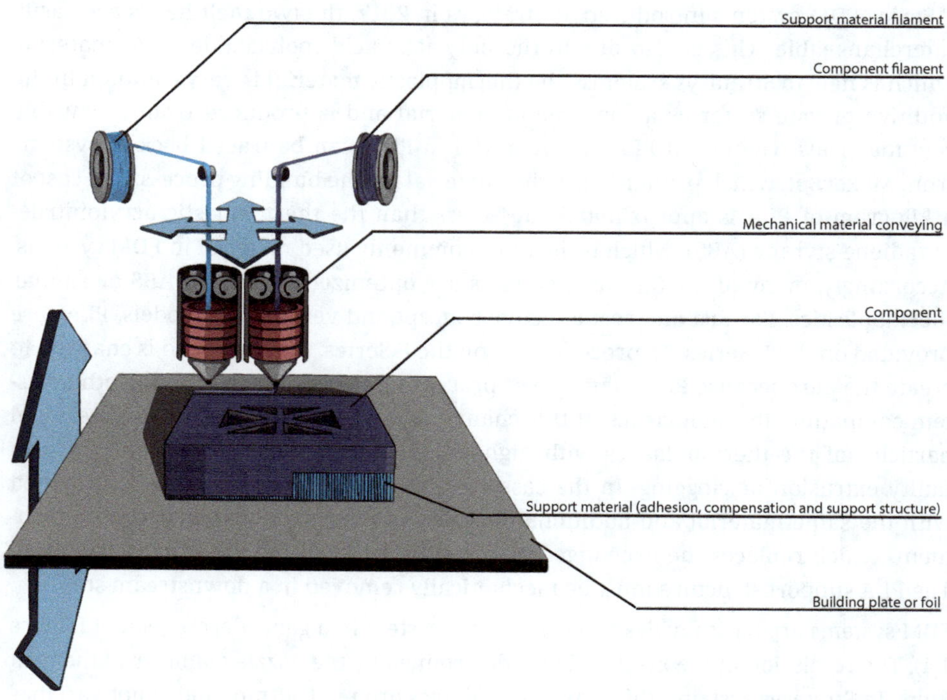

Figure 1.4 Functional principle of the FDM process from Stratasys (© Leonie Fensterle, from [Fischer 2020])

At the beginning of the build process with the z-resolution of 0,25 mm, approximately nine layers of support material are created, which ensures optimum perpendicular alignment from the last support material layer to the melting nozzles. This support material substructure is generated wherever the first component layers or support structures are created in the further course of the building process. If the nozzle is not optimally aligned with the layer, this can lead to clogging of the nozzle on the one hand and loss of adhesion between the layers on the other. If these support material layers are created from dissolvable thermoplastic, the detachment of the component from the support substrate can take place automatically. The support material substructure in FDM is the result of the calibration process that takes place prior to the build process. During this process approximately nine points on the substrate base are captured using styluses.

1.3.2 Characteristics of Fused Deposition Modeling

Components created via FDM are characterized by the thermoplastic ABS, which is often used in Stratasys systems, and the distinctive extrusion beads. Depending on the nozzle diameter and z-resolution used, the stacked extrusion beads are haptically perceptible and clearly visible. When filling the layers via the 45° hatching, the tracks are rotated by 90° in each layer. This strategy is intended to prevent too many extrusion beads from being deposited directly on top of each other, and thus leads to better layer adhesion within the components. As a result of the diameter of the extrusion beads and the 45° hatching, certain geometric features can lead to defects without material being deposited in the layers. These flaws can have negative effects on intended functionalities such as load-bearing capacity, tightness and surface quality.

1.3.2.1 Surfaces

The surfaces of components produced by additive systems exhibit a stair-step effect due to the layered structure, especially in the z-direction; an effect that is more or less pronounced. Additive systems that primarily process liquids have the lowest stair-step effect due to a high z-resolution and thus the smoothest surfaces in the z-direction. Surfaces in the xy-direction have in the majority of cases a lower and process-related resolution. With FDM, the z-resolution is generally higher than the resolution in the xy-direction. The following z-resolutions are provided (in different combinations) on the various Stratasys systems:

- 0,127 mm
- 0,178 mm
- 0,254 mm
- 0,330 mm.

The most common z-resolution is 0,254 mm, which is available for the majority of systems.

> In FDM, components with visible surfaces should be created in such a way that these surfaces are largely embossed via the z-resolution. The possible surface finishes in FDM are much more effective for surfaces created with z-resolution. In parallel, however, possible increases in cost and build time due to the specific orientation of the part should also be evaluated. Furthermore, it should be noted that in particular the possible tensile load for components in the z-direction is lower than that in the xy-direction.

The extrusion beads typical of FDM are more clearly visible in the xy-direction and reflect the typical nozzle diameter of the systems. The nozzle diameter for Stratasys systems is approximately 0,45 mm. Defects can also occur, especially in the xy-direction, which result from the interaction of the nozzle diameter, the geometry of the part and the corresponding path planning (Figure 1.5). In some cases, it is physically impossible to place another extrusion bead between the component contour and the component filling. This ultimately results in a defect, which can occur both on the surface in the xy-direction and inside the component. The surfaces produced in this way can usually only be optimized with mechanical finishing steps to produce smooth surfaces.

Figure 1.5 Here you can see the path planning of a layer with 45° hatching in FDM. This is the fully filled mode with voids inside the component. The green lines represent the component and the purple lines the necessary support material in the layer (© Andreas Fischer, from [Fischer 2021])

Chemical finishing operations such as solvent dissolving of the thermoplastic surface, as is done in the Stratasys "Finishing Touch Smoothing Station", cannot produce the desired surface finish on these special surfaces. The solvent used in the Finishing Touch Smoothing Station penetrates a maximum of 32–63 mm deep and does not change the part surface by more than 0,023 mm. In the z-direction, the extrusion beads protrude by approximately 0,051 mm during FDM. The extrusion beads are basically smooth and glossy after deposition. Along the beads in the z-direction, this results in a smooth haptic impression. If the component is chemically post-processed, the surface remains glossy. If the components are mechanically reworked, such as by manual grinding, sandblasting, or vibratory finishing, the surface of the machined components becomes matt. If the components have sharp edges or fine details, both chemical and mechanical post-processing of the components can lead to negative results.

Another factor that can lead to negative effects both in the component and on the surface are waste products that form at the die during extrusion. These include burnt thermoplastic residues and parts of extrusion beads which, if not removed, may fall off the die when the layers are produced and thus be incorporated into the component. Burnt thermoplastic residues often lead to deformations on the surface of the components that are difficult to repair and thus to AM rejects (Figure 1.6). Residues of extrusion beads that have been incorporated into the component, can usually be easily removed mechanically, although minor defects of the surface can result. To counteract this specific problem of extrusion and thus increase the quality of the components, FDM systems have a mechanical nozzle cleaning unit. The cleaning unit consists of a metal brush with an upstream heat-resistant rubber lip. In Stratasys' F123 systems, in addition to the mechanical nozzle cleaning, the purge tower is generated in parallel with the build process. In addition to cleaning, this separated object also ensures that the nozzle is optimally filled with material. Below the mechanical nozzle cleaning, the removed waste products and the extrusion beads, which are generated there in intermediate steps, are collected in a replaceable container. The extrusion beads generated in the intermediate steps are also used for optimum material filling of the nozzles. Each time the model material and the support material are changed, the nozzles are run over the mechanical cleaning combination with a rapid back-and-forth movement, and thus cleaned.

Figure 1.6
FDM component with surface defects in the xy- as well as z-direction, caused by the inclusion of waste particles from the nozzle in the component (© Andreas Fischer, from [Fischer 2022])

After a layer has been created, in FDM the z-axis is lowered by the value of the z-resolution and the printing unit is moved to the cleaning unit. There, the last nozzle used, in this case the nozzle for generating the component, is cleaned. The nozzle for the support structure is prepared for the generation of the following layer. It is cleaned, extrudes briefly, and is cleaned again. Then, at the starting point defined by the path planning, the generation of the support structure begins, followed (in the same way) by the preparation of the nozzle for the component generation and the start of the layer generation of the component at the planned starting point. If the planned start and end points of the component layers lie exactly on top of each other several times, clearly visible material accumulations can occur on the surface (Figure 1.7). These seams often occur in components generated by rotational axes, especially if such a component is generated such that the rotational axis is perpendicular to the XY orientation of the system. The problem can basically be solved by changing the starting points in the path planning, but this intervention in the path planning is not possible with all Stratasys systems. However, the seams can be easily removed by mechanically reworking the components. In principle, the specific FDM surface in the z-direction is particularly suitable for increasing the grip of a component.

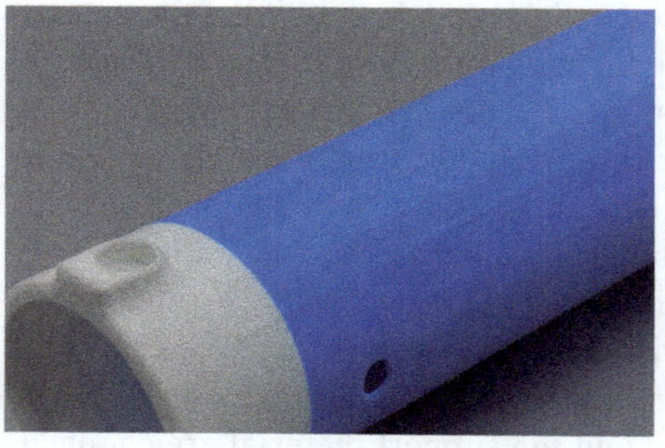

Figure 1.7 Two-color cylindrical FDM component with seams on the surface (© Andreas Fischer, from [Fischer 2022])

1.3.2.2 Support Structures and Building Chamber

In FDM, the necessary support structures are generated via the second nozzle of the printing unit. A support structure is required to generate overhangs that exceed 45°. Here, the z-axis of the FDM system represents the zero angle. This support structure stabilizes part contours and prevents specific parts from deforming during layer generation, or falling off during the build process. On some Stratasys systems, there are three selectable strategies for calculating the support structure:

- Basic
- SMART
- Surround Support.

The "Basic" strategy represents necessary support structures, which are filled with a 45° hatching. This structure is particularly suitable for component segments to be supported, which represent a high volume and thus weight. Due to its extended volume and dense filling, the Basic strategy leads to increased support material consumption and longer manufacturing times. With "SMART", the volume of the necessary support structures is reduced by about half of the basic strategy. In addition, the filling of the support structures is also halved, which leads to a reduction in support material and manufacturing time. The surround support encloses the entire component, and this is filled to the same density as in the "Basic" variant. This support structure is particularly suitable for filigree components, which are thus additionally protected by the structure. The consumption of support material and the manufacturing time are extremely pronounced with this strategy.

After completion of the build process, the support material must be removed from the component. This can be done mechanically or chemically. However, manual removal of the support material is possible for all geometric conditions. Support structures, which are located in cavities with little access, for example, can hardly be removed mechanically and must therefore remain in the component. However, this can be planned as a stabilizing function when designing the component. Furthermore, filigree segments of the component can also be affected during mechanical removal. The surround support strategy is not suitable for mechanical support structure removal (Figure 1.8).

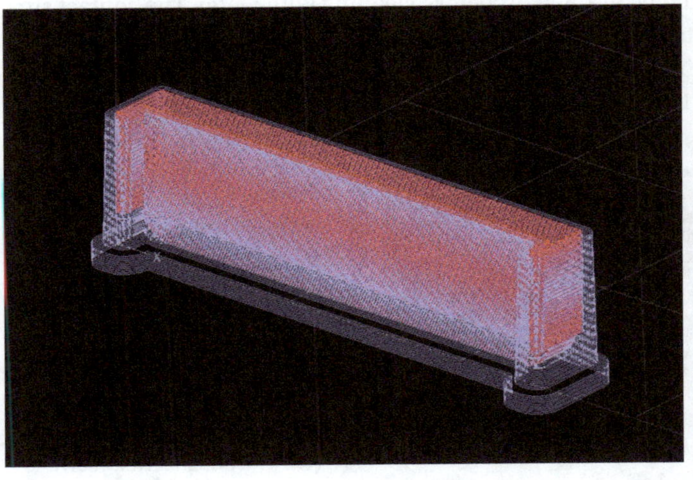

Figure 1.8 Surround support during path planning of a FDM component: The component is shown in red and the support structure in purple (© Andreas Fischer, from [Fischer 2021])

Depending on which thermoplastic is used for the component, the suitable thermoplastic support material can be removed mechanically or dissolved chemically. In the case of mechanically removable support material, it should be noted that cavities can sometimes be difficult to free from the support structure, and filigree component segments can be deformed or damaged by the removal of the structure. The chemically dissolvable thermoplastic support material is removed in a circulating liquid at a temperature of approx. 60 °C; care must be taken to ensure that the liquid is not too hot. Also it must be ensured that liquid can flow in and out of intended cavities through openings. If these openings are not optimally selected, the dissolution of the support structure may require longer processing times. Components with thin and large-area component segments can be easily deformed by the circulation and the influence of heat.

In principle, several components can be processed in the specially designed systems for chemical support structure removal (Figure 1.9). Components with sensitive geometric segments should be processed individually, since the components can damage each other, especially due to the circulation of the liquid. If the interior of the components was created by means of a hollow structure, these can fill with the liquid during the chemical removal of the support structure. Furthermore, the surrounded air in the hollow structures ensures that the components float to the surface of the liquid. As a result, parts of the support structure, which are thus not in proper contact with the liquid, cannot be dissolved. FDM components with hollow structures or cavities should, if possible, be positioned so that the liquid can completely cover the component. After dissolving the support material, the components are cleaned with clear water and then dried. In the case of components with cavities or hollow structure, any residual liquid should be drained off before cleaning with water. The contact surfaces of the components to the support structure have a greater tendency to show defects on the surface. In the case of mechanical removal, defects on the surface can additionally be created by the removal process.

Figure 1.9 Recirculation tank of the HP Designjet 3D Removal System for the removal of support structures of FDM components made of ABS with maximum dimensions of 200 × 200 × 150 mm (© Andreas Fischer, from [Fischer 2021])

Combining ABS as the material for the component, Stratasys uses P400-SR thermoplastic support material. Fourier-transform IR spectrometry of the ABS support material shows the typical absorption bands of an acrylic copolymer based on methacrylic acid methyl ester (methyl methacrylate, MMA) and acrylic acid butyl ester (butyl methacrylate, BMA). The glass transition temperature is 103 °C (midpoint). Rheological analysis also showed that the support material is thermoplastic and that the viscosity does not decrease noticeably with time.

In the Stratasys HP systems, the SR30L support material was used in combination with ABS. IR shows the typical absorption bands of an acrylic thermoplastic based on polystyrene-*co*-butyl acrylate-*co*-acrylic acid for this material. The glass transition temperature is 124 °C (midpoint). The two Stratasys support materials can be dissolved using sodium hydroxide powder dissolved in water, by agitation and application of heat. The mixing ratio should be 950 g of the P400SC powder and 42 L of water. Removal of the support material has no detectable negative effects specifically on the ABS components. If large quantities of the support structure are removed using the chemical process, special storage and disposal of the resulting liquid may be required. The volume of the material to be disposed of can be drastically reduced by the targeted evaporation of the water content of the liquid.

Another special feature of the FDM systems are the build chambers. The size of the components that can be produced ranges from the former entry-level model of Stratasys (the Mojo) with 127 × 127 × 127 mm, to the top model (the Fortus 900mc) with 914 × 610 × 914 mm. Inbetween are the two "Idea"-series systems and the three "Design"-series with three systems. In the "Idea" series, parts can be produced on the µPrint SE with 203 × 152 × 152 mm and on the µprint SE Plus with 203 × 203 × 152 mm. In the "Design" series, components of 203 × 203 × 305 mm can be produced on the Dimension Elite and components of up to 253 × 254 × 305 mm on the Dimension 1200es or Fortus 250mc. The "Production" series incorporates Stratasys' highest-performance FDM systems. Here, the size of the components that can be generated ranges from 355 × 254 × 254 mm on the Fortus 360mc to 914 × 610 × 914 mm on the Fortus 900mc. Inbetween are the Fortus 380mc with 355 × 305 × 305 mm, the Fortus 400mc with 406 × 355 × 406 mm, and the Fortus 450mc also with 406 × 355 × 406 mm. In terms of performance and purchase price, the current F123 series is positioned between the Fortus 250mc and the Fortus 380mc. The build chamber capacities here range from 254 × 254 × 254 mm for the F170, to 305 × 254 × 305 mm for the F270, and 356 × 254 × 356 mm for the F370. The "Idea"-series and "Design"-series systems in particular were widely adopted and are often still in use – despite discontinued service from Stratasys. Specialized companies such as iSQUARED AG from Lengwil in Switzerland continue to offer service and consumables for these systems.

All the build chambers of these FDM systems can be heated uniformly by fans to around 80 °C and are self-contained. This property is very beneficial in additive component production via thermoplastics, as it prevents thermally induced residual stresses in the component in particular. These residual stresses can lead to plastic deformation of the component, which in turn negatively affects adhesion to the substrate material and the application of further thermoplastic. Furthermore, the higher and more uniform temperature in the chamber ensures that the additive extrusion process is more effective, thus also reducing scrap during part creation. If, for example, errors occur in the material feed or in the printing unit during the creation of a component, the component continues to be kept at a constant specific temperature. Troubleshooting can then take place in parallel and the building process can be continued afterwards without any problems. When the production of a component is started, the build chamber door locks automatically and mechanically. This ensures that operating personnel are not endangered during part creation, and that the ingress of foreign bodies or particles that could negatively influence the process is prevented. In order to prevent errors when threading or removing the thermoplastic filament into or out of the printing unit, which is located in the build chamber, and also to protect the operating personnel at the same time, this process is solved by automatic conveying of the filament. The process of threading and removing the filament takes place in the temperature-controlled build chamber. The printing unit is positioned in the build chamber via the axes in such a way that bends through which the filament has to be conveyed are minimized. If the filament to be threaded is in front of the printing unit, it is placed again above the cleaning unit via the axis system. Once the processing temperature of the nozzle is reached, the filament is mechanically conveyed into it and an extrusion bead of a specific length is extruded into the waste container. The nozzle is then cleaned mechanically. When the filament is removed, the die is heated to the process temperature to break the adhesion between the cooled thermoplastic and the metal of the die. Once this is done, the filament is mechanically conveyed out of the printing unit and can then be pulled out through the feed channel.

1.4 Functional Principle of Fused Layer Modeling

Fused layer modeling (FLM) belongs to the extrusion process group and uses the same process principle as described in Section 1.3. The emergence of FLM systems is due to the expiration of the maximum protection period of early Stratasys patents. This now made it possible for other manufacturers to use the process principle developed by Stratasys. Additionally, conducive to this development was the fact that the basic design of a system based on additive extrusion is simple and can be achieved with lower-cost components than other process principles.

The simplest variant here are the 3D printing pens such as the 3Doodler or the Play 3D Pens from Polaroid. The pens basically contain a mechanical conveyor for the thermoplastic material and a heatable nozzle. The nozzle diameter is around 0.4 mm, and the material is fed in by means of rods or filaments with a diameter of 1,75 mm. The thermoplastics acrylonitrile-butadiene-styrene (ABS) and polylactic acid (polylactide, PLA) are primarily used here. These 3D printing pens are very similar in function to a mechanical hot glue gun, which in turn was used as a source of inspiration in the early development phase of fused deposition modeling (FDM). In very simplified terms, additive extrusion is a hot glue gun that is moved in a targeted manner via an axis system or manipulators. In the case of 3D printing pens, the user's locomotor system takes over the role of the manipulator. The focus of use for these devices is the creative hobby sector. In some cases, professional designers also use the systems to create sculptures – for example, Grace Du Prez and her team, who were commissioned by Japanese car manufacturer Nissan to model the Nissan Qashqai in black using a 3D printing pen on a scale of 1:1. In around 800 h of work, the 4400 mm long and 1600 mm high sculpture was created by the design team from around 14 kilometers of thermoplastic filament.

FLM is particularly characterized by the variation in the structure of the different systems. In addition to the number of melting nozzles per printing unit, worthy of mention are the manipulation system for the printing unit, the design of the building chamber, and the material feed method. The most widespread FLM system variant is the combination of one melting nozzle, a mechanical filament feeder, a three-axis gantry system, and an open building chamber (Figure 1.10).

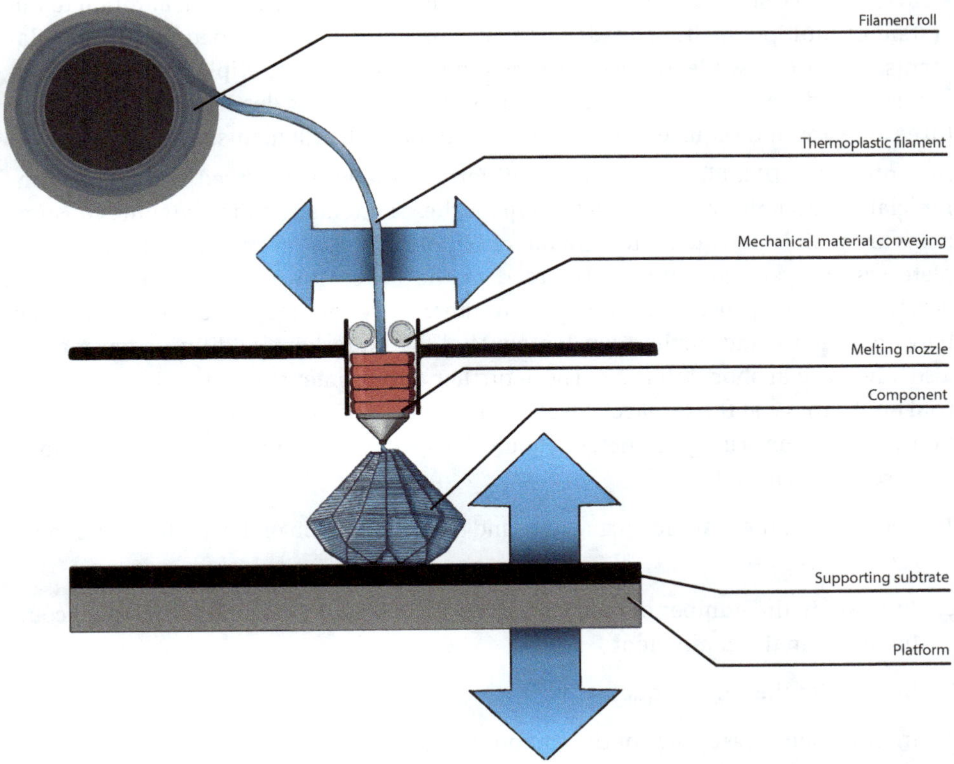

Figure 1.10 Functional principle of the FLM process with one melting nozzle (© Leonie Fensterle, from [Fischer 2020])

An important distinguishing feature of FLM from other additive processes is the possibility of using different thermoplastic materials in one component. A prerequisite for this, however, is that the materials adhere to each other. To create such multi-material components using FLM, it is possible to equip the system with several melting nozzles and respective material feeds, or to weld together pre-calculated lengths of different filaments to form a multi-material filament, which then ensures that the material change takes place at specific points in the component. FLM systems are equipped with a maximum of up to four melting nozzles. Additional melt nozzles are theoretically feasible depending on the type of manipulation system used for the printing unit. With four melting heads, three different materials and a suitable support material are possible. In addition to the adhesion between the materials for the component, the support material must be designed to provide adhesion to the component materials.

A very widespread feature of FLM is the possibility of using the material in different colors in order to produce multicolored components based on one material. It is possible to produce specific parts of a component in the required color. In addition to the system of multiple melting heads or the use of adapted and welded multicolor filaments, it is also possible to use special melting nozzles with multiple accesses for filament to produce multicolor components. These special nozzles have approximately three accesses and require coordinated conveying of the filaments.

In contrast to FDM, FLM systems with different designs are offered by various commercial manufacturers. The target group for these systems and the matching consumables are mainly private customers but also industrial customers. In addition to complete systems, kits are also widespread in the lower price segment. In FLM, the variability of the process and the large number of system manufacturers mean that new approaches and further developments of the underlying additive extrusion are also advanced at short intervals. These further developments can enable new industrial applications at the research level as a first step. In order to advance such applications, additive process parameters should be specifically optimized. These parameters are graded as follows:

1. Reduction of the manufacturing time and thus elimination of the limited component size
2. Increase in the number of materials that can be processed and their possible combinations in the component
3. Increase of the surface quality
4. In-line quality assurance of the component
5. Integration into serial production.

The parameters (1) and (2) have been further developed by various systematics in FLM. Parameters (3) and (4) have not yet been solved acceptably for industrial purposes. Parameter (5) is partly possible through the use of industrial robots as a manipulation system for the printing unit (see Section 1.4.2.2).

1.4.1 Material Feed

In FLM, the thermoplastic material can be fed via various methods to the die that applies the thermoplastic extrusion beads for the component or support structure. Depending on the method, the material is available in the following forms:

- Filament
- Foil
- Granules.

As with FDM, the most common method is to feed the thermoplastics in filament form. In addition to the filament diameter of 1,75 mm, diameters of 3 mm are also occasionally available. A larger filament diameter also results in a larger die diameter, with the consequence of a larger cross-section of the extrusion beads. As a result, the achievable resolution and the manufacturing time for components are reduced. The principle of reducing the manufacturing time for components by means of a larger diameter of the extrusion beads is used in particular for feeding the material in granular form. Depending on the material, the manufacturing costs for 1 kg of thermoplastic filament is around €13, with the conversion of the granules into filament accounting for around €8.

Thermoplastic foils are another efficient way of supplying the material Compared to filament-based print heads, the throughput of a foil-based FLM printing unit is significantly higher. This results in reduced component manufacturing time while maintaining the same resolution. Another advantage of foil is that it requires less space than filament during storage. Furthermore, deviations in thickness or diameter are smaller with foil than with filament. This achievable higher precision with the foils leads to higher component quality. The printing units for thermoplastic foils do not differ significantly from filament-based ones. The foil is conveyed via a cooled area into the melting unit, where it is converted into melt under the influence of the supplied thermal energy and discharged via a defined nozzle geometry [New Media Publisher 2019].

The third method of material feeding in FLM is via thermoplastic granules. In traditional plastics processing, granules are used for injection molding or extrusion. Granules are produced by cutting extruded and cooled strands after polymerization or compounding. The pourability of granules makes them easy to convey and transport, which can be particularly advantageous for high-volume AM. By using granules directly, the intermediate step of filament or foil extrusion can be bypassed and a wider range of materials can be accessed. In addition, the direct use of granules can reduce the manufacturing costs of the components. Due to a significantly higher achievable throughput with the direct use of granules, higher processing speeds can be achieved in order to manufacture large-volume components (in particular) in an acceptable time.

When feeding the material as pellets, there are two options for placing the extruder that produces the thermoplastic melt. Direct placement of the extruder as a pressure unit is one variant. It should be noted that either the extruder must be moved by the axes or, in the case of rigid placement, the build platform must be moved accordingly. Using an extruder directly as a printing unit can limit the advantages of granule-based material feeding, such as reducing manufacturing time. In addition, this method of placing the extruder may compromise any increase in axes that is possible with FLM (see Section 1.4.2.2). To take full advantage of granule-based material feeding, it is necessary to separate the printing unit and the extruder. The thermoplastic melt generated by the extruder is fed to the material-applying die of the printing unit via a heated hose system.

Example systems for granule-based FLM systems are the KamerMaker and "big additive area manufacturing" (BAAM). The KamerMaker was originally developed for a house construction project by the architecture group DUS from the Netherlands. The design of the FLM system is based on a three-axis gantry arrangement and was developed in cooperation with Ultimaker BV, an FLM system manufacturer from the Netherlands. The KamerMaker is 6 m high and occupies 6 m^2 of floor space. The components, which were created for a house construction project using the KamerMaker, weigh approximately 180 kg and have dimensions of 2000 × 2000 × 3500 mm. The material used was the thermoplastic Macromelt 6900 E from Henkel. This transparent material is a polyamide (PA) based on vegetable oil, and consists of 77 % biologically renewable raw materials. The processing temperature is a comparatively low 170 °C.

The BAAM system was developed in cooperation between Cincinnati, a US-based manufacturer of metalworking systems, and Oak Ridge National Laboratory. The FLM system has a building chamber with dimensions of 6100 × 2360 × 1830 mm. Materials that can be processed include polyethylene terephthalate (PET), polyethylene terephthalate glycol (PETG), polylactic acid (polylactide, PLA), polycarbonates (PCs), thermoplastic polyurethane (TPU), polyamide (PA), acrylonitrile-butadiene-styrene (ABS, acrylonitrile-styrene-acrylate (ASA), polyetherimides (PEIs), polyether ether ketones (PEEKs) and polyphenylsulfones (PPSUs). The possibility of adding carbon and glass fibers can improve the thermal resistance as well as the strength of the components. Printing is performed at a mass flow rate of 17.24 kg/h with a nozzle diameter of 7,62 mm. Additionally, nozzle diameters of 5,08 mm and 10.16 mm are available. The relatively large nozzle diameters can shorten the manufacturing time for components, but the extrusion beads typical of FLM are very noticeable on the surface of the components. The BAAM system, however, also offers the option of subsequently smoothing the surface by means of milling. The positioning speed of the system can reach up to 250 m per minute. At the same time, the printing unit is subjected to acceleration forces of over 2*g*. The system's applications range from cutting and drilling jigs to concrete formwork, resin transfer molds, sand casting molds, stretch forming jigs, furniture, and vehicles. Parts of the Strati electric vehicle from Local Motors were additively manufactured using BAAM. In addition to the chassis, frame, and exterior trim, parts of the interior were made in one piece. The vehicle's mechanics come from the Renault Twizy. The additively manufactured components of the vehicle were produced within 44 h, which can be reduced to around 24 h by using a larger nozzle diameter.

Another thermoplastic additive system that supplies the material in granulate form is the ARBURG freeformer [ARBURG 2021]. However, this system cannot be classified as FLM, as it does not generate extrusion beads for building up the components. Rather, this system applies thermoplastic droplets via a piezo nozzle shutter. ARBURG, a German engineering company, calls this process "plastic freeforming". In the freeformer, it is not the printing unit that is moved via axes, but the building platform on which the component is produced. In contrast to the KamerMaker and BAAM, the freeformer 200-3X, which is based on three axes, features two thermoplastic extruders. This means that the freeformer can also be used to create components from two materials or to use a water-soluble support material. With a building chamber of 200 cm^3 and a maximum component size of 154 × 134 × 250 mm, the freeformer is suitable for office use. The nozzle diameter of the 200-3X is 0.2 mm, and the build chamber can be heated to a maximum of 120 °C. The freeformer 300-3X, with a build area of approximately 300 cm^3 and with up to three thermoplastic extruders, enables the creation of additive components consisting of two different thermoplastics and the additional use of support material. The possibility of using standard granulates is made possible by the necessary predefined parameter sets provided by ARBURG. These parameter sets act as reference materials. The freeformer can process thermoplastic elastomers (TPE), polyamides (PAs), polycarbonates (PCs), polymethyl methacrylate (PMMA), polypropylene (PP), polylactic acid (polylactides, PLAs) and acrylonitrile-butadiene-styrene (ABS).

1.4.2 Characteristics of Fused Layer Modeling

FLM systems are for the most part open systems. The systems basically offer the possibility of making changes to the process parameters as well as adaptations to the hardware. In contrast, FDM systems are primarily closed. The ability to change the process parameters means that there is a greater variance in the materials that can be processed, including the color bandwidth. Furthermore, the open system structure means that consumables in FLM are not tied to specific manufacturers. As a result, consumables are approximately one-third less expensive than with FDM. With FLM, the open structure makes it possible to produce components in series materials.

The open system structure in FLM is made possible by the necessary software, which is provided in part by third-party suppliers, but also by the system manufacturer (and branded as such). It is divided into two categories:

- Layer generation and path planning (slicer)
- Machine control.

The software solutions resulting from the categories can be used separately or via a uniform graphical user interface. In particular, the solutions regarding layer generation and path planning of third-party suppliers are also used as a basis for in-house developments. Since FLM uses a wide variety of system variations, including swarm arrangements with industrial robots, there is a large number of very different software solutions in the machine control category. In contrast to FDM, complete specific software developments for FLM systems are rarely found (Figure 1.11).

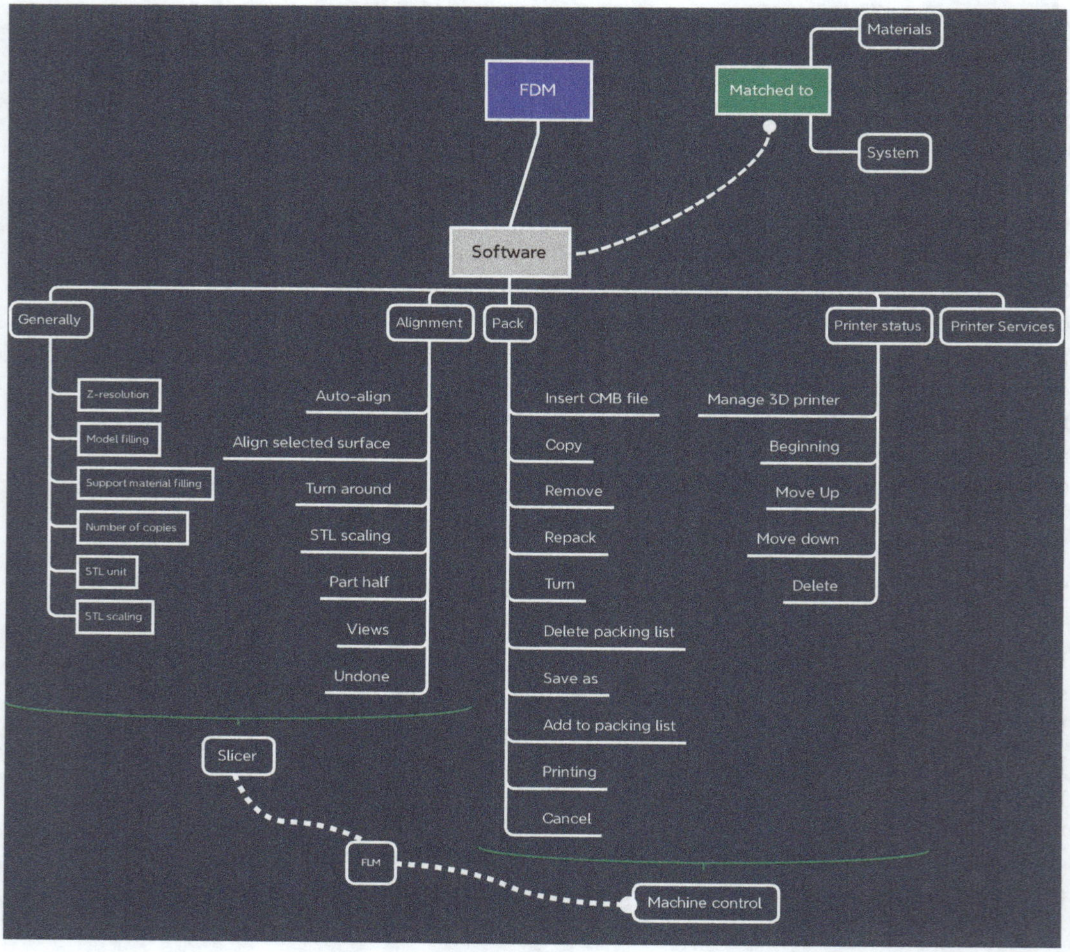

Figure 1.11 Comparison of the software functionality of FDM and FLM (© Andreas Fischer, from [Fischer 2021])

The high variation of the systems and the open architecture of the FLM systems can lead to a reduction in the quality of the components. Due to the open architecture, thermoplastic filaments from third-party suppliers can also be used, which in many cases represent a price advantage. These filaments can have deviations in diameter as well as voids inside, and thus represent variations in the quality of the filament. Defects in the filament cannot be detected or compensated by FLM printing units, and variations in the diameter of filaments can lead to conveying problems. Voids and variance in nozzle diameter result in increased or decreased thermoplastic feeding, which can lead to defects in the component as well as on the surface of the components.

Furthermore, the high variance of the systems leads to different characteristics of machine-specific quality-promoting system components such as automated mechanical nozzle cleaning. On the one hand, the absence of such a cleaning unit leads to shorter manufacturing times for components, since the necessary movements to the cleaning unit and the cleaning itself are saved. On the other hand, the omission of the unit can lead to unintentional flowing thermoplastic, contamination of the nozzle, or the incorporation of short extrusion beads into the component, thus reducing the quality of the component and having a negative impact on the build process.

1.4.2.1 Support Structures and Building Chamber

In FLM systems, water-soluble thermoplastic materials or component material is used as support material. Sometimes thermoplastic materials are also used, which have a low adhesion to the component material and have to be removed mechanically afterwards. However, there are also approaches to dispensing with support material by means of machine-side strategies. If FLM systems with five instead of three axes are used, for example, the component can be positioned relative to the printing unit when the current layer is applied, so that no support is required.

For systems with three axes, the USC Viterbi School of Engineering in the USA is developing a variable build platform that provides the necessary support for the components via individually extendable platforms. Each mechanical support is moved in the z-direction by its own motor to support component-specific areas. This support strategy reduces production times by around 40 % and material usage by 30 %. The use of thermoplastics as support materials is a mandatory selection criterion for support materials in additive extrusion systems due to the process. When using the same material for component and support structure, the number of direct contact points between component and support structure is kept as low as possible, since this support structure must be mechanically removed from the component in a downstream step. Due to the open software architecture of FLM systems, the parameters of the support structure can be adapted depending on the material and component geometry. Support structures in FLM usually have a negative effect on the surface quality of the affected areas of the component at the direct contact points. In the case of direct support with the material of the component, depending on the size of the contact area, local destruction of the surface of the component can also occur when the support structure is removed (Figure 1.12). The melting point of the dissolvable support material should be between 150 °C and 280 °C, depending on the process.

Figure 1.12
Egg cup made of acrylonitrile-butadiene-styrene (ABS) and manufactured by means of FLM with defective local surface caused by the mechanical removal of the support material (© Andreas Fischer, from [Fischer 2022])

The thermoplastic and water-soluble materials that can be used for dissolvable support structures are essentially limited to the following five thermoplastics:

- Polyvinyl alcohol (PVAL)
- Polyvinyl methyl ether (PVME)
- Polyacrylic acid (PAA)
- Polyvinylpyrrolidone (PVP)
- Cellulose ether (CE).

With a melting point of –122 °C, PME is not suitable for additive extrusion. PAA and PVP, with melting points of 12.5 °C and 13.9 °C, respectively, also are in a temperature range that is not compatible with that of the materials used for components. CE, with a melting point between 244 °C and 255 °C, can be used as a dissolvable support material with positive adhesion to the component material. The material is obtained from renewable raw materials and is mainly used in construction products. With a melting point of 230 °C, PVAL can also be used as a dissolvable support material in FLM. The material is often used as an adhesion and thickening agent in coatings and adhesives. PVAL is established as a water-soluble support material for FLM systems and is available as filament. When using PVAL, the systems should have either a temperature-controlled build chamber or a heatable build platform.

Furthermore, the use of an adhesion promoter may be necessary depending on the support substrate of the build platform. An adhesion promoter based on PVAL and water is suitable for the component materials polylactic acid (polylactide, PLA), acrylonitrile-butadiene-styrene (ABS), and polystyrene (PS). If an adhesion promoter based on acetone and ABS is used, ABS, PC, or a blend of the two can be used as component materials.

PVAL cannot be processed too dry or too wet. Therefore, a defined amount of water of approximately 2–2.5 % must be absorbed by the thermoplastic beforehand. If the filament is too dry, it becomes brittle and hinders the production process by frequently breaking in the feed. If PVAL is processed too moist, gas bubbles form when the filament melts, leading to foam formation in the melting nozzle. This foaming can lead to defects in the component or abortions of the manufacturing process. PVAL filament should be stored at a humidity of 20–30 %. PVAL is not suitable as a support material for geometries with cavities if adhesion to the material of the component is not given. In this combination, defects may occur in the component produced. If different support substrates are required for the component material and PVAL, this support material is not suitable in the combination.

The function of heatable build chambers in additive extrusion systems is primarily to counteract thermally induced residual stress in components. This stress is caused by different rates of cooling of the edge and core areas of components; the faster cooling and shrinkage of the component edge zones can lead to tensile stresses and plastic deformation there. These plastic deformations can lead to partial or total detachment of the component from the substrate and thus to the abortion of the AM process. Thermally induced residual stress is additionally promoted by specific component geometries and processing temperatures of the component materials. Components which present a large-area geometry in the first layers to be generated tend to have thermally induced residual stress and are more likely to be generated in closed and heatable build chambers. This also applies to component materials whose melting points are in higher ranges and thus require a higher processing temperature. Another influencing factor for thermally induced residual stress, especially in open build chambers, can be temperature fluctuations at the installation site of the FLM system.

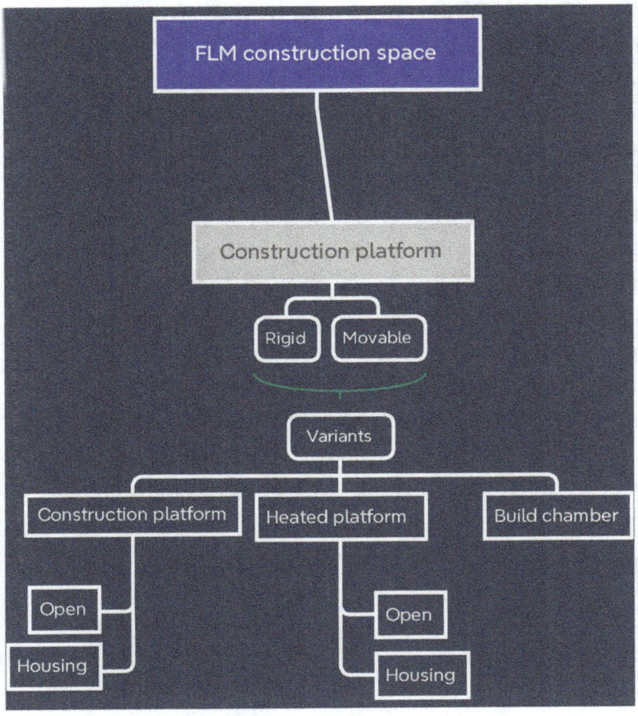

Figure 1.13
Variants of build areas for FLM systems (© Andreas Fischer, from [Fischer 2021])

The simplest form of a build area design for FLM systems is a rigid or movable build platform that can be loaded with different support substrates (Figure 1.13). This type of build area design is often found in systems based on industrial robots. Materials that have been modified with regard to shrinkage are advantageous for such installation spaces. The installation site for such systems should be chosen so that homogeneous ambient conditions can be created.

A variation of this basic design for FLM systems are build chambers in which the build platform is equipped with a controllable surface heating system. These heatable build platforms are again used with and without enclosures. The advantage of these build spaces is that the thermally induced residual stress in the lower layers of the component can be minimized and thus better adhesion of the component to the substrate can also be achieved. Enclosures of build spaces can minimize the negative effects of ambient conditions on the build process. In combination with integrated surface heating systems, their efficiency can be increased. Build chambers, which can be homogeneously tempered, are another variation of designing build areas. In principle, these are very similar to the build chambers used in FDM (see Section 1.3.2.2). The thermally induced residual stress can be counteracted most effectively in such build chambers.

1.4.2.2 Robot-based Fused Layer Modeling

In FLM, the necessary axis system, which is responsible for the precise guidance of the print head and the build platform, can be represented by industrial robots. AM using industrial robots is classified (like milling, riveting, drilling, or measuring with robots) as a precision application. This combination of AM and industrial robots follows the general trend of replacing machine tools or special machines with industrial robots for specific manufacturing tasks. Industrial robots have a high potential to reduce costs in production due to large workspaces and comparably low investment costs. However, applications such as FLM can only be implemented by taking measures to increase accuracy. Since in FLM, similar to milling, the basis of motion programming is based on digital 3D data and off-line programming, the repeat accuracy, as opposed to absolute accuracy, is not relevant for the process. The repeat accuracy of industrial robots is the accuracy that can be achieved when repeatedly moving to identical points in space, and is usually around 0.1 mm. The absolute accuracy varies depending on the manufacturer, age, and stress of the robot system, and can therefore be several millimeters. Through calibration measures, an absolute accuracy of approx. 0.5 mm can be achieved, and if the working area of the system is also reduced, values of around 0.1 mm can also be achieved.

The following factors can negatively influence the absolute accuracy:

- Axial zero positions of the robot links
- Length and angle error of the robot links

- Length changes due to temperature fluctuations (heating of the robot during operation)
- Variable load on the flange of the robot.

The load on the flange of an industrial robot used for FLM is largely fixed and not variable. In filament-based systems, the load results from the weight of the printing unit. The weight of a single printing unit of the robot-based 3D Fibre PrinteR is around 800 g. The FLM build area variants (see Section 1.4.2.1) can also be implemented with a robotic axis system. If an enclosure of the build chamber is necessary, it should be of variable design and only include the flange together with the printing unit of the FLM system. The open variants with build platform or surface heating are the most widespread for robotic FLM systems, as they are easier to implement than the closed variants.

Dirk van der Kooij, a Dutch product designer, can be seen as a pioneer of FLM using industrial robots. Faced with the difficulty of creating affordable large-volume components via AM, he began adapting a decommissioned industrial robot from the automotive industry in 2010. Van der Kooij's FLM system has an open build platform as its build area and a small granular extruder attached directly to the flange. The system uses recycled granules from old refrigerators as its component material, primarily to create furniture. The industrial robot can only operate in three axes due to the direct attachment of the granule extruder to the flange. The relatively large diameter of the extrusion beads can reduce the manufacturing time of large components, but the van der Kooij system cannot generate a support structure. Instead, the components are placed on differently arranged building platforms, which then serve as a support structure and (in places) are also give shape to the components.

Another robotic FLM system inspired by van der Kooij's work is the 3D Fibre PrinteR. Development began in 2012 at the Fraunhofer Institute for Manufacturing Engineering and Automation in Stuttgart under the direction of Andreas Fischer. The prototype of the filament-based system uses a KUKA KR60 HA (Figure 1.14), where the abbreviation "HA" stands for high accuracy, because of the improved repeat accuracy of the industrial robot used. The payload of the KR 60 industrial robot is 60 kg, which means that the prototype is readily expandable in terms of the printing unit and the number of filament extruders. As implied by the naming of the FLM system, the focus of the first prototype is the integration of continuous fiber according to the principle of FLM (see Section 1.6.1). The printing unit is equipped with three filament-based extruders, whereby the outer ones are intended for fiber integration and the middle one for materials used to generate a support structure. The thermoplastic material for the robotic system is supplied as filament with a diameter of 3 mm from an external climatic cabinet via an encapsulated roller system. The installation area of the 3D Fibre PrinteR is variable and results from the industrial robot used. For the KUKA KR60 HA used, this is approximately 2000 × 1000 × 1000 mm. In the prototype, an open build platform was used, which can take in carrier substrates with maximum dimensions of 2000 × 1000 mm completely or partially via a matrix vacuum plate.

Figure 1.14 Concept of the 3D Fibre PrinteR with three melting heads and double fiber integration (© Andreas Fischer, from [Fischer et al. 2013])

The controller of the 3D Fibre PrinteR is equipped with the additional KUKA.CNC control option and is thus able to process G-code. This CNC core was developed for robot-based milling and thus also contains all the necessary M commands required for milling. In order to extend the range of applications to the specific FLM G-code, the additionally required equipment and standard M commands have been developed and integrated (Figure 1.15). By using this control strategy, the 3D Fibre PrinteR prototype could only be used for three-axis based FLM.

From 2015 to 2017, robotic FLM based on the 3D Fibre PrinteR was further developed at AGOR GmbH's site in Bietigheim-Bissingen. In addition to the generation of multi-material components and the improvement of the printing units with regard to flexible materials such as thermoplastic polyurethane (TPU), the focus of the further developed systems was also on a new control approach that enables independent layer generation. This new approach enables components to be applied to given geometric and free-form workpieces by AM using the axis degrees of freedom of the industrial robot (Figure 1.16). If the robot kinematics itself is not sufficient for the application, it can be extended by integrating additional systems, such as a rotary or tilting table or another industrial robot. Furthermore, the use of axis degrees of freedom makes it possible to minimize necessary support structures, and facilitates the integration of semi-finished products into components.

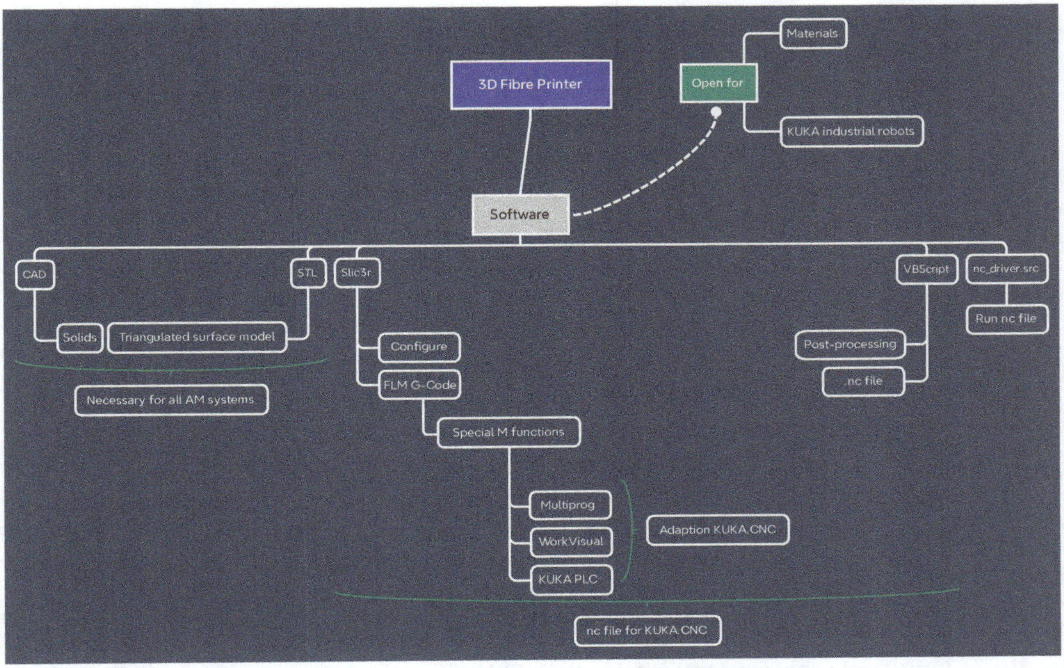

Figure 1.15 Process mapping from CAD to the NC data file of the components to be printed on the 3D Fibre PrinteR (© Andreas Fischer, from [Fischer et al. 2015])

Figure 1.16
Top: Three-axis FLM demonstrator with two printing units on a KUKA KR 6 processing thermoplastic polyurethane (TPU). Bottom: Direct application of TPU to a metal cylinder with a KUKA KR 60 HA (© Andreas Fischer)

Since 2018, the know-how around the 3D Fibre PrinteR has been used by Syncree UG in Blumberg to develop AM applications. The focus was placed on time-saving and cost-reducing AM of large-volume components. Furthermore the laminating process group (see Section 1.2.2) was focused. In addition to approaches to personalizable and lightweight vehicle bodies made of closed-cell rigid foam on a polymethacrylimide (PMI) base, the company mainly focuses on a limited series of design furniture produced from a wide variety of sheet materials.

One way of producing large-volume components using FLM in combination with industrial robots in shorter periods of time without increasing the diameter of the extrusion beads is the cooperative robot approach. AMBOTS Inc., based in Fayetteville, is launching the patent-pending Autonomous Mobile roBOTS and Advanced Manufacturing roBOTS. The robots, which operate in a swarm, can perform cooperative three-axis FLM in addition to traditional assembly tasks. In this process, individual robots each produce sub-segments of the overall component. The 3D printing robots can be autonomously transported to the individual sub-segments to be manufactured by the mobile platforms integrated in the system. The system has been developed since 2015 in the AM3 Lab at the University of Arkansas Department of Mechanical Engineering under the direction of Dr. Wenchao Zhou. The development was supported by the SiSi lab led by Dr. Zhenghui Sha.

Industrial robots in combination with FLM offer a high variation of possible applications. Existing industrial robots can be converted into a highly adaptive FLM system using printing units and adapted software. In particular, three-axis FLM is the simplest variant to implement. However, commercial FLM systems based on industrial robots are rather rare, and many of the systems are tailored solutions or demonstrators. In FDM, the Stratasys company has also developed a demonstrator based on industrial robots in collaboration with Siemens. The eight-axis system Robotic Composite 3D focuses in particular on the AM of fiber-plastic composite components (see Section 1.6.1).

A commercial FLM system based on robots is offered by Variobotic GmbH in Neu-Ulm. The DOBOT M1 is used as the axis system, which is a selective compliance assembly robot arm (SCARA) with four axes. Three of the axes of the horizontal jointed-arm robot are rotary, and the fourth axis is linear. The system has a semi-cylindrical installation space. The repeatability is ±0,02 mm with a maximum reach of 400 mm, and a payload of 1.5 kg. The DOBOT M1 becomes a FLM system with the filament-based maker kit available as an accessory. This allows the use of a cylindrical printing area with a diameter of 80 mm and a height of 130 mm. The software used is the open-source slicing application for the Cura 3D printer, which is now available from the Ultimaker company.

1.5 Functional Principle of Selective Laser Sintering

Selective Laser Sintering (SLS) belongs to the sintering process group and uses the principle of local melting of thermoplastic powder materials. The classic SLS process consists of the following phases:

1. Preheating phase
2. Printing process with build phases
3. Cooling phase.

In the preheating phase, the powder material is heated close to the melting point or to the glass transition temperature of the thermoplastic material, in order to generate optimum process conditions. This is ensured by controllable heating systems above or directly on the powder supply chamber. These systems allow heat to be supplied and dissipated. The actual build chamber is heated in parallel to the necessary process temperature and inverted by the addition of the gaseous inert nitrogen. This serves not only to avoid a powder explosion or material oxidation, but also to reduce negative process influences due to the inhomogeneous preheating of the powder material. Depending on the material, the nitrogen atmosphere generated in the build chamber has a residual oxygen content of 0.1–3.5 %. After the process temperature has been reached and the powder supply has been preheated, the build process begins in the build chamber at a slight positive pressure. The preheating phase in SLS can take several hours, and is therefore considerably longer than in additive extrusion systems. The build phases of SLS are divided into four individual steps:

1. Lowering the build platform
2. Application of the powder layer
3. Preheating of the powder layer
4. Melting of the component layer.

In the process step of lowering the build platform, the platform is lowered by a layer thickness via a lifting mechanism. This layer thickness, consisting of thermoplastic powder, is approximately 60–180 µm. The thermoplastic powder used has particle diameters of 20–100 µm. Depending on the arrangement of the powder supply chamber, the powder is supplied to the coating system either by lifting or in bulk. Depending on the system or manufacturer, the powder supply chamber is located below or above the building platform. During the subsequent application of the powder layer, the thermoplastic powder is applied to the building platform via a coating system to form a level, compacted powder layer. Depending on the system, this is done by a roller or wiper system (Figure 1.17). In the following process step, the powder layer produced by the coating system is preheated. This is achieved by means of an IR heater located above the building platform. Preheating is continued until the desired temperature distribution is as homogeneous as possible. In the final step of the build phase, the current layer representation of the component is melted by means of a laser. For this purpose, most systems use a focused CO_2 laser, which is deflected by an XY scanner system and melts line by line. Depending on the energy input and material properties, the thermoplastic powder in the layer currently being processed is melted by the temperature increase and bonded to the underlying layer.

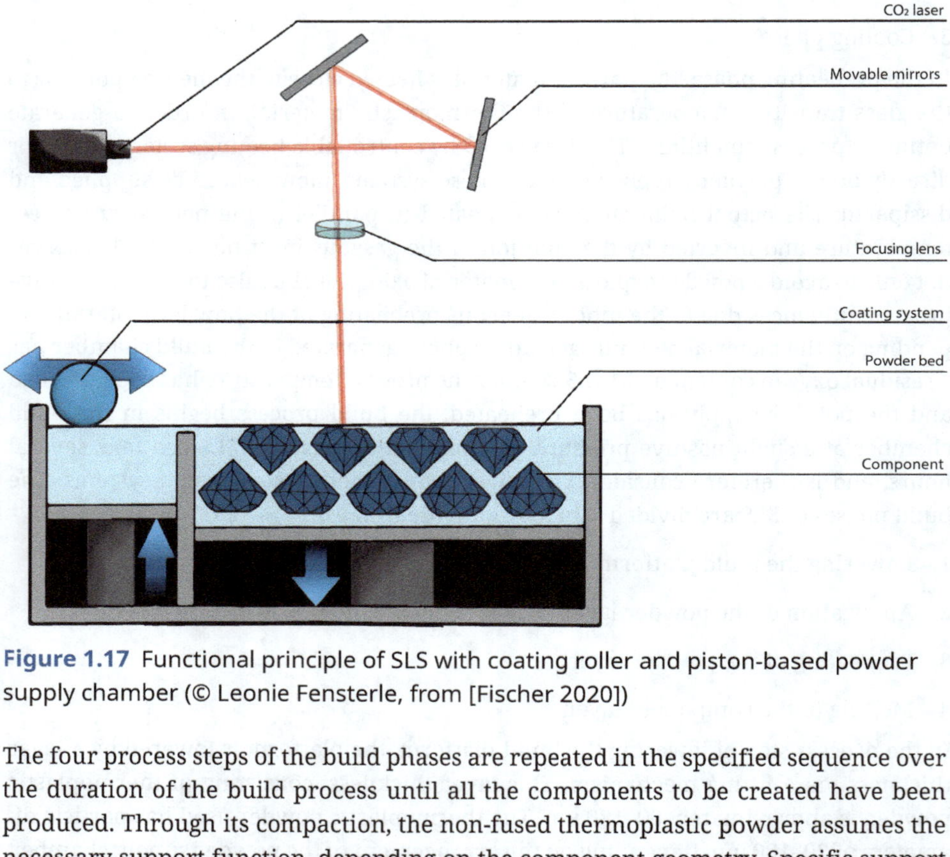

Figure 1.17 Functional principle of SLS with coating roller and piston-based powder supply chamber (© Leonie Fensterle, from [Fischer 2020])

The four process steps of the build phases are repeated in the specified sequence over the duration of the build process until all the components to be created have been produced. Through its compaction, the non-fused thermoplastic powder assumes the necessary support function, depending on the component geometry. Specific support structures, such as those used in FDM (see Section 1.3.2.2) are therefore not necessary. After completion of the build process and generation of the components, several millimeters of additional powder layers are applied via the coating system, which no longer have any fused segments. These powder layers serve as thermal insulation in the upper area of the unprotected powder cake, which consists of components and powder bed. After these final protective layers have been produced, the heating elements of the systems are switched off and the cooling phase is initiated. During this process, the component geometries solidify due to heat conduction.

After the build process is complete, the components are in the powder cake, which has a temperature of between 170 °C and 210 °C, depending on the material used. The generation of a layer in SLS can take between 30 and 40 s. As a result, a build process can take hours or days. The cooling phase also takes several hours. After the cooling phase, the components are freed from the powder bed. This downstream process is also called unpacking and is performed by workers with the use of tools such as brushes and compressed air. The components and the non-fused powder of the powder bed have an identical color after the building process. Therefore, especially in the case of filigree geometries of the components, the removal of powder close to the component must be carried out carefully. In the case of non-filigree geometries, adhering powder can be removed by blasting with glass beads. Due to the high process temperatures, which also affect the material over longer periods of time, aging processes take place in the powder. The accumulating, non-fused powder from the build process can be reused after reprocessing.

This so-called residual powder from the build process can be reintroduced after cleaning processes and mixing with 50–75 % unused (virgin) powder. The problem with this procedure is that there is no traceability with regard to the runs of the powder particles. The inconstant quality of the residual powder, in particular, results in different mixing qualities. These can lead to fluctuating component qualities during processing. In the worst case, the low mixing quality can lead to the formation of an uneven surface ("orange peel") or to rejected components. Components produced by SLS have a comparatively high surface roughness and porosity. If this is not modified by downstream finishing and post-processing, dirt particles can be deposited and liquids can penetrate into the outer layers of the component. The surface roughness results from the stair-stepping effect common in AM and from the use of a relatively coarse material powder.

Components made of semi-crystalline plastics exhibit good mechanical properties. Thus, SLS is also suitable for the generation of functional components and end products. The generation of fine structures and the integration of functions are also very easily achieved. For example, the generation of a pneumatic bellows can be realized as an applicable actuator in SLS. The use of the non-fused powder as a support function and the problem of reusing the residual powder, in combination with the associated costs, lead to highly efficient use of the possible build volume in SLS. The components to be produced are placed in the build volume in such a way that as little residual powder as possible is produced per build process. As a result of this placement strategy, the duration of a build process tends to be in the order of one day. Compared to additive extrusion, the relationship between costs and residual powder means that it is not possible to produce a small component quickly and effectively. The necessary preheating of the systems due to the process and the use of lasers for the melting process leads to high energy consumption in SLS. For this reason, the systems in professional use are restarted immediately after the produced powder cake is removed.

Another problem of SLS is the insufficient reproducibility of the component properties. The reasons for this are specific effects in the manufacturing process and special effects in the melting of the thermoplastic powder by the laser energy input. Compared to injection molding, SLS can be advantageous in terms of economy and energy efficiency, especially for small batches. Depending on the component volume, the limiting number of pieces for small components can be in the four-digit or low five-digit range. For relatively large components, this limit number can drop to several hundred parts. The most widely used material in SLS is polyamide 12 (PA12). Approximately 90 % of all components produced by SLS are made from this thermoplastic. Other materials such as PA6, PA11, polypropylene (PP), polyether ketone (PEK), and thermoplastic polyurethane (TPU) are commercially available. However, these represent a niche product due to their more problematic processing compared to PA12. The limited range of materials compared to additive extrusion thus restricts the area of application of SLS [Wegner 2015].

1.6 Hybrid Additive Manufacturing with Thermoplastics

The term hybrid additive manufacturing covers a variety of approaches in which AM processes are used in combination to produce components. One of these approaches is the production of basic components or semi-finished products in a conventional manner, and the further processing of these using AM. A variation of this approach is the creation of the basic component by means of AM and further processing, for example, by means of conventional subtractive manufacturing methods. However, the creation of casting molds by means of AM can also be classified as hybrid AM.

Very strongly anchored in hybrid AM is the combination with industrial robots (see Section 1.4.2.2). The focus here is on fully automated, flexible and hybrid production lines. In this context, assembly or handling approaches then also play an important role. Another broad approach is the combination of different materials in an additively produced component. In addition to multi-material components (see Section 1.4), these can also be fiber-plastic composite components or components in which the additively generated material combination represents an intermediate step. The background to the more complex approaches in particular is to exploit the advantages of additive processes as well as to reduce the disadvantages and thus increase integration into existing production processes [Landshut 2021].

1.6.1 Fiber-reinforced Additive Components

Composites made of plastic matrix and fibers are used in various products, particularly with a view to saving weight and thus conserving resources. Such composites are used in larger quantities in the aerospace industry, but lighter and stronger materials are also being used more and more frequently in automotive engineering. Composites with fibers require a complex automation solution or manual handwork in addition to molding tools and, in some places, preforming tools for production. This increases the cost of composite components in addition to the actual fibers. The material costs for fiber-reinforced plastic components can be as high as 60 %, with offcuts taking up to 30 % in this case. The matrix material of the composites can consist of both thermoset material as well as thermoplastics. In contrast to a thermoset material matrix, a thermoplastic matrix has the advantage that it can be subsequently formed. However, complete infiltration of the fibers is significantly more complex with viscous thermoplastics than with low-viscosity reactive resins.

Three types of fibers are primarily used to produce the composites. Carbon fibers offer the best mechanical properties at a very low weight. A high level of energy is used to produce the fibers, which both increases costs and reduces its sustainability. The range of applications is primarily in the aerospace, automotive, and racing industries. In contrast to carbon fibers, glass fibers can be produced comparatively cheaply, and offer good mechanical properties at moderate weight. Glass fibers are used in aircraft construction, vehicle construction, shipbuilding, and in the manufacture of rotor blades for wind turbines. Since glass fibers are electrically insulating, they are also used in the manufacture of printed circuit boards. The third type of fiber are aramid fibers, which offer good mechanical properties that lie between carbon fibers and glass fibers. The elongation at break and impact strength are twice as high as glass fibers, and with a further weight reduction. Aramid fibers are mainly used in the production of bulletproof vests or safety ropes.

For the production of fiber-plastic composite components with thermoplastic matrix by means of AM, the process groups sintering and extrusion are particularly suitable (see Section 1.2.1). However, it is also possible to manufacture composite components using the laminating process group. In this process, contoured sheets or mats are arranged on top of each other and bonded together to produce composite components. Commercial systems are available for the sintering and extrusion process groups. The length of the fibers used is assigned to specific terms.

In SLS, various polyamide (PA) powders with fibers are used to produce fiber–plastic composite components. The thermoplastic powders used have particle diameters of 0,02–0.1 mm. The average particle diameter is therefore around 0,06 mm. For processing reasons, the fiber length of the carbon fibers in these special PApowders cannot be significantly longer than the average particle diameter. Thus, the fiber length of these PApowders is below the smallest length value of the short fibers (Table 1.2). Components manufactured with these special PApowders exhibit increased stiffness and strength as well as lower weight. Furthermore, the components are electrically conductive due to their carbon fiber content. However, the mechanical properties differ in the three spatial axes, which is due to the process-related XY alignment of the carbon fibers. The use of PApowders with carbon fibers can partially replace additively produced metal components, but due to the higher costs of these special PApowders, SLS with fibers is more suitable for smaller quantities. SLS system manufacturer 3D Systems calls its carbon fiber-reinforced engineering plastic powder the Dura Form HST Composite. At systems manufacturer EOS, the fiber-reinforced powder is listed as Carbon Mide PA12-CF, which has a modulus of elasticity of 7800 MPa. Compared to components made of PA12, the integration of the short fibers increases tensile strength by 60 %, elastic modulus by 323 %, flexural strength by 108 %, and flexural stiffness by 298 %.

Table 1.2 Designations of the Different Fiber Lengths for Fiber–plastic Composite Components

Designation	Length in millimeters
Short fiber	0.1–1.0
Long fiber	1.0–50.0
Continuous fiber	> 50.0

With classic FDM or FLM printing units, it is not possible to process filaments filled with long fibers. Filaments with short fibers and high fiber content can lead to accumulations of fibers at the nozzle outlet, which cannot be dissolved with the low pressure provided by the printing units, and thus leads to clogging and abortion of the printing process. Two carbon-fiber-reinforced materials are available in FDM. The carbon-fiber-reinforced acrylonitrile-butadiene-styrene (ABS) from system manufacturer Stratasys, called ABS-CF10, has a fiber content of 10 % [Stratasys 2021]. The components produced with it are more insensitive to moisture, 50 % stiffer, and 15 % stronger than components made from Stratasys' standard ABS material. Typical applications of the material in combination with additive extrusion include manufacturing tools, fixtures, and gripping elements, which also benefit from reduced part weight. The ABS-CF10 material can be used with the Stratasys F170, F270, and F370 systems, and can be used in combination with soluble support material.

The second carbon-fiber-reinforced material used in FDM is FDM Nylon 12CF. This is PA12 with a carbon fiber content of around 35 % and a short fiber length of 0,15 mm. The reinforced material has the highest flexural strength of all thermoplastic FDM materials and therefore also has the best stiffness-to-weight ratio. Applications for FDM Nylon 12CF include high-performance tooling solutions as well as end products with low weight and functional prototypes for applications, for example in the aerospace or automotive sectors. This carbon fiber-reinforced PAis available for the Stratasys Fortus 450mc and F900 AM systems and is compatible with the SR-110 soluble support material. The elastic modulus of FDM Nylon 12CF ranges from 7515 MPAto 2300 MPa.

The company Stratasys is also implementing the ability to additively create carbon-fiber-reinforced components in combination with industrial robots. The Robotic Composite 3D Demonstrator presented at the International Manufacturing Technology Show (IMTS) in 2016 was developed together with Siemens, which was primarily responsible for the motion control and CNC automation of the demonstrator (Figure 1.18). The main components of the system are a KUKA industrial robot in combination with a rotary tilt table. Thus, the Stratasys demonstrator has an eight-axis motion system, which also makes it possible to dispense with support structures [Stratasys 2016]. The system can be classified as hybrid AM in two respects (see Section 1.6). In addition to fiber integration, the demonstrator is able to apply directly to conventionally produced components, for example to apply lightweight and stable reinforcements. The aim of the development is to better integrate additive extrusion into industrial production lines.

Figure 1.18
The Robotic Composite 3D Demonstrator from Stratasys and Siemens applying reinforcements in multiple axes (© Katharina Juschkat, Vogel Communications Group GmbH & Co. KG)

In FLM, solutions are available for short fiber as well as continuous fiber. Numerous variations are available for short-fiber-reinforced filaments, which can be processed on a wide variety of FLM systems. However, processing these filaments is more difficult compared to classic materials and requires specific configurations of the FLM systems, depending on the material combination. In addition to hardened nozzles for extrusion and thus counteracting abrasive wear, at least heated enclosures to counteract material distortion are also recommended (see Section 1.4.2.1). For the thermoplastic matrix of the reinforced materials, acrylonitrile-butadiene-styrene (ABS), polyamide (PA), polypropylene (PP), polyethylene terephthalate (PET), and polyether ether ketone (PEEK) are primarily used. Fiber reinforcement in FLM ranges from carbon fiber to glass fiber to aramid. The fiber content of short-fiber-reinforced filaments ranges from about 10–30 %. In principle, the use of short-fiber-reinforced filament achieves significantly better mechanical properties along the discharged material than across it. This is due to the fact that the fibers are oriented along the direction of flow when they are discharged from the die in the plastic stream. As a result, the tensile properties in the tensile stress direction can be greatly improved. However, only slight or no improvement can be produced transverse to the tensile stress direction.

Components reinforced by continuous fibers and produced additively can currently only be manufactured by means of additive extrusion. The continuous fiber can be integrated into the thermoplastic matrix in three different ways:

- Before processing
- Inside the nozzle
- On the building platform.

In the first and simplest variant of integration, the continuous fiber is impregnated by the matrix material in the separated and upstream step of filament production. In this continuous fiber-reinforced material, the fiber lies within the FLM filament. The advantage of this variant is that additive components reinforced by continuous fiber can be produced without major adjustments to the FLM system. However, the first variant has disadvantages in terms of material flexibility compared to the second. The very high reinforcement effect can be as high as 12.73 %. In the second variant, the continuous fiber is impregnated by the thermoplastic matrix inside a special FLM nozzle. In this process, the continuous fiber is fed to the nozzle in a separated manner, comes into contact with the plasticized matrix material within the nozzle, and is generally conveyed through the nozzle by the flow of the liquid matrix. In the second variant, there is a complexity with regard to handling the pure continuous fiber, as well as the parallel control of the additive process and the impregnation of the fiber. The reinforcing effect and the fiber volume fraction are lower compared to the first variant. The third variant of integrating continuous fibers into additively produced components splits the output of matrix material and fiber. The necessary impregnation of the continuous fiber with matrix material takes place on the building platform, whereby the continuous fiber is first set and then impregnated with matrix material.

In addition to the two thermoplastics polylactic acid (polylactide, PLA) and polyamide (PA), the Mark One FLM system from the company MarkForged can also process reinforced filament with continuous fibers. Similar to short fibers, carbon fibers, glass fibers and aramid are used here. PAis used as the matrix material for the filament reinforced with continuous fibers. The system for processing the filament reinforced with continuous fibers is referred to by the manufacturer as composite filament fabrication (CFF). The z-resolution of the purely thermoplastic filament is 0.1 mm and that of the fiber-reinforced filament is 0.2 mm. The Mark One is equipped with a classic FLM printing unit and a specific one for continuous filament processing. This specific printing unit is equipped with a separating device to separate the continuous fiber from the component after the build process. The system's build chamber is 305 × 160 × 160 mm, and the diameter of the thermoplastic filament is 1,75 mm. The system can process two materials in parallel. Complex geometries can only be produced to a limited extent, as no detachable support material is provided. The components that can be produced on the Mark One consist of unreinforced PAsegments and segments reinforced by continuous fibers. The segments reinforced by continuous fibers are sheathed by the unreinforced PAsegments to give the part a smoother surface. To additionally reduce the component weight, hollow structures (for example in honeycomb form) can be created inside the component.

Another approach to additively manufacture components reinforced by continuous fibers is to introduce the fiber into the melt of the thermoplastics through a specially designed FLM nozzle. Under the name 3D Fibre PrinteR, this strategy was developed in 2012 at the Fraunhofer Institute for Manufacturing Engineering and Automation (IPA) in Stuttgart (see also Section 1.4.2.2). Similar to the Mark One, the specific continuous fiber printing unit can be combined with classic FLM printing units due to its modularity, for example to create necessary support structures. The fiber must be introduced into the special nozzle of the continuous fiber system in the correct ratio to the thermoplastic to ensure complete encapsulation of the fiber. This sheathing is very important to create the best possible adhesion between the fiber and the thermoplastic. The thermoplastic for the matrix is fed to the 3D Fibre PrinteR as a filament with a diameter of 3 mm. It is fed to the nozzle by a classic feed device. The thermoplastic filament is melted in the special continuous fiber nozzle. The fiber is fed to the side of the melt in the lower section of the nozzle. The fiber is conveyed here by the continuous movement of the melt, which makes it easy to switch the fiber integration on and off. To take advantage of this effect, a cutting device is used to separate the continuous fiber, similar to the Mark One. The specific nozzle of the 3D Fibre PrinteR has a diameter of 1 mm and can be operated with a maximum of 1K (67 tex) fiber roving consisting of 1000 individual filaments with a diameter of 7 µm. The fiber volume fraction of the components produced is around 10 %. In addition to carbon fiber, glass fiber and aramid can also be processed with this system. The fiber is coated by the thermoplastic melt and conveyed through the nozzle: otherwise the function of the casing and the conveyance through the thermoplastic melt cannot be guaranteed.

With the thermoplastic-based, additive ARBURG plastic freeforming process (AKF, see Section 1.4.1), continuous fibers can also be integrated into AM parts. In order to generate greater stability of the parts, the continuous fiber is applied simultaneously to the droplet-applied thermoplastic during the production of the parts. The continuous fiber embeds itself in the thermoplastic material and bonds with it. Furthermore, the tensioning of the continuous fiber eliminates the need for support structures The fiber feed unit is separate from the thermoplastic printing unit, which has the advantage that the fiber structures can be better influenced and thus, for example, the fiber strands can be laid in a different direction from layer to layer. In this system, the necessary separation of the continuous fiber is not achieved by a cutting device. Rather, the continuous fiber is moved against the discharge direction as well as bent around the edge of the cannula and broken in the process. This type of separation is only successful if the continuous fiber is fixed both within the fiber feed device and in the component.

The mechanical properties of additively produced components can be improved a little by integrating short fibers and significantly with continuous fibers. The material selection is much higher with short fibers than with continuous fibers. Furthermore, the opportunities for material selection is much more restricted with reinforced materials than with unreinforced ones. In addition, different fiber volumes in the component cannot be achieved with SLS and FDM, but this is possible to a degree with FLM using multi-material systematics. Furthermore, the data preparation systems of the various methods are not able to take into account the position and orientation of the fibers in a way that is appropriate for the load. As a result, the efficiency of the reinforcement decreases. In addition, the geometric complexity of the components is limited by minimum bending radii, especially in additive continuous fiber integration.

1.6.2 Atomic Diffusion Additive Manufacturing

Atomic diffusion additive manufacturing (ADAM) can create metal components through hybrid additive manufacturing. The Metal X was presented by Markforged at the Consumer Electronics Show (CES) in 2017. In ADAM, as in additive extrusion, components are created by the principle of spraying molten polymers through a nozzle (see Section 1.2.1). The materials used are a bound-metal powder filament, namely a thermoplastic polypropylene (PP) filament with metal particles, and the ceramic release filament with ceramic particles. The filament with metal particles is also used to create the necessary support structure. In order to be able to detach this support structure from the component more easily later on, the filament with the ceramic particles is applied in thin layers between the component and the support structure. The support structures are removed from the component after the sintering process. In ADAM, the metal part generation is performed in three steps on three different devices (Figure 1.19):

- Additive extrusion (Metal X 3D printer)
- Washing
- Sintering.

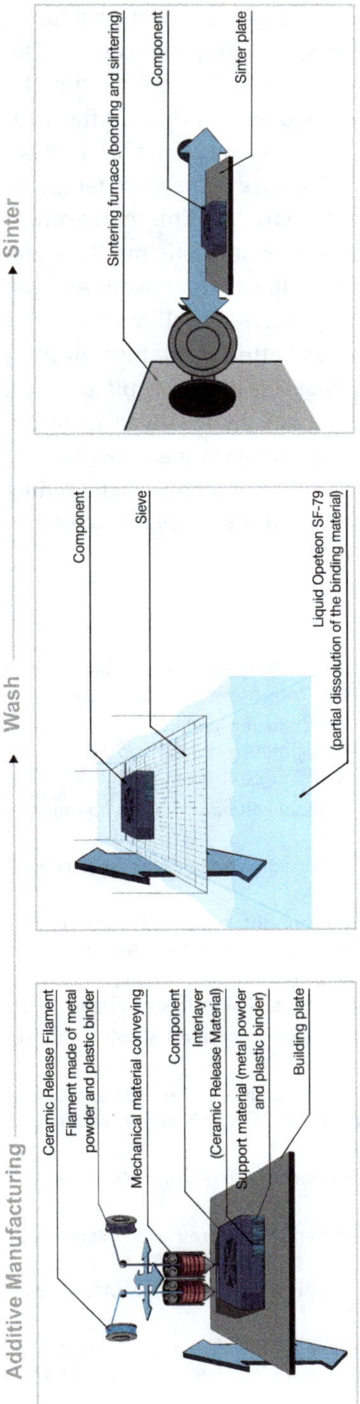

Figure 1.19
The three process steps of ADAM to the metal component (© Chao Hui Lin)

After the additive generation of the component and the necessary support structures, both are subjected to a washing process. In this process, a large proportion of the thermoplastic binder material is dissolved using the Opteon SF-79 liquid. After the washing process, the component including the support structure is dried. After drying, the component is positioned on a sintering plate. The sintering plate with component and support structure is then heated in a sintering furnace from room temperature to approximately 85 % of the melting point of the metal particles. This temperature increase in the sintering process dissolves the remaining thermoplastic binder material. At up to approx. 1300 °C, the ceramic release material turns into powder and the metal particles become a solid component. The sintering process also allows the metal particles to bond through the usual AM layers, resulting in better mechanical properties of the additive metal components. After a cooling phase, the sintering plate can be removed from the sintering furnace and the component can be removed from the support structure by lightly tapping it. In addition to the comparatively low system costs and thus low manufacturing costs, especially for complex geometries, the reduction of toxicity and flammability risks due to the integration of the metal particles into the thermoplastic filament is another advantage of ADAM.

References

3Druck 2016: Die HP Multi Jet Fusion 3D Drucker sind da! 3Druck.com, 17 May 2016. *https://3druck.com/drucker-und-produkte/die-hp-multijet-fusion-3d-drucker-sind-da-3244851/* (retrieved 27 May 2024)

3Druck 2017: Markforged introduces Metal X 3D printer with Atomic Diffusion additive manufacturing technology. 3Druck.com, 9 January 2017. *https://3druck.com/drucker-und-produkte/markforgedstellt-metal-x-3d-drucker-mit-atomic-diffusion-additive-manufacturing-technologie-vor-1653202* (retrieved 15 February 2021)

3Faktur 2020: HP Multi Jet Fusion. *https://3faktur.com/3d-druck-materialien-und-verfahren/hp-multi-jet-fusion-das-verfahren* (retrieved 26 August 2020)

AMBOTS 2021: AMBOTS – Swarm 3D Printing for Autonomous Manufacturing. *https://ambots.net* (retrieved 18 August 2021)

ARBURG 2021: ARBURG freeformer. Additive manufacturing – Free thought. *https://www.arburg.com/fileadmin/redaktion/mediathek/prospekte/arburg_freeformer_680835_de.pdf* (retrieved 29 March 2021)

Balzereit 2020: S. Balzereit, Structures and properties of selectively laser-sintered polyamide and polyamide-copper systems for applications in three-dimensional circuit carriers. University of Bayreuth, 2020

Baumann 2020: F. Baumann, Additive manufacturing of continuous fiber reinforced plastics with the ARBURG plastic free-form process. Shaker Verlag, 2020

Campbell & Wohlers 2017: I. Campbell & T. Wohlers, Markforged: Taking a different approach to metal Additive Manufacturing. Metal AM, vol. 3, 2017 *https://www.metal-am.com/articles/markforged-taking-a-different-approach-to-metal-3d-printing* (retrieved 15 February 2021)

Cincinnati 2021: Additive Solutions. Fact Sheet. *https://wwwassets.e-ci.com/PDF/Products/Additive-Fact-Sheet.pdf* (retrieved 26 March 2021)

Cirtes 2020: Stratoconception®. Process. *https://www.stratoconception.com/procede* (retrieved 7 September 2020)

Fischer 2015: A. Fischer, Thermoplastics: 3D Printing for Production. In: Kunststoffe, 11/2015. Special: Rapid Manufacturing. Carl Hanser Verlag, 2015

Fischer 2020: A. Fischer, 3D-Druck & Co. – Potenziale der additiven Fertigung. In: Lang, Michael & Müller, Michaela: From Augmented Reality to AI. The most important IT topics you need to know for your business. pp. 203–229. Carl Hanser Verlag, 2020

Fischer et al. 2013: A. Fischer, S. Rommel & T. Bauernhansl, New Fiber Matrix Process with 3D Fiber Printer. A Strategic In-process Integration of Endless Fibers Using FDM (FDM). In: International Conference on Digital Product and Process Development Systems (IFIP), Springer Verlag, 2013, pp. 167–175

Fischer et al. 2015: A. Fischer, S. Rommel and T. Bauernhansl, 3D Printed Objects and Components Enabling Next Generation of True Soft Robotics. In: Verl, Alexander et al.: Soft Robotics. Transferring Theory to Application. Springer Verlag, 2015, pp. 198–208

Fischer et al. 2018: A. Fischer, S. Gebauer & E. Khavkin, 3D Printing in the Enterprise: Decision models, best practices and application examples. Taking Fused Layer Modeling (FLM) as an example. Carl Hanser Verlag, 2018

Funding Universe 2020: Stratasys, Inc. history. *http://www.fundinguniverse.com/company-histories/stratasys-inc-history* (retrieved 19 August 2020)

Galati & Minetola 2019: M. Galati & P. Minetola, Analysis of Density, Roughness, and Accuracy of the Atomic Diffusion Additive Manufacturing (ADAM) Process for Metal Parts. In: Materials 2019, 12(24), 4122. *https://www.ncbi.nlm.nih.gov/pmc/articles/PMC6947362* (retrieved 15 February 2021)

Gebhardt 2000: A. Gebhardt, Rapid Prototyping. Tools for rapid product development (2nd edn), Carl Hanser Verlag, 2000

Gebhardt 2016: A. Gebhardt, Additive Manufacturing. Additive manufacturing and 3D printing for prototyping – tooling – production (5th edn), Carl Hanser Verlag, 2016

Gebhardt et al. 2016: A. Gebhardt, J. Kessler & L. Thurn, 3D Printing. Fundamentals and applications of additive manufacturing (AM) (2nd edn), Carl Hanser Verlag, 2016

Gibson et al. 2015: I. Gibson, D. Rosen & B. Stucker, Additive Manufacturing Technologies. 3D Printing, Rapid Prototyping, and Direct Digital Manufacturing, Springer Verlag, 2015

Jäger & Hauke 2010: H. Jäger & T. Hauke, Carbon fibers and their composites. Manufacturing processes, applications and market development. The Library of Technology, Volume 326. Verlag Moderne Industrie, 2010

Landshut 2021: Hybrid additive manufacturing creates new potential for industrial 3D printing. Landshut University of Applied Sciences, 19 July 2021. *https://www.haw-landshut.de/kooperationen/transfer-und-zusammenarbeit/news/detailansicht/article/hybride-additive-fertigung-schafft-neues-potenzial-fuer-den-industriellen-3d-druck.html* (retrieved 9 July 2021)

Mark Allen Engineering 2021: Dynamic additive platform prints using movable supports. The Engineer, 6 April, 2021. *https://www.theengineer.co.uk/dynamic-additive-platform-prints-using-movable-supports* (retrieved 14 April 2021)

Markforged 2021a: A Revolutionary New Way to Manufacture Metal Parts. *https://markforged.com/resources/blog/adam* (retrieved 15 February 2021)

Markforged 2021b: Metal-X system. *https://markforged.com/3d-printers/metal-x* (retrieved 15 February 2021)

Mittweida 2020: Laser Microsintering. Mittweida University of Applied Sciences, Laser Institute *http://www.laser.hs-mittweida.de/forschung-entwicklung/materialbearbeitung-beschichtung/additive-fertigung/lasermikrosintern.html* (retrieved 18 August 2020)

New Media Publisher 2019: Covestro: 3D printing by film extrusion – More efficient production than with filaments. Plasticker, 18 November 2019. *https://plasticker.de/news/shownews.php?nr=36332* (retrieved 18 March 2021)

Peters 2018: S. Peters, Additive Manufacturing. The path to customized production. Hessen Trade & Invest GmbH, 2018

PROTIQ 2020: Selective laser sintering: History of selective laser sintering. PROTIQ GmbH *https://www.selektiveslasersintern.de/historie-selektives-lasersintern* (retrieved 19 August 2020)

Radi & El Hami 2016: B. Radi & A. El Hami, Material Forming Processes. Simulation, drawing, hydroforming and additive manufacturing. Wiley/ISTE, 2016

Schmid 2015: M. Schmidt, Additive Manufacturing with Selective Laser Sintering (SLS). Process and material overview. Springer Verlag, 2015

Sehrt 2016: J. T. Sehrt, Lecture Additive Manufacturing Processes. Classification and process flow. University of Duisburg-Essen, 2016

Stratasys 2021: Stratasys ABS-CF10: ABS-CF10. Material data sheet. *https://www.alphacam.de/fileadmin/user_upload/products/fdm-materialien/fdm-datenblaetter/Materialdatenblatt-ABS-CF10-2021-02-11.pdf* (retrieved 13 September 2021)

Stratasys 2016: Stratasys Infinite-Build 3D Demonstrator. 3D demonstrators designed for bigger, lighter auto and aerospace parts. *https://www.stratasys.com/de/explore/blog/2016/infinite-build-robotic-composite-3d-demonstrator* (retrieved 14 September 2021)

Stratasys 2018: Stratasys FDM Nylon 12CF. Material data sheet. https://*www.alphacam.de/fileadmin/user_upload/products/fdm-materialien/fdm-datenblaetter/Materialdatenblatt-Nylon12CF-DE-2018-10-25.pdf* (retrieved 13 September 2021)

TechnikNeuheiten 2016: Largest 3D pen sculpture ever – Nissan Qashqai traced life-size with 3Doodler. TechnikNeuheiten.com, 13 September 2016. *https://www.technikneuheiten.com/groesste-3d-stift-skulptur-aller-zeiten-nissan-qashqai-mit-3doodler-lebensgross-nachgezeichnet* (retrieved 4 March 2021)

Variobotic 2021: DOBOT M1 – Multiflexible industrial robot. *https://variobotic.de/industrieloesungen/dobot-m1* (retrieved 20 August 2021)

Voxeljet 2022: HP Multi Jet Fusion and Voxeljet High Speed Sintering in comparison. *https://www.voxeljet.de/case-studies/konsumgueter/hp-multi-jet-fusion-und-voxeljet-high-speed-sintering-im-vergleich* (retrieved 11 January 2022)

Wegner 2015: A. Wegner, Theory on the continuation of melting processes as a basic requirement for robust process control in laser sintering of thermoplastics. University of Duisburg-Essen, 2015

Zäh 2014: M. F. Zäh, Economic Manufacturing with Rapid Technologies. User guide for the selection of suitable processes. Carl Hanser Verlag, 2014

2 Polymers in Additive Manufacturing

Polymers are chemical compounds that are built up (polymerized) from a chain of molecules. They are composed of identical or similar units, known as monomers [Bonnet 2009]. Synthetically produced polymers, colloquially also often referred to as plastics, play a major role in additive manufacturing (AM, also known as 3D printing), and in fact plastics represent the largest and fastest growing market for 3D printing materials in terms of volume consumed (Figure 2.1).

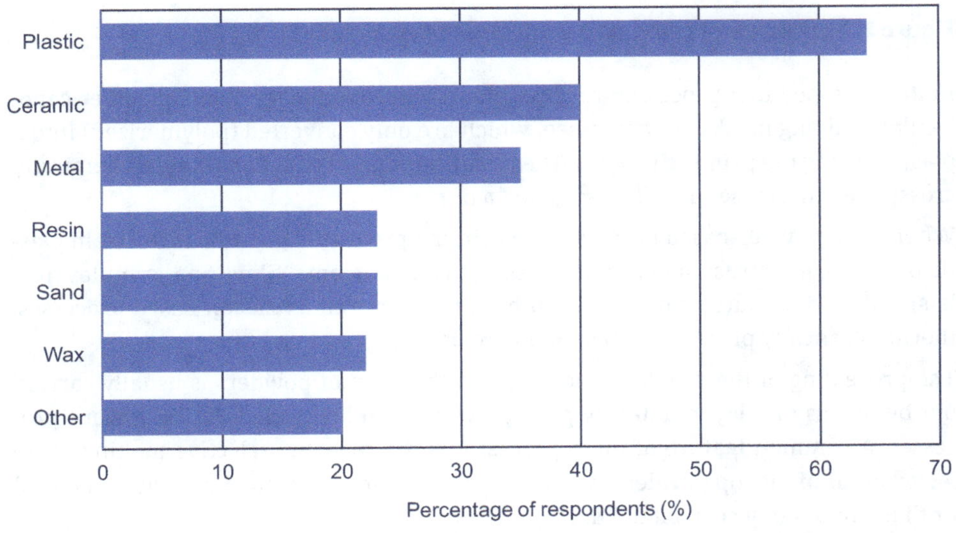

Figure 2.1 Survey of the most-used materials in 3D printing in 2020 [Statista 2021]

The term "plastics" includes all polymers that can be processed as thermoplastics (via melting), while the term "resins" include all monomers that are processed via light-induced polymerization into polymers. The widespread use of polymers in a variety of technical and non-technical applications, as well as their ease of processing, has given rise to a number of different additive manufacturing methods. Commercially used additive process technologies are divided into photopolymer processes (resins), powder-based processes (powders), and "material extrusion" (MEX) (filaments, films, or pellets) (Figure 2.2).

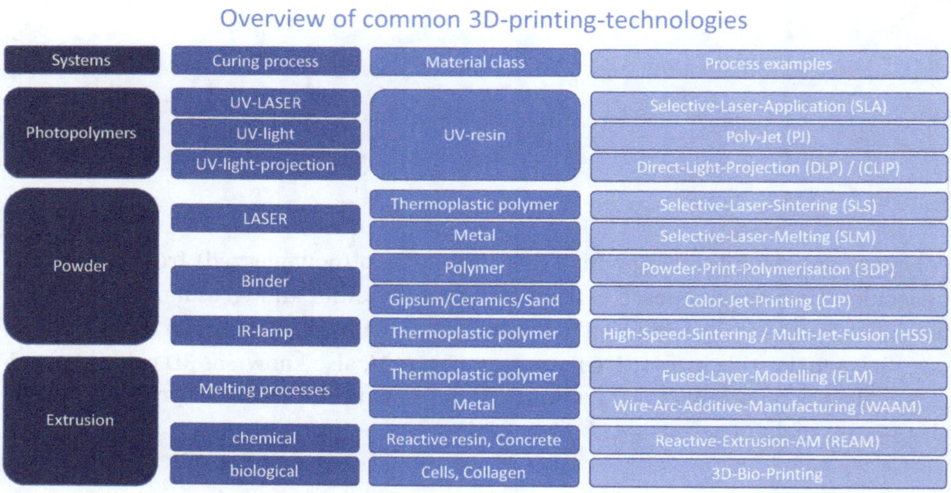

Figure 2.2 Overview of common 3D printing technologies

In the photopolymer processes, reactive monomeric and longer-chain oligomeric molecular building blocks are processed, which are only converted (polymerized) into a plastic during the printing process. The resulting plastic is a non-meltable, chemically crosslinked thermoset plastic (also called a duromer).

Whereas in powder-based processes the printing product is manufactured from plastic powders, in extrusion-based processes, plastic filaments, films, and granules (pellets) are used as starting materials. In both powder- and extrusion-based processes, thermoplastically processable polymers are used.

The processing of thermoplastic polymers in the form of powders is usually carried out by means of a layer sintering process, which can be carried out by various processes. A common feature of these processes is the repeated selective melting of selected areas of the top powder layer, which is repeatedly refreshed with new material until the target object is obtained.

When designing complex components, such as hollow bodies, that are to be manufactured using powder printing, outlets must always be included so that unused powder can flow off. A particular advantage of powder printing is that overhangs do not have to be supported separately: the non-sintered powder ideally serves as a recyclable printing or support material. After printing, the sintered objects are freed from non-sintered powder. Common methods of introducing the necessary energy for melting the powder layer are the use of lasers and IR lamps.

In addition to the powder-based processes for processing thermoplastic polymers in 3D printing, there are also extrusion-based processes. In the DIY sector, most widespread is the feeding of the thermoplastic polymers to the heating and extruder unit – the so-called "hotend" – by means of filament strands. The transport to the hotend can be achieved either by a direct drive extruder mounted above or by a spatially offset conveying unit, the so-called Bowden extruder [Dragon 2020]. In the professional sector and for printing large components, the material feed is also solved via pellets in granulate or pellet extruders [Alexandre et al. 2020] and films in a film or ribbon extruder [Pockat et al. 2010].

In the extrusion-based processes – fused layer modeling (FLM)/fused filament fabrication (FFF) or fused deposition modeling (FDM) – the build-up takes place in layers by selectively depositing the polymer melt emerging from the hotend in the form of a polymer bead or as touching drops, known as "freeforming", onto a build platform [ARBURG 2021; Kloke 2018]. Overhangs >45° must be stabilized with mechanically or chemically removable support structures.

The term FDM refers to extrusion-based processes developed by the Stratasys company, and protected under this name, but the terms FLM, FFF, FDM are effectively equivalent. FLM and FDM refer to all extrusion-based processes based on thermoplastics, and the differences between these processes are largely negligible. In the following discussion, the umbrella term "MEX process" is used for all extrusion-based processes, and the umbrella term "FLM process" for filament-based processes.

In addition to the processing of "finished" thermoplastic polymers in the melt, polymers can also be produced in situ in special 3D printing processes through light-induced photopolymerization of liquid reactive resins. These processes start from liquid monomeric or oligomeric resins. Monomers as the precursors of polymers are applied in complete layers, analogous to powder sintering processes, and are only selectively polymerized or cured in one plane in selected areas. Typical methods like direct light processing (DLP) or stereolithography (SLA) produce highly crosslinked polymer objects, so-called thermosets (also called duromers) or elastomers [Formlabs 2020].

The ease and versatility of processing plastics is the main reason for their popularity and prevalence as a build material in various additive manufacturing technologies. In a 2018 survey, 46 % of respondents said they used FLM/FFF, which is an extrusion-based printing process (Figure 2.3). After FLM/FFF, the processing of thermoplastic powders by selective laser sintering (SLS) is the most widely used 3D printing technology. Thus, among the three most widely used 3D printing technologies, two are based on the thermal processing of thermoplastic polymers via the melt state, and one is based on the generation of chemically crosslinked polymers (thermosets or elastomers) in the SLA process by selective photopolymerization of reactive resins.

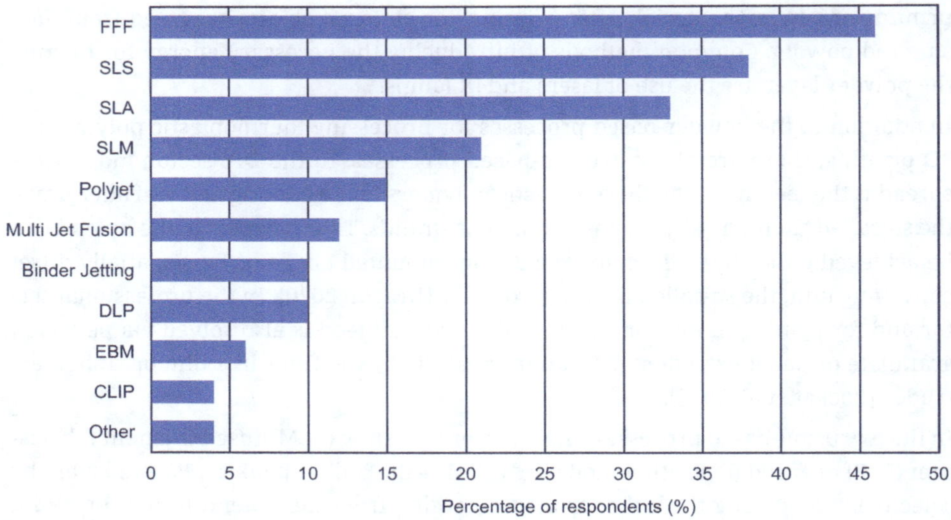

Figure 2.3 Survey of the most widely used 3D printing technologies in 2018 [Statista 2018]

Thermosets (duromers) and elastomers differ mainly in the glass transition temperature and the degree of their crosslinking. The glass transition temperature describes the temperatures below which molecular chains in a polymer freeze and lose their mobility, or above which mobility is regained. The material is below the glass transition point in a "solidified" glassy state.

Duromers are tightly crosslinked materials with a high glass transition temperature significantly above service temperature, while elastomers tend to be wide-meshed crosslinked materials with a glass transition temperature significantly below service temperature. Due to their chemical crosslinking, neither can be melted and thus they cannot be thermally deformed after their manufacture. Thermosets with a glass transition temperature above the service temperature are typically hard and rather brittle, while elastomers with a glass transition temperature below the service temperature can be tough and elastic.

Thermoplastics, which are easily processed by powder sintering and extrusion in the melt, currently form the basis for the largest proportion of additively manufactured products in the industry. They combine the advantages of a wide range of well-known and readily available materials with simple and cost-effective processing.

The expiry of important basic 3D printing patents has led to very rapid growth in 3D printing applications since 2010. This has significantly increased the number of production devices (3D printers) active in the market [Wohlers 2020]. Almost daily, new applications for 3D printing are published and improved printer generations in terms of quality, reliability, and throughput are introduced.

2.1 History of Artificial Polymers (Plastics)

The term "plastics" originally referred to artificially produced materials that were used as substitutes for natural products such as horn, ivory, plant resins, or natural rubber. Particularly between the First and Second World Wars, plastics were specifically developed as alternatives for scarce conventional materials and building materials, such as rubber, silk, wood, and many more [Braun 2017].

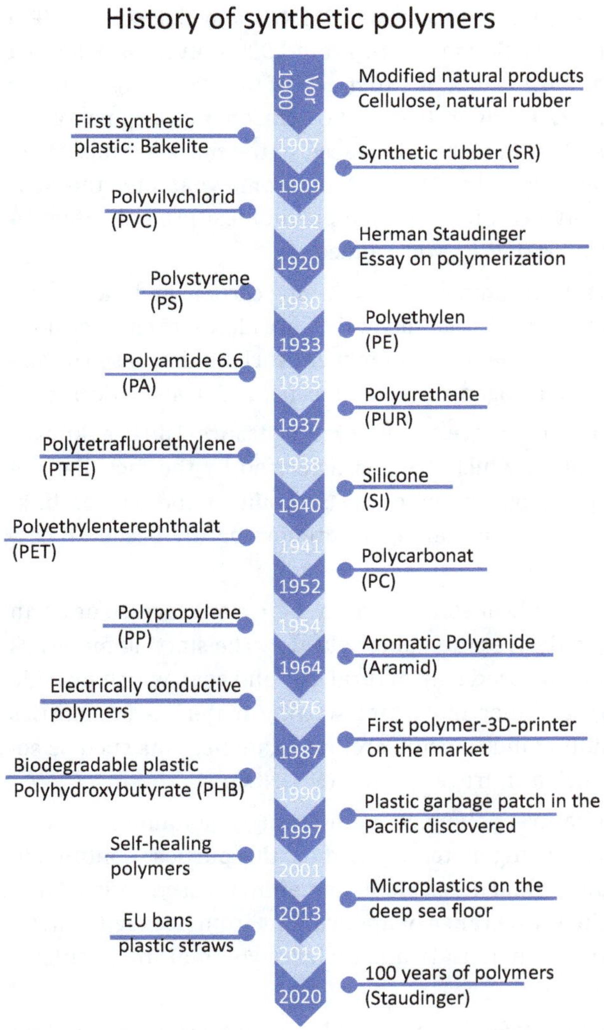

Figure 2.4 Timeline of important development steps in the history of development of plastics [GDCh 2020]

The first plastics produced on an industrial scale were based on polymeric natural products that were chemically modified (Figure 2.4). Prominent examples are vulcanized rubber after the patent of the Goodyear company in 1839 [Goodyear 1844] and products made from chemically treated cellulose, for example nitrocellulose in 1845, Celluloid® in 1871 and artificial silk in 1880 [Endres & Siebert-Raths 2009]. In 1872, the German chemist A. von Baeyer first succeeded in polymerizing small molecules like phenol and formaldehyde to produce a resin-like plastic. In 1907, the Dutch chemist Baekeland produced Bakelite, the first high-performance, fully synthetic phenolic resin, named after him [Kossmehl 2010].

After the First World War, polymethyl methacrylate (PMMA) and polystyrene (PS) were added to the growing list of plastic materials. From 1930 onwards, polyvinyl chloride (PVC), polyvinyl acetate (PVA) and, in the first half of the 1940s, high-density polyethylene (HDPE), polyamides (PA), polyurethanes (PU), silicones (SI), epoxy resins (EP), and polytetrafluoroethylene (PTFE) were introduced to the market. Since then, more engineering plastics such as polycarbonates (PC) and, somewhat later, the class of aromatic high-performance polymers, for example polyimides (PI), polyetherimides (PEI), and polyaryl ether ketones (PAEK), were added.

Plastics are mainly polymeric chemical compounds containing carbon, and they therefore belong to the family of "organic" chemicals. Today, they are almost exclusively produced synthetically, and in addition to the main component carbon, they also contain significant proportions of hydrogen, nitrogen, oxygen, fluorine, sulfur, and chlorine.

In addition to these organic plastics, there are also inorganic plastics. The best-known representatives are the polysilicones, which are characterized by the fact that the main chain of the polymer (polymer backbone) consists of silicon and oxygen linkages $(Si–O)_x$. However, the side chains are then again predominantly based on carbon.

Most plastics are plastically processable at some point in their manufacture or as an intermediate product, from which derives the name "plastic". The starting materials for plastics production are still mainly crude oil, natural gas, and coal, which provide the carbon, hydrogen, and oxygen. Other important starting materials for plastics containing nitrogen, chlorine, sulfur, and fluorine are, for example, salts such as sodium chloride (NaCl) and gases such as nitrogen oxides (NO, NO_x).

In addition to plastics, modified natural polymers can nowadays be found in a number of technical applications. The starting materials used for this purpose continue to be cellulose, proteins, natural rubber, and other plant- and animal-based resins. Modified natural polymers are used in a wide range of applications from packaging materials to adhesives, coatings, composite materials, and (in large volume) still in rubber tires in the transport sector.

There is currently great interest in developing renewable raw materials as a carbon source instead of crude oil, natural gas, or coal for the production of plastics. However, in order to be able to use these alternative raw materials to produce suitable starting materials for plastic production, major chemical or biological refining and processing steps are typically necessary. Subsequently, they can be easily integrated into existing production plants. The final properties of the polymers produced in this way are identical to the respective originals synthesized from fossil raw materials. Such products are also called drop-in products. Examples of these drop-in products based on renewable raw material sources are “bio-PET” (polyethylene terephthalate), “bio-PE” (polyethylene), “bio-PP” (polypropylene) and “bio-PVC” (polyvinyl chloride).

Another area of research is the development of novel biopolymers, based on starting materials for polymerization that are not readily available via the classic refinery processes fed by crude oil and natural gas. These novel biopolymers can provide new and specific properties that, in the best scenarios, open up new areas of application for plastics. Developing new biopolymers involves high R&D investment, and it can take years before the desired material can be technically tested, approved, and produced cost-effectively according to industrial requirements. These new “biopolymers” include, for example, polylactides (PLA), polyhydroxyalkanoates (PHA), and polyethylene furanoate (PEF).

The physical properties of all plastics depend not only on their elemental composition (content of carbon, hydrogen, oxygen, nitrogen, etc.) but also on the way in which these are linked, the length, structure, and arrangement of individual polymer chains, and the interaction of the polymer chains with one another. Accordingly, plastic materials are divided into so-called crosslinked polymers (elastomers and duromers) and non-crosslinked meltable (thermally plasticizable) polymers (thermoplastics).

One significant advantage of plastics compared to other materials is their high strength at low material density and thus their low mass for a given functional part. This explains their widespread use in vehicle construction and aviation. In addition, their typically low water absorption and often good resistance to weathering, as well as their good properties as electrical and thermal insulators, make them important building materials in the construction industry.

Many plastics are not or hardly attacked by bacteria, fungi, acids, alkalis, and solvents. Others are impermeable to air or absorb UV radiation. They can be very well combined in layers and their properties modified by blending, and they are therefore particularly desired as high-performance transport and packaging materials in applications in the chemical and food industries. They are also extensively used for many consumer goods, and not least as textiles.

Pure plastics are – contrary to popular belief – largely inert to interactions with the environment and are non-toxic as the material itself. The extreme diversity of available properties, the ease of processing and shaping, and the fact that most plastics can be produced very inexpensively explain their widespread use in our everyday lives.

A deep understanding of the fundamental product and process properties of plastics (polymers) is an essential prerequisite for developing new applications for plastics in AM. Only then can the associated AM processes and the products obtainable from them be meaningfully optimized in terms of process costs and product properties.

The following sections therefore systematically deal with the basic classification principles of polymeric materials (plastics) with special attention to the class of thermoplastics. For this purpose, manufacturing processes, materials, and processing properties, as well as typical areas of application of plastic materials, are repeatedly related to the resulting process and product properties in AM.

Depending on the number, type, and composition of the basic building blocks (so-called monomers) of polymers (plastics), they are classified into homopolymers and copolymers, and into organic and inorganic polymers, which are further described in Section 2.1.1–Section 2.1.3.

2.1.1 Homopolymers

A homopolymer consists of a long chain of connected monomers of one type. This also includes the largest and most cost-effective polymer class in total, that of standard plastics, with an annual production volume of >300 Mt [Statista 2021]. Important representatives are polyethylene (PE), polypropylene (PP), polyvinyl chloride (PVC), polystyrene (PS), polyamide 6 (PA6), and the poly(meth)acrylates (PAC).

2.1.2 Copolymers

Copolymers are composed of at least two types of monomers, and they are often used as engineering plastics. Important representatives of this class are polyesters (PES), polyurethanes (PU), polycarbonates (PC), and also many polyamides (PA6.6, PA4.6, PA4.10, etc.). Copolymers are characterized by a wide variety of possible properties due to the variability of the monomer composition. According to this classification, copolymers also include many biopolymers such as proteins (= copolyamides) or polynucleic acids (DNA, RNA = polyphosphoric acid esters).

2.1.3 Organic and Inorganic Polymers

Organic polymers are often subdivided into plastics (artificial polymers) and biopolymers. Plastics are mostly polymers of carbon produced on the basis of fossil or renewable raw materials in predominantly industrial processes, and have no analog in nature. Biopolymers are, for example, the building blocks or ingredients of living organisms such as proteins (e.g. silk), DNA, cellulose, starch, chitin, or natural rubber. Biopolymers are often additionally chemically modified for technical use, with examples being refined cellulose such as viscose (artificial silk), cotton acetate, nitrocellulose, and celluloid. Colloquially, polymers based on naturally renewable raw materials such as polylactide (PLA), polyhydroxyalkanoates (PHA), or polyethylene furanoate (PEF) are also frequently referred to as biopolymers, but they also belong to the class of organic plastics (artificial polymers).

A simple and widespread example of naturally occurring inorganic polymers are glassy rock types such as the volcanic glasses obsidian and pitchstone, asbestos and in (in the living world) the skeleton of diatoms. Artificial inorganic polymers include window glass, glass fibers, and (as an example of elastic, soft materials) polysilicones.

2.2 Polymer Chemistry

Polymer chemistry summarizes all chemical reactions that lead to the formation, modification, or degradation of polymers. After an introduction at this point, the subject is dealt with in greater depth in Section 2.4. Chain formation, i.e. the linking of individual monomers to form polymers, occurs through so-called polymerization reaction (polyreactions). A distinction is made between chain-growth reactions (involving radical, anionic, cationic, or coordinative polymerization) and step-growth reactions (involving polycondensation or polyaddition polymerization). In this context, polymerization is a chemical reaction that is repeated many times and describes the formation of a polymer chain from monomeric building blocks. If these reactions take place in a technical reactor, they are described with the added term "process".

The most widely used synthetic polymers are industrially produced by chain-growth-reactions and include polyethylene (PE), polypropylene (PP), polyvinyl chloride (PVC), and polystyrene (PS). In chain-growth reactions, chain growth occurs via an active chain end. The common feature of the monomers is an unsaturated carbon–carbon bond (a C=C double bond). This is converted by the polymerization reaction into a saturated carbon–carbon bond (a C–C single bond). This links the monomers together to form stable long carbon chains and thus the backbone of the polymer. Depending on the choice of monomers used, various side groups can branch off at regular intervals from the backbone.

In the step-growth reaction, chain growth takes place in steps. Monomers with at least two reactive groups are required for this. These first react to form dimers and can then react again with other monomers or dimers to form higher homologs. These reactions take place either by an addition reaction or a condensation reaction. While condensation releases a molecule such as water or hydrochloric acid (HCl) after each reaction step, no byproducts are formed in the addition reaction.

Well-known examples of polymers produced by polycondensation processes are polyesters (PES) and polyamides (PA), but also the polycarbonates (PC). Important representatives of polymers produced by polyaddition processes are polyurethanes (PU) and polyepoxides (EP).

Since no molecules have to be removed in polyaddition processes, the material reaction volume shows a significantly lower shrinkage during polymerization. Polyaddition processes are therefore frequently used in the in-situ production of semi-finished and finished products. The polymers are formed by mixing the components at the point or in the form in which they are to be used, e.g. as foams, coatings, adhesives, potting compounds, or composites.

Polycondensation reactions also differ from the products of chain-growth reactions obtained by polymerization of C=C double bonds, in that heteroatoms (i.e. atoms other than carbon atoms) are incorporated into the polymer backbone. Typical heteroatoms are oxygen and nitrogen but other heteroatoms such as sulfur, silicon, or phosphorus can also be part of the polymer backbone.

> Natural polymers such as cellulose, proteins, chitin, and DNA are structurally and biosynthetically most similar to polymers obtainable from polycondensation processes.

> Fully polymerized thermoplastic materials are used in powder- and extrusion-based AM processes because they can be molded and processed layer by layer from the melt by powder sintering or extrusion-based processes. For AM based on liquid reactive resins (photopolymer systems), almost exclusively monomeric or oligomeric compounds are used, which are polymerized or cured in the desired spatial form by subsequent polymerization or addition processes (avoiding the shrinkage during polymerization as much as possible).

2.3 Polymers as Materials

Engineering polymers (plastics) can be divided into four groups based on their physical behavior: thermoplastics, thermoplastic elastomers, duromers, and elastomers (see Figure 2.5).

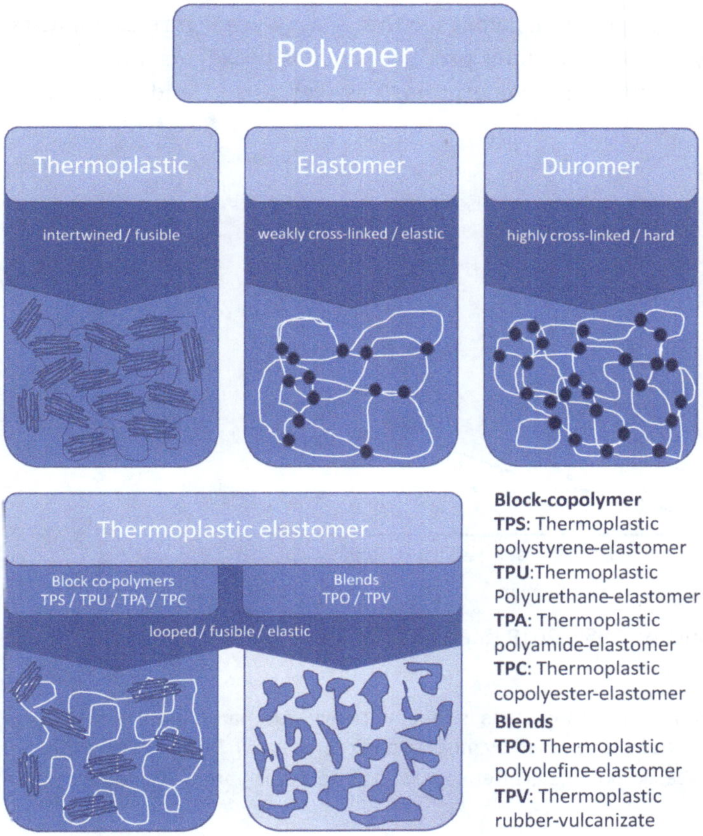

Figure 2.5 Overview of the various polymer classes with focus on their morphology

2.3.1 Thermoplastics

Thermoplastics are plastics that soften when heated to temperatures above their melting point and then become malleable or flowable. This behavior is used technically to transform them into their useful form by melt processes such as extrusion, injection molding, or thermoforming. The melting point or the softening temperature of thermoplastics is by definition well above their service temperature. In amorphous polymers the melting point is given by the so-called glass transition temperature. Below this temperature, the molecular chains are frozen in their movement like glass. In semi-crystalline polymers, in addition to the glass transition temperature, crystalline structures with an even higher melting point exist due to the organization of polymer chains, which are only broken up at their melting point. By far the largest proportion of polymers manufactured industrially today, accounting for approx. >80 % of total production, belongs to this group – see Figure 2.6 [Statista 2021]. They are widely used in consumer products such as packaging, automotive, electronic housings, and pipes, and come in hard as well as soft and viscoplastic variants.

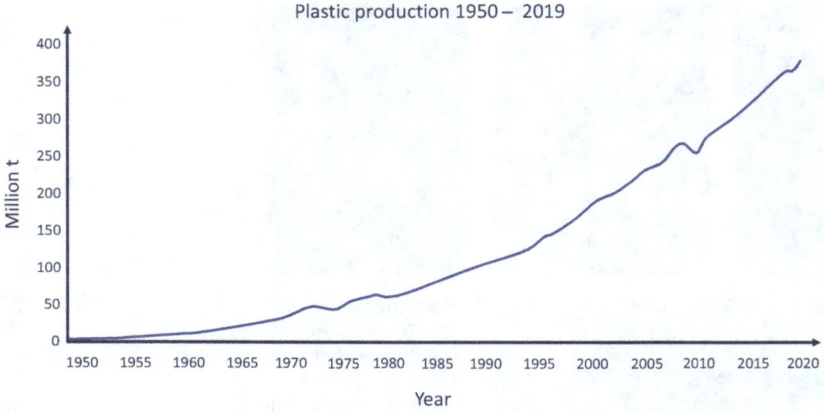

Figure 2.6 Plastic production 1950–2019 [Statista 2021]

A major advantage of thermoplastics is their suitability for material reuse (recycling), which can often be achieved by melting and reprocessing in injection molding or extrusion-based processes after separation into single-variety products.

In AM, thermoplastics represent the largest polymer product class. They are used in both powder-based and extrusion-based processes [Wohlers 2020].

2.3.2 Duromers

Duromers consist of highly crosslinked polymer molecules with a glass transition temperature (T_g) above the service temperature. They are often used in products where high hardness and temperature resistance are important, such as coatings, adhesives, rigid foams, and composites. Duromers are often brittle because they are in the glass state.

Both duromers and their "soft relatives" (the elastomers) are crosslinked materials and cannot be liquefied by simple melting processes. They can therefore not be directly recycled as a material but must be broken down beforehand by chemical or thermal decomposition into their building blocks.

The photopolymer systems in AM span the range from soft elastomers (low T_g, low crosslinking density) to hard duromers (high T_g, high crosslinking density) in their polymerized (cured) form. In their initial form, they consist as liquid reactive resins of various monomeric building blocks that are selectively polymerized in the presence of light and photoinitiators. The product is thus formed as an elastomer or duromer only during the printing (build) process, by polymerization of the reactive resin to a polymer with varying degree of crosslinking. The large variety of suitable and readily available monomeric building blocks and the high component production speed and component resolution make photopolymerization an important AM technology. The elastomers and duromers available in this way are developing into a strongly growing product class of their own in AM [Wohlers 2020].

2.3.3 Elastomers (Rubbers)

Elastomers (often named rubbers after natural rubber) are characterized by rubber-elastic properties and consist of long polymer chains that are only weakly crosslinked (chemically bonded with a T_g well below their service temperature. The glass transition temperature is the temperature at which the mobility of the molecules freezes to such an extent that the material appears hard and sometimes brittle (glassy). Chemical crosslinking of rubber is achieved during a process called "vulcanization". In this process, the raw rubber polymer, which is a highly viscous material at room temperature, is crosslinked by heating in presence of crosslinkers, for example sulfur or peroxides.

Elastomers differ from thermoplastics in that the chemical crosslinking of the polymer chains prevents melting, although the number of crosslinking points is lower in elastomers than in duromers. The best-known and also largest application of rubber-like materials is their use in vehicle tires of all kinds.

Elastomers can be reused after their service life by grinding as a filler for the elastification of, for example, floor coverings such as road asphalt, playground flooring and sports fields. This is what is known as "downcycling". Alternatively, the reprocessed rubber components of tires, for example, are considered a high-grade fuel (via certified scrap tire disposal companies) [Zertifizierte-altreifenentsorger 2021].

Typical elastomers are only slowly finding their way into AM because they are difficult to process using the additive processes relevant to polymers and require special process modifications. In order to successfully process elastomers in powder- and extrusion-based processes (such as FLM, SLS, and binder jetting), they can only be crosslinked or vulcanized after the build process. For this to work, the processing temperature during shaping in the process must be well below the vulcanization temperature. Vulcanization is advantageously carried out in a subsequent annealing process under pressure.

In contrast, elastomers based on liquid reactive resins (e.g. photopolymer systems) can be polymerized using additive processes such as SLA, digital light processing (DLP) or polyjetting. However, due to the high degree of crosslinking required by the process to achieve a sufficient build stability, in combination with the demand for a soft product with a low glass transition temperature, the process is limited by the availability of suitable reactive resins. The properties achieved by the printing of these resins are more similar to soft coatings and potting compounds than to the comparatively high-molecular weight, highly filled and slightly crosslinked technical rubber elastomers used in the rubber industry.

AM based on liquid reactive resins currently only works with liquids that have a low processing viscosity in the range of approx. 5 mPa·s to 5000 mPa·s. The viscosity of rubber formulations before crosslinking (vulcanization), on the other hand, is typically much greater than 10,000 mPa·s.

2.3.4 Thermoplastic Elastomers (TPEs)

Thermoplastic elastomers (TPEs) are a subtype of thermoplastic, and combine the mechanical properties of crosslinked (vulcanized) elastomers (rubber) at room temperature with the processability of thermoplastics in the melt.

Instead of chemical crosslinking points between the polymer chains, TPEs have physical crosslinking points. They consist of components with a T_g below the service temperature (soft phase) and at least one "separate" component with a softening temperature (melting temperature/second glass transition temperature) above the service temperature (hard phase). There are always at least two separate polymer phases present.

TPEs behave like physically reversible (temperature-dependent) crosslinked polymers. After heating to a temperature above the softening temperature (known as the melting temperature or second glass transition temperature), they can be processed in the melt just like other thermoplastic polymers. After cooling to the service temperature, they reversibly solidify by forming physical crosslinking points (hard phase) and thus revert to an elastic material.

Important examples of this product class are the thermoplastic polyurethanes (TPU) with a crystalline polyurethane hard phase and an amorphous polyester or polyether soft phase, and the thermoplastic polystyrene block copolymers (SBCs) with an amorphous high-T_g hard phase and a second amorphous low-T_g soft phase.

Like thermoplastics, TPEs can often be reused after careful material separation, by melting and reprocessing.

A special form of TPEs are thermoplastic blends. Here, the amorphous soft phase with a low T_g is finely distributed, but phase-separated, in an amorphous or crystalline hard phase with a high T_g or high melting point. Elastic/ductile properties can be adjusted via the amount and composition of the soft phase. The soft phase may be present in a low crosslinked form, as in thermoplastic vulcanizates (TPVs), or in a non-crosslinked form, as in most thermoplastic polyolefin elastomers (TPOs).

In the case of a thermoplastic elastomer (TPE), the polymer is thus melted when the melting temperature/second glass transition temperature is reached and analogous to normal thermoplastics, can be processed in the melt. After the melt has cooled, the typical properties of the TPE regenerate through the reformation of the hard phase.

As a rule, physically crosslinked TPEs have a higher initial modulus than chemically crosslinked elastomers with a low glass transition temperature (see rubber). The modulus of TPEs can be specifically lowered by increasing the percentage of the soft phase and by blending with plasticizers that are selectively soluble only in the soft phase. However, the mechanical properties of the overall product often suffer more than in a similarly crosslinked rubber.

> The material properties of TPEs depend decisively on the processing conditions ("product by process" [Ritzkowski 2019]). If the melt is sheared heavily in the process, this can lead to a different distribution of the phases in the product. The resulting changes in the phase structure (morphology) in turn control the mechanical properties. In addition the cooling rate in processing and the presence of crystallization nuclei also have an important influence on the expression of the phase morphology of the product and thus on the mechanical properties.
>
> Especially in AM, the process and cooling conditions differ significantly from the otherwise typical thermoplastic processing methods. Due to the layer-by-layer build process, good interlayer adhesion in the component is crucial. To enable processing of TPE in the FLM or SLS process, it is therefore important that the last layer is still liquid or is melted again by the new layer. The melting/solidification times that occur in this process depend decisively on the component size and the layer build speed. Depending on the component geometry, the thermoplastic elastomer can thus exhibit different properties from component to component, as well as within a component, simply by specifically controlling the process conditions.

Industrial component manufacturers using processes based on additive melting of thermoplastics (and in particular TPEs) must be aware of these relationships between process conditions and the resulting product properties. These relationships can be used specifically to vary material properties within a component to a certain extent. Thus, by selecting process parameters such as build speed, build temperature, activation temperature, and build direction, it is possible to adjust not only density gradients in the component but also the material modulus and surface effects with local precision.

This variability of properties is both a curse and a blessing. On the one hand, compared to conventional manufacturing technologies, additional degrees of freedom in the spatial resolution of component properties are accessible in addition to the new geometric degrees of freedom. On the other hand, however, meaningful conventional component inspections and quality analyses are made more difficult if they are based on a false assumption that material properties are always homogeneous.

In AM, TPEs are always used where high toughness and elasticity of the products are required. Examples include the orthopedic sector (shoe insoles, prostheses, and body supports), the body protection sector (helmets and protectors) and comfort applications (cushions, seats, and mattresses). TPEs are used in AM in powder- and extrusion-based processes. Commercially available are thermoplastic polyurethanes (TPU), thermoplastic copolyesteramides (COPE), polyether block copolyamides (PEBA), styrene-block copolymers (TPE-S/SBC) and thermoplastic polyolefin-elastomers (TPO).

2.4 Polymer Manufacturing Process

Polymerization is a general collective term for synthesis reactions that convert similar or dissimilar monomers into polymers. Technical polymerization reactions are mostly used for the synthesis of plastics. They can be divided into chain-growth reactions and step-growth reactions (Figure 2.7).

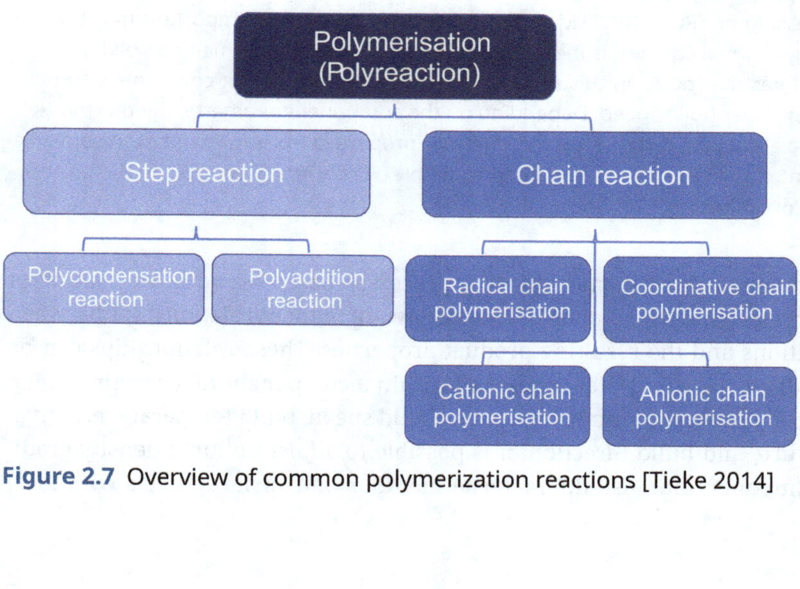

Figure 2.7 Overview of common polymerization reactions [Tieke 2014]

Chain-growth reactions occur via one active chain end at a time, until chain termination. They can be divided into radical, cationic, anionic, and coordinative chain-growth reactions. Step-growth reactions occur via the simultaneous parallel polycondensation or polyaddition of all monomers to form chains, continuously producing longer and longer chains until no monomer is present. Biological polymerization reactions proceed according to other mechanisms and are much more complex in comparison, but highly selective.

2.4.1 Chain-growth Reactions

In chain growth, a start reaction (initiation) is followed by a continuous linkage of monomers (M) to the growing polymer chain (P_i), which is formed from i monomer units (M). In this process, the initiation reaction is often triggered by a "change of state" – for example, temperature, light, pH, or oxidation state. In this process, a starter molecule (initiator) is usually generated in the reaction mixture, which reacts with one of the aforementioned stimuli and thus initiates the polymerization reaction.

$$P_i + M \longrightarrow P_{i+1} \tag{2.1}$$

In this equation, i can take all integer values ≥2. Such growth occurs when only the growing polymer chain carries a reaction-active functionality. It can be a radical (radical chain polymerization), a cation (cationic chain polymerization), an anion (anionic chain polymerization), or a coordinative complex/catalyst (coordinative chain polymerization). The formation of the polymers proceeds as a chain reaction. Thus, step by step, only one monomer is added at a time. This growth results in the formation of large and long macromolecules even at very low monomer conversion rates, since the number of growing chains is always very small compared to the absolute number of monomers. The molecular weight thus depends on the effective initiator concentration and the probability of a chain termination reaction.

In free-radical polymerization, this chain termination reaction occurs through a so-called transfer reaction (one chain stops, another chain takes over the growth reaction, often helped by a chain regulator), recombination (two chains add to a single chain, and growth stops), or disproportionation (two chains react, transferring a molecular building block from one chain to the other, and growth stops in both).

A special form of chain-growth reaction is the living polymerization reaction. Here, only a fixed number of growing chains is formed, which grow continuously, statistically simultaneously, and ideally without a termination reaction, until all monomers are consumed. The molecular weight increases linearly with the conversion. The slope is controlled by the number of simultaneously growing chains.

In 3D printing, the chain-growth reaction is used above all in photopolymer systems in the form of radical chain polymerization. The products of this reaction process are usually strongly crosslinked duromers due to the use of mono- and multi-functional monomers (chain-branching agents). The trick with 3D printing based on photoactivatable starter molecules (photoinitiators) is to trigger only local polymerization. Polymerization only takes place where the photoinitiator breaks down into active starter molecules as a result of excitation by light/laser energy. By using a controlled amount of oxygen or other polymerization inhibitors that can effectively stop the free-radical polymerization reaction, the reaction can be carried out in a narrowly localized manner. The resolution is limited only by the resolution of the laser/light pattern. Thus, depending on the technology, high-resolution parts/components consisting of crosslinked polymers (elastomers/duromers) can be produced down to a size of 0.001 mm.

The high speed at which chain polymerizations take place makes this technology particularly suitable for in-situ polymerization processes such as those that take place in 3D printing based on reactive resins (SLA, DLP, etc.). Since the processes take place under normal pressure, the choice of suitable monomers is limited to compounds with low vapor pressure, such as (meth)acrylates and relatively robust polymerization processes such as free-radical polymerization. This in turn limits the properties as well as the possible applications of the products obtainable from them.

Alternative processes based on the cationic polymerization of epoxides and, in particular, hybrid polymerizations in which different polymerization processes are combined, provide partial relief here. However, this is often done at the expense of the build speed and/or by increasing the process complexity.

For 3D printing processes based on thermoplastics, the development focus is still on process adaptation to make even more existing materials processable. In the case of photopolymer systems, a particularly large amount is being invested in the product development of new resin formulations, in order to expand the range of applications away from pure prototyping towards functional products with improved mechanical properties.

2.4.2 Step-growth Reactions

Step-growth reactions include polycondensation and polyaddition reactions. Step growth can only occur if the monomers carry at least two functional groups, both of which are reactive. The growth of polymer chains proceeds in steps by reaction of two functional groups with each other. If these groups are of different natures, then the ratio of the respective groups (stoichiometry) to each other determines the degree of polymerization that can be achieved. Ideally, the ratio of the reactive groups (for example, A and B) to each other is exactly 1:1. The groups can be distributed over two monomers (AA + BB) or be present in one molecule simultaneously (AB + AB). In special cases, the reactive groups are also the same and can react with themselves (AA + AA). Since all monomers have the ability to react directly with each other, initially only dimers, trimers, and oligomers are formed at low and medium conversions. Only when the functional groups are almost completely converted do high molecular weight polymers form.

2.4.2.1 Polyaddition Reactions

In polyaddition, growth occurs via addition reactions. First, dimers (P_2) and trimers (P_3) are formed from the monomers (M):

$$\begin{aligned} M + M &\rightarrow P_2 \\ P_2 + M &\rightarrow P_3 \\ P_i + P_j &\rightarrow P_{i+j} \end{aligned} \tag{2.2}$$

As growth progresses, adducts with any degree of polymerization react with each other. In the reaction equation, *i* and *j* can therefore assume all integer values ≥2.

2.4.2.2 Polycondensation Reactions

In polycondensation reactions, chain growth proceeds in steps analogous to the polyaddition reaction, but with the elimination of one reaction molecule (L) at a time, for example water. The progress of the chain growth is otherwise analogous to the polyaddition reaction:

$$\begin{aligned} M + M &\rightarrow P_2 + L \\ P_2 + M &\rightarrow P_3 + L \\ P_i + P_j &\rightarrow P_{i+j} + L \end{aligned} \tag{2.3}$$

In Table 2.1 the basic properties of the different polymerization reactions are listed and compared.

Table 2.1 Comparison of the Properties of the Different Polymerization Reactions

Radical polymerization	Living polymerization	Polyaddition polymerization
A growing chain reacts with both monomers and other active chains under transfer, recombination, and disproportionation.	A growing chain reacts only with monomers.	Monomers, dimers, and oligomers react with each other.
The degree of polymerization is almost independent of conversion.	The degree of polymerization increases linearly with conversion until all monomers are consumed.	The degree of polymerization increases exponentially with conversion.

Table 2.1 Comparison of the Properties of the Different Polymerization Reactions (*continued*)

Radical polymerization	Living polymerization	Polyaddition polymerization
In addition to growing chains, the reaction mixture also contains polymers that no longer grow, and monomers. After consumption of the initiators, the reaction stops due to the continuously occurring termination reactions.	All chains grow statistically uniformly until the monomer is completely consumed.	The mixture contains monomers, oligomers, and polymers with different chain lengths. All grow simultaneously until the monomers are completely consumed.
At low conversion, polymers with long chains are already present.	At low conversion, only short chains are present, depending on the monomer to initiator ratio.	Polymers with long chains do not form until turnover is almost complete.
An initiator or a special catalyst is required to start a chain.	An initiator or a special catalyst is required to start a chain.	No special initiator is needed to start a growth reaction, but temperature and catalysts accelerate the reaction.
The growth of a chain is terminated by a termination reaction.	No termination reactions occur.	No termination reactions occur.
These are often very fast reactions (seconds).	These are often very fast reactions (seconds).	These are often slow reactions (minutes to hours).

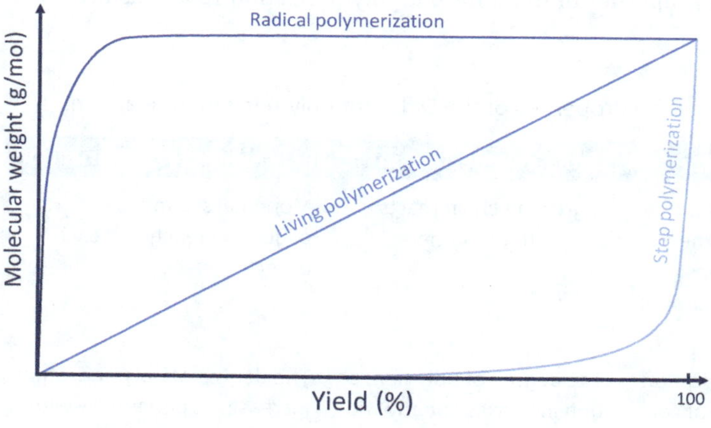

Figure 2.8 Molecular weight as a function of conversion

Many products generated by step-growth-reactions, in the form of thermoplastics such as polyamide (PA), polyester (PES), polycarbonate (PC), polyether ether ketone (PEEK), and thermoplastic polyurethanes (TPU), find application in AM using extrusion-based and/or powder-based processes. There is currently no commercial 3D printing process in which thermoplastics are first formed in situ through step reactions. The input products for AM with thermoplastics therefore originate from classic industrial manufacturing processes. This is also the reason for a certain difficulty in developing thermoplastics specifically for 3D printing. The industrial manufacturing processes for thermoplastics are designed for high throughputs and large production volumes (typically 100–10,000 t per production run).

The initially low sales volumes in 3D printing have so far rarely been sufficient to justify the development of products tailored for 3D printing processes based on standard thermoplastics and engineering thermoplastics. PA12, PA11, and PLA represent exceptions to a certain extent. Accordingly, at the beginning of AM based on thermoplastics, there were only a few products that could be processed by the 3D printing processes available at the time, either by extrusion or as powder. In particular, polylactide (PLA) and polyacrylonitrile-butadiene-styrene (ABS) for the extrusion-based processes, and PA12 and PA11 for the powder-based processes, should be mentioned here.

In addition to the engineering thermoplastics PLA, ABS, and PA12, which have long been used in 3D printing, other important standard thermoplastics and engineering thermoplastics produced by chain polymerization on an industrial scale can now also be used in AM. High-impact polystyrene (HIPS), polyoxymethylene (POM), and thermoplastic elastomers (TPEs), such as the block copolymers of styrene (SBS/SEBS), are particularly worthy of mention here, and increasingly also polyethylene (PE) and even polyvinyl chloride (PVC).

Today, the first products are being developed specifically for extrusion-based and powder-based AM processes. Examples include special polyaryl ether ketone (PAEK) variants such as polyether ether ketones (PEEK), polyimide (PEI) variants, polyamides (PA), polyamide copolyesters (COPE), polyether block copolyamides (PEBA), polyesters (PETG) and, in particular, thermoplastic polyurethanes (TPU). Development is often limited to blending existing products (polymer blends, polymer alloys) and mixing in (compounding) suitable fillers such as glass fibers, carbon fibers, basalt fibers or inactive fillers such as talc, chalk, glass beads, and (of course) pigments, dyes, and plasticizers.

At the same time, processes are increasingly being adapted to make specific AM applications possible in the first place, for example in medical technology or aerospace using high-quality but difficult-to-process materials such as PEI and PEEK.

The large number of existing thermoplastic products and the possibility of further modifying and optimizing them by mixing and blending processes for use in AM explains the interest in thermoplastic build materials for AM. In addition, intensive work is being done on the adaptation and specialization of AM processes to the ideal processing conditions of individual selected high-quality thermoplastics.

Overall, as processable thermoplastics and thermoplastic processing methods become used more and more widely, they are in many cases already allowing additive processes to be used to create functional components. In this way, AM can thus develop into a "new" standard in manufacturing technology.

2.4.3 Biological Polymerization

In living organisms, polymerization reactions serve not only for the synthesis of DNA and proteins, but also for the production of cellulose and starch. They often proceed according to different and much more complex mechanisms than those previously listed. In particular, DNA and protein syntheses involve temporary complex formation with a special matrix. For example, in protein biosynthesis, mRNA serves as a template that enters into a complex with ribosomes. Subsequently, the sequence of the template is transferred to the newly formed polymer. The extremely complex mechanisms allow a high degree of control over the final polymer, not only in chain length but also in the chain sequence and the resulting folding of the proteins, which provides the polymer with a controlled three-dimensional superstructure and the desired properties.

Biological polymerization reactions, despite their great potential in terms of controlling product properties through perfect chain and monomer sequence control, have so far only been made use of in a few selected cases. There are very few examples to date in which enzymatic polymerization or refinement has been used in the production of commercial engineering polymers. Similarly, despite major efforts, there are few examples to date of the commercial use of enzymes for the targeted degradation of engineering polymers.

Processes in which a new 3D shape is created from an initial 3D-printed shape by external stimuli such as temperature and humidity are known as "4D printing". 3D printing is also finding more and more applications in the biomedical field, for example when it comes to printing self-regenerating skin, organ precursors, or bone templates. Special 3D printing processes are often used for this purpose, which allow biological materials or biologically compatible materials to be applied, for example, via a gentle (low-temperature and low-shear) extrusion process such as cell plotting or bioprinting [Jose et al. 2016].

2.5 Physical Properties of Polymers

Measurement and knowledge of the physical properties of thermoplastic polymers is an important prerequisite for the correct selection of polymers for a desired application. The measurement of physical properties helps to determine both the application properties and the processing parameters. The following two statements can be made:

- The properties of the solid state determine the application properties.
- The properties of the melt, as well as in the transition to the solid state by crystallization (or after falling below the glass transition temperature), determine the possible processing methods and processing conditions.

2.5.1 Viscoelasticity

Viscoelasticity is an important parameter both for understanding the mechanical properties of polymer materials and for their processing. Viscoelasticity is understood as a partially elastic, partially viscous material behavior. Viscoelastic materials thus combine characteristics of energy-elastic solids and viscous liquids. The property of viscoelasticity is time-, temperature- and frequency-dependent, and can be observed both in polymer solids and in polymer melts.

The viscoelastic behavior of polymers can be simulated by a combination of springs and damping cylinders, coupled in parallel and/or series (Figure 2.9).

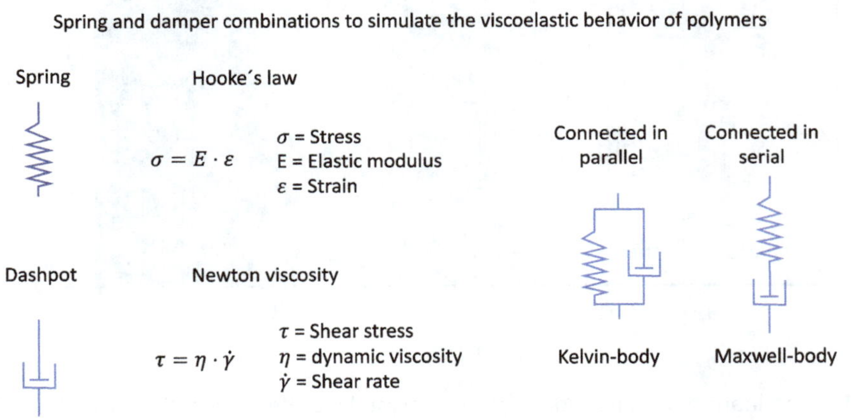

Figure 2.9 Simple models of viscoelasticity

The spring contains the Young's modulus of the material (in the fully elastic case, the Young's modulus corresponds to the spring constant D of Hooke's law), and describes the tensile stress σ at a given deformation ε.

A damping cylinder is an element which shows a constant strain rate $\dot{\gamma}$ controlled by the viscous behavior of a material at a constant applied stress τ.

For a quantitative description of the viscoelastic behavior of polymer solids and polymer melts, the modulus of elasticity E or the shear modulus G are plotted against temperature or frequency, depending on the measurement method. In the glass region below the glass transition temperature (T_g), the elastic material behavior dominates. Under linear-elastic behavior, it is assumed that, according to Hooke's law, a specific strain value (stress–strain diagram) is assigned to each stress value, regardless of the duration of the load.

In the glass transition region, the material becomes "anelastic" and can be approximately described by the series connection of a spring and a Kelvin body (see Figure 2.10). Anelasticity describes the property of solid bodies, under tensile, shear, or compressive stress, to resume their original shape only after a certain time has elapsed since the stress was removed. In the rubber plateau the elastic material behavior (spring) dominates, while in the melt transition the behavior can be described by a Maxwell body. Plotting the Young's modulus against frequency leads to a mirror-image curve in comparison to plotting against temperature. The Young's modulus then increases with increasing frequency. In the case of polymers that are not ideally crosslinked, creep of the material must be expected at a constant load over the entire temperature range.

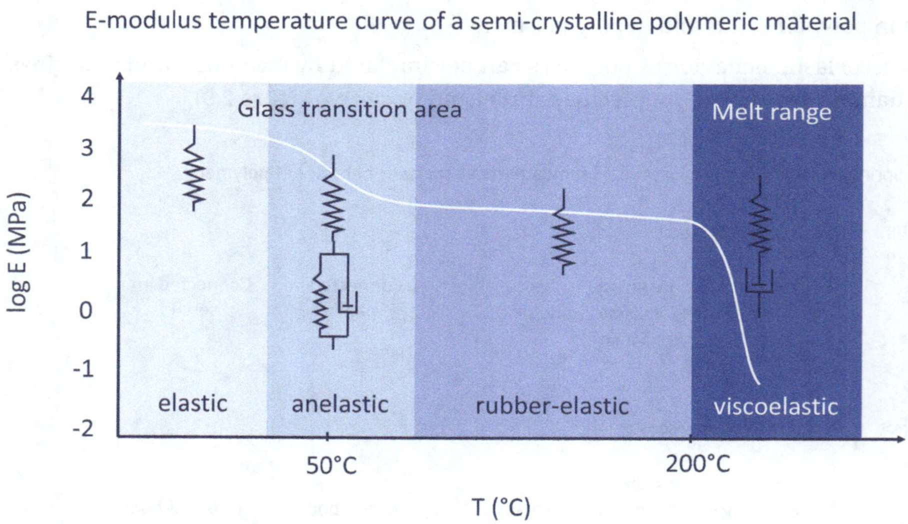

Figure 2.10 Dominant material models for semi-crystalline thermoplastics as a function of temperature

2.5.2 Dynamic Mechanical Analysis (DMA)

The properties that are decisive for the application of polymers can be determined to a large extent by dynamic mechanical analysis (DMA) of a solid specimen as a dependence of the modulus on temperature (Figure 2.12, cf. [Lüpke 2015b]). Depending on the measurement method (oscillatory loading of the specimen in shear/torsion or tension), the storage modulus G' (for shear/torsion loading) or E' (for tension loading) and the loss modulus G'' or E'' of a material are plotted against temperature at a given measurement frequency and deflection (Figure 2.11). The storage modulus is also colloquially called elastic modulus. Analogous to an ideal spring, it indicates the part of the response of a material to a deformation that is returned without loss. The loss modulus or viscous modulus, on the other hand, corresponds to the part of a material's response to a deformation that is irreversibly retained. The loss modulus thus corresponds to the behavior of a damping cylinder, whereby fluid is displaced by a piston in a cylinder and the energy required for this is completely converted into heat by the resulting friction.

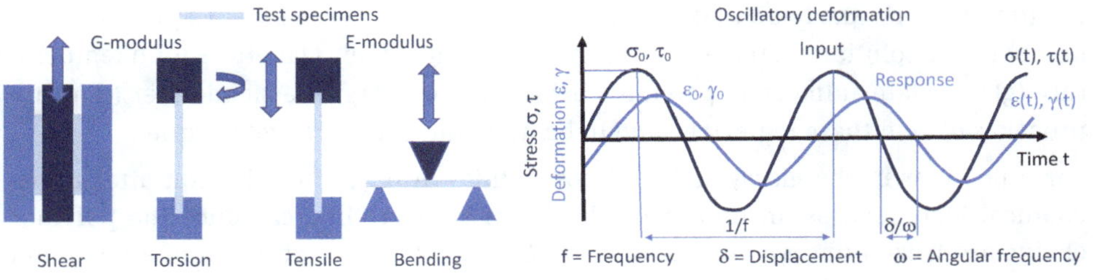

Figure 2.11 DMA measurement: Test setup

In the following, the terms elastic and viscous modulus are used for reasons of clarity. From the course of the elastic and viscous modulus curves obtained in a DMA measurement, it is possible to read off not only the temperature dependence of the moduli but also basic polymer properties such as the glass transition temperature and, in the case of semi-crystalline polymers, the melting point.

The quotient of the viscous modulus and the elastic modulus is called loss factor or $\tan\delta$ and is a measure of the damping or the energy dissipation during a mechanical load/deformation. The $\tan\delta$ takes the value 0 for an ideally elastic (energy elastic) body ($G'' = 0$ and $G''/G' = \tan\delta = 0$). For an ideally viscous body, the damping approaches infinity accordingly ($G' = 0$ and $G''/G' = \tan\delta = \infty$).

If G' and G'' are determined in an oscillatory measurement at small deformations, then the deformation and the resulting stress are linearly related, corresponding to the modulus. The magnitude of the complex viscosity $|\eta^*|$ is obtained via the angular frequency ω according to Equation 2.4. The angular frequency $\omega = 2\pi f$, with f being the measurement frequency.

$$|\eta^*| = \sqrt{\left(\frac{G'}{\omega}\right)^2 + \left(\frac{G''}{\omega}\right)^2} = \frac{|G^*|}{\omega} \tag{2.4}$$

The shear modulus G^* and the tensile modulus E^* are related to each other via the transverse contraction number ν (Poisson's ratio), where Equation 2.5 applies strictly only to isotropic bodies, i.e. for bodies that have the same properties in all directions independent of the angle of rotation.

$$G^* = \frac{E^*}{2 + 2\nu} \tag{2.5}$$

For elastomers and melts, approximations are $\nu = \frac{1}{2}$ and $E = 3G$.

In the typical use case, that is, for thermoplastic polymers in the solid state below the softening temperature and for crosslinked polymers below the decomposition temperature, tanδ is usually $\ll$1.

In the range of the glass transition temperature and in the range of the melting point of a thermoplastic polymer, tanδ *increases* significantly. In the glass transition temperature region, tanδ has a maximum. When tanδ is $\gg$1, one speaks of a polymer melt or a polymer solution. In the temperature range below their glass transition temperature (glassy state), almost all polymers behave in an energy-elastic manner (tanδ approaches 0), and these materials tend to be hard and comparatively brittle.

For a polymer in the glassy state, all possibilities of energy dissipation after a mechanical load such as an impact or shock are severely limited, since the polymer chains are frozen in their translational mobility – with the exception of a few bond vibrations and possibly bond rotations.

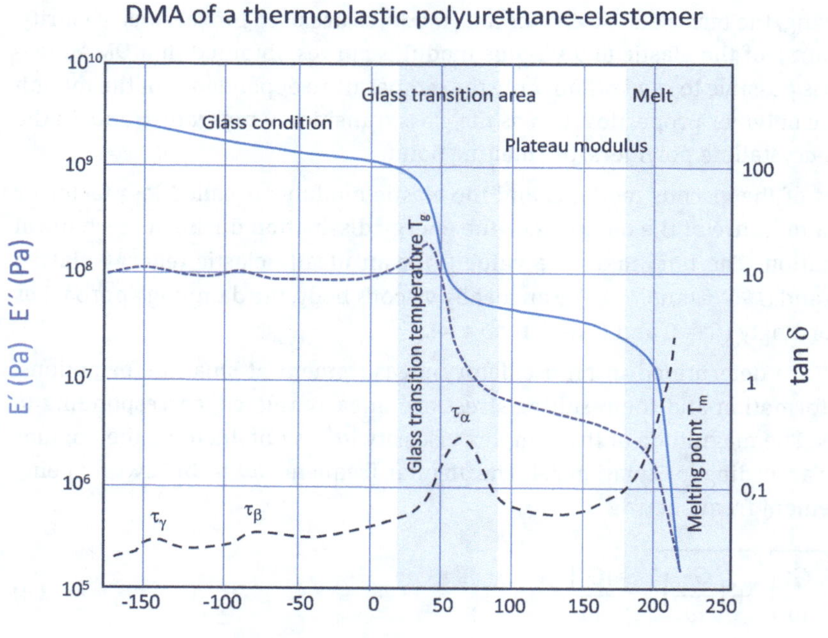

Figure 2.12 DMA curve of a hard semi-crystalline TPU elastomer

A brittle fracture, often glass-like in appearance, occurs when the excitation energy is too high and can no longer be returned in a fully elastic manner or dissipated as heat.

Shortly before and shortly after the glass transition temperature, individual molecular regions of the polymer chains thaw, and initial movements are possible (rotation of atoms/molecular units, sliding of molecular units or chains against each other). As a result, the material undergoes an initial transition to a tougher property behavior, and the tanδ, as a quotient of the viscous and elastic modulus, increases (Figure 2.12). The energy introduced from the outside can now be partially dissipated as heat (friction due to molecule/chain movement during irreversible deformation). Since both the elastic and viscous moduli remain high in the glass transition temperature range for the time being, the material is still hard, but now also impact-resistant [Menge et al. 2022].

Above the glass transition point, the modulus typically continues to drop steeply by several powers of magnitude, and the material softens due to an increasingly high mobility of the polymer chains. In this transition region, semi-crystalline or cross-linked polymeric materials are often particularly impact-resistant, since the impact energy can be converted into deformation of the still-high modulus material and thus into significant heating. This behavior is expressed tanδ being a maximum.

If there is no melting point (as in the case of semi-crystalline polymers) or crosslinking (as in the case of elastomers and duromers), then the modulus of a polymer decreases continuously with temperature after the glass transition point. The polymer starts to flow and a polymer melt results, with the viscosity of the melt continuously decreasing with a further temperature increase (Figure 2.13).

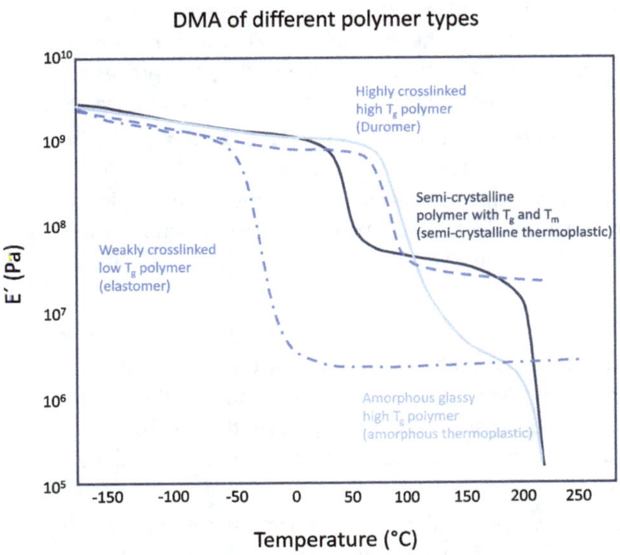

Figure 2.13 DMA curve of different polymer types

In the case of linear high-molecular-weight polymers, if the polymer chains are prevented from flowing by sliding off each other, then a so-called "rubber plateau" is established. From the rubber plateau, the molecular weight of a polymer can be inferred if the inherent chain mobility is known. If the chains are prevented from further sliding by chemical or physical crosslinking, then a crosslinking modulus is obtained, which in turn (if the chain mobility is known) provides information about the chain length between two crosslinking points and thus about the crosslink density.

In a classical DMA measurement, the loss modulus, the storage modulus, the quotient of loss modulus, and the storage modulus are plotted against temperature. The measurement takes place in a sinusoidal manner, with the strain signal running repeatedly from maximum to minimum at a given frequency. The strain range is selected in such a way that mainly linear dependencies between excitation and result can be assumed.

The DMA measurement is extremely versatile with respect to the measurable specimen geometry. In addition to films in tension and shear, different test bars can also be measured in tension, torsion, or under 3-point bending load. This makes the method particularly suitable for load-direction-dependent measurement of heterogeneous materials as well as composites.

Unlike other thermo-analytical methods, DMA allows simultaneous analysis of thermal, frequency-dependent, and mechanical properties. DMA measurement is particularly important for understanding polymers in application. The change of moduli with temperature allows direct estimation of the temperature-dependent properties of a material, such as elasticity and damping, and of softening behavior in the glass transition and melting transition. Comparative DMA measurements provide information about the phase distribution in polymer blends or the influence of fillers, for example in fiber-filled polymers.

The shape of the transition from the glassy state to the crosslinked or rubber plateau state, and from the solid state to the melt, also provides important information about quality-relevant properties that are influenced by the processing conditions, such as the degree of crystallinity of semi-crystalline thermoplastics.

In addition to measurement in the linear range, a special DMA design (hydropulser with temperature-controlled test rig, cf. [Kohl & Pešík 2016]) can also be used for measurement in the non-linear load range, whereby, for example, the fatigue behavior of materials under dynamic loading at different temperatures, frequencies, and deformations can be investigated.

For 3D printing, the DMA method is particularly important because it provides an understanding of the interrelationships between the measured quantities, and thus enables the developer of products from photopolymerization in the SLA and DLP processes to design materials specifically with regards to their temperature-dependent properties (glass transition temperature) and crosslinking density (crosslinking modulus).

The variability of the possible test specimen geometries and measurement conditions is particularly important for 3D printing processes using melt-based methods (FLM, SLS), since material properties are typically highly dependent on the local thickness and orientation of the part in the printing process. DMA usually works for small deformations in the linear and non-destructive range. For the intended application of polymer materials, this range is decisive. Measurements in the non-linear range and in the range of non-reversible deformations rather describe the safety and fatigue behavior of a material under loads that are not appropriate for the application, or hint to performance under incorrect component design.

The shape and position of the glass transition as measured by DMA is not only an important piece of information for the selection of the correct process conditions in printing but also for the production of polymer powders by the cryo-milling process for the various powder sintering processes, such as selective laser sintering (SLS) and high-speed sintering (HSS). Furthermore, DMA provides information on the temperature at which a material softens and starts to flow. This information is important for the design of extrusion heads (cooling and heating zone) and the selection of the right temperature for hot bed and build chamber.

2.5.2.1 Networking Density

The crosslink density describes the average chain length between two crosslink points of a polymer chain, analogous to the thread length between two linkage points (mesh density) in a fishing net. A crosslinked polymer above the glass point can only be reversibly stretched until the densest part of a network is fully stretched and taut. Oscillatory loading at a constant frequency in the linear measuring regime results in a typical modulus that is controlled by the crosslink density of the polymer chains.

The crosslink density v_e as a measure of the degree of curing (Equation 2.7) gives the number (in moles) of effective crosslinking points per cm^3 in the rubber-elastic range at temperatures of $T > T_g$ [Ehrenstein & Bittmann 1997]. It can be obtained by determining first the mean chain length M_C between the crosslinking sites from the measured elastic modulus E' according to Equation 2.6, at a given material density and temperature. Looking at the structure of a network, it is obvious that the higher the chain length M_C between the crosslinking sites, the more flexible a material is [Gu et al. 2020]. M_C has thus been routinely experimentally determined from measurements of the elastic modulus G' or E' in the rubber-elastic state using the theory of rubber elasticity [Schäfer & Rehage 1969].

$$\begin{aligned} G' &= n \cdot k \cdot T \\ E' &= 3 \cdot G' \\ E' &= 3 \cdot n \cdot k \cdot T \\ \rho &= \frac{n \cdot M_C}{N_A} \\ R &= k \cdot N_A \\ \rightarrow M_C &= \frac{3 \cdot \rho \cdot R \cdot T}{E'} \end{aligned} \tag{2.6}$$

G': Shear modulus in the rubber-elastic range (MPa)

E': Modulus of elasticity in the rubber-elastic range (MPa)

n: Number of molecular chains per unit volume ($1/mm^3$)

k: Boltzmann constant (J/K)

T: Temperature at *E′* (K)

M_C: Average molar mass between the crosslinking sites (in g/mol)

N_A: Avogadro number (1/mol)

p: Density (g/cm^3)

R: General gas constant (J/(mol·K))

$$M_C = \frac{\rho}{\nu_e} \rightarrow \nu_e = \frac{\rho}{M_C} = \frac{E'}{3 \cdot R \cdot T} \tag{2.7}$$

ν_e: Crosslink density (mol/cm^3)

2.5.2.2 Entropy Elasticity

Above the glass transition temperature, crosslinked polymers exhibit a network modulus, which is called the plateau modulus. Linear high molecular weight and only lightly crosslinked elastomers have a so-called rubber plateau modulus. The rubber plateau modulus is a temporary (transient) network of chemical and physical network points, which is mainly formed by entanglements of polymer chains and polymer–polymer interactions. Under long-term constant stress, the entanglements and polymer–polymer interactions can become unhooked. The chains then flow past each other and the stress is relieved. In this case, the network modulus is time-dependent.

$$\nu = \nu_e + \nu_c \tag{2.8}$$

ν_e: Physical crosslink density (for example by entanglements) (in mol/cm^3)

ν_c: Chemical crosslink density (in mol/cm^3)

Contrary to expectations, the modulus of a polymer in the rubber plateau increases slightly with temperature. This behavior is called rubber elasticity or entropy elasticity [Gordon 1976]. The entropy is a measure of the disorder of a system. The lower the order, the greater the entropy. Nature strives towards the state of greatest disorder (respectively the greatest degree of freedom).

Entropy elasticity expresses the effort of macromolecules to return to the most entropically favorable state, the disordered tangle state, after deformation. When a flexible polymer material is subjected to mechanical stress, the macromolecules align themselves in the applied stress field. The resulting higher molecular state of order is accompanied by a reduction in the entropy (disorder) of the system.

If the flow of chain segments under load is prevented, for example, by chemical or physical crosslinking, and the network of polymer chains become more oriented, then the macromolecules strive to maximize entropy when the load is removed and return to the disordered initial state. For a polymer network, this means that it opposes the imposed order in the direction of a mechanical load it is subjected to.

The standard Gibbs free energy change, ΔG, indicates the thermodynamic favorability of a physical or chemical process. When $\Delta G < 0$, the process is thermodynamically favored.

The change of the Gibbs free energy ΔG in a materially closed system (no material exchange) is a summation of the change of the enthalpy ΔH and the product of the temperature (in K) and the change in entropy ΔS (see Equation 2.9).

$$\Delta G = \Delta H - T\Delta S \tag{2.9}$$

While the enthalpy ΔH or internal energy of a polymer chain changes hardly at all during macroscopic deformations (highly energetic atomic bond lengths and bond angles are not significantly deformed under normal load conditions), the entropy decreases significantly due to the reduction of disorder of the system when the chains are oriented mainly by low-energy rotational deformations in the direction of the applied stress field (load). Hence the term $-T\Delta S$ becomes positive and the Gibbs free energy increases, meaning the process of deformation increases the energy of the system.

Since ΔS is multiplied by T, the buildup energy of the system (and thus also the resulting restoring force) increases with increasing temperature and decreases with decreasing temperature. Thermodynamically, the temperature-dependent effect of rubber elasticity is thus essentially based on a decrease in entropy ΔS under deformation.

In reality, this purely entropic effect, which increases the modulus of the polymer with increasing temperature, is often superimposed by other interactions within polymeric materials, which contribute to the physical network density and thus the modulus, and usually decrease with temperature. Examples of this are the interactions of polymers with fillers or the interactions of polymers with each other (van der Waals interactions, dipole–dipole interactions, hydrogen bond interactions etc., cf. [Minch 1999]).

2.5.2.3 Relaxation Times and Temperature

The DMA measurement clearly shows the temperature dependence of the modulus of polymeric materials. The modulus can be understood as a measure of the sum of physical and chemical interactions between individual chain segments. In addition to a temperature dependence, the strength of the interactions also shows a frequency dependence via the so-called relaxation time τ of an "interaction". If the relaxation time at a given temperature is much smaller than the measurement frequency, its effect on the modulus is small. If the relaxation time τ is equal to or greater than the measurement frequency, the polymer chains cannot respond to the applied stress field by relaxing (flowing), and the interactions have an effect similar to that of a non-transient chemical crosslinking site. The modulus is therefore frequency-dependent, because the strength or effectiveness of the described interactions between individual chain segments, which in sum influence the modulus, are all influenced by the measurement frequency and temperature via so-called relaxation times τ.

The relationships can be illustrated in the stress relaxation test where the decrease in stress σ with time is measured after a defined deformation (Figure 2.14, see also [Elias 2002]). Equation 2.10 describes the dependence of the stress σ on the relaxation time τ_{max} of a specific interaction, here described by a spring connected in series with a damping cylinder (Maxwell element). In reality, polymers have a large number of interactions, which are expressed in a large number of individual relaxation times, and are summarized in the temperature- and frequency-dependent properties of polymers, such as the modulus.

$$\sigma = \sigma_0 \cdot \exp\left(-\frac{t}{\tau_{max}}\right) \tag{2.10}$$

The temperature and frequency dependence of the modulus above the glass transition temperature and in the melt can be determined experimentally by a superposition of measurements at different temperatures and frequencies and mapped in a so-called master curve [Lüpke 2015a].

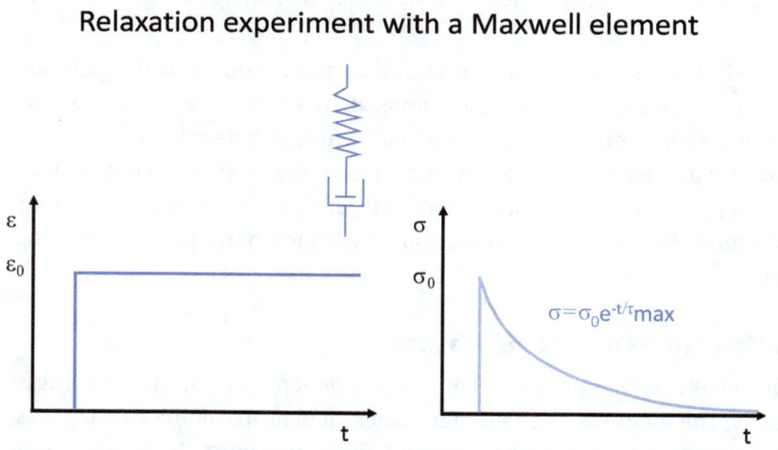

Figure 2.14 Relaxation test with a Maxwell element measured in strain

Figure 2.15 shows the composite master curve for a polycarbonate, which covers a frequency range of about 15 orders of magnitude via the temperature–time superposition principle. The flow region, the rubber plateau, and the glass transition region can be clearly read in the master curve.

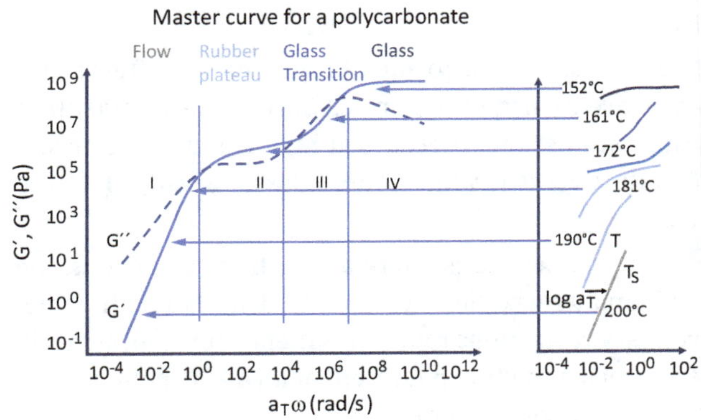

Figure 2.15 Master curve for a polycarbonate measured in shear (composed of several frequency sweeps at different temperatures)

Knowing the critical degree of polymerization (the critical chain length), a knowledge of the master curve can be used to estimate not only the viscosity or modulus at arbitrary temperatures and frequencies, but also the molecular weight for linear normally distributed polymers. The critical degree of polymerization describes a chain length specific to each polymer at which individual polymer chains begin to interpenetrate and effectively entangle with each other. The entanglements act as reversible physical network points whose effect depends on time, frequency, and temperature. The critical degree of polymerization n^* can be obtained from a plot of the logarithm of the viscosity η against the degree of polymerization n (number of repeat units of a polymer chain) as the point at which the slope changes abruptly (Figure 2.16). The critical chain length (better critical chain weight) results from the product of the degree of polymerization n with the molecular weight of a chain repeat unit (chain segment).

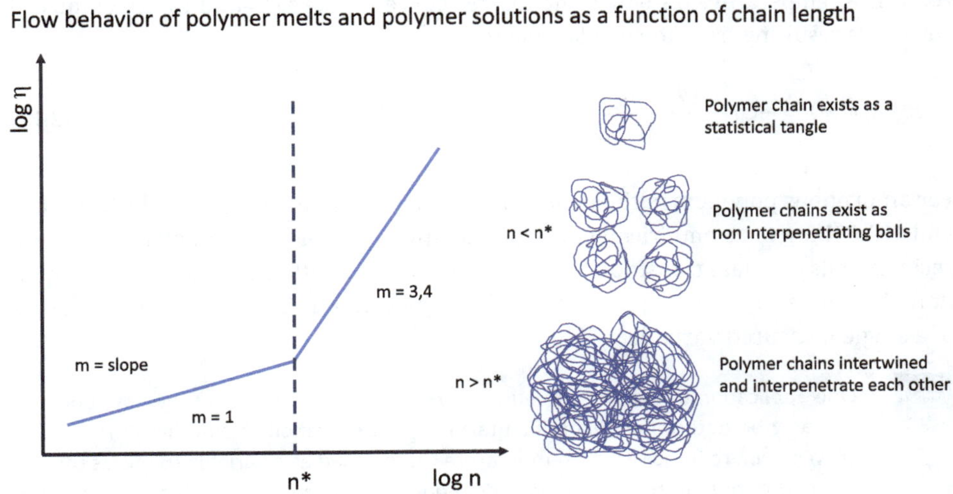

Figure 2.16 Viscosity as a function of the chain length of polymers

2.5.2.4 Observable Relaxations

When a dynamic mechanical experiment is performed on polymers, different relaxations at different frequencies are recorded (Figure 2.12). The main relaxation (α-relaxation) shows the largest modulus step. It occurs at high frequencies or low temperatures. From here on, collective rearrangements (glass transition) of the polymer main chains can take place.

At even higher frequencies or lower temperatures, there are further relaxations, the so-called secondary relaxations (β- and γ-relaxations, cf. [Doi & Edwards 1988]). A molecular interpretation of the secondary relaxations represents them as only local movements of molecule/chain segments, such as a kink movement of an alkyl chain, a "phenyl flip", or a rotation of a dipole (e.g., an ester group).

The frequencies of all polymer relaxations, especially those in the melt, are temperature-dependent, and shift to higher frequencies at higher temperatures (relaxation times *decrease*). Instead of plotting the curves for different temperatures against frequency (see Figure 2.15), it is therefore also possible to plot curves for different frequencies against temperature.

The so-called Williams–Landel–Ferry equation (WLF equation, or time–temperature superposition equation) developed by Malcolm L. Williams, Robert F. Landel, and John D. Ferry (see Equation 2.11, see also [Williams et al. 1955]) describes this temperature dependence of the relaxation mechanisms of amorphous polymers and other materials above the glass transition temperature in the polymer melt.

The influence of crystallization in crystallizing materials, as well as the relaxations below the T_g, are not covered by the equation. The $\log_{10}(a_T)$ is the decadic logarithm of the displacement factor a_T, where T is the measurement temperature, and T_s is the reference temperature on which the master curve is based. C_1 and C_2 are empirical constants resulting from the displacement.

$$\log_{10}(a_T) = -C_1 \frac{(T - T_s)}{C_2 + (T - T_s)} \tag{2.11}$$

For amorphous polymers in the melt, van Krevelen recommends the value 8.86 for C_1 and 101.6 K for C_2 as empirically determined, quasi-universal parameters. For T_s, he recommends the glass transition temperature T_g + 43 K. With the aid of the WLF equation, the temperature dependence of the viscosity can thus be estimated on the basis of a single measured variable.

> One application of the WLF equation (the essential parameters of the equation can only be determined experimentally) is the generation of so-called "master curves", where frequency or temperature can be used as a variable to represent a broad temperature or frequency dependence of complex modules, as required. The method is used for predictions of polymer properties, such as viscosity

above the glass transition temperature at high frequencies, since these are often difficult to determine experimentally. As a rule, a frequency jump by a factor of ≥10 corresponds graphically to a decrease in measurement temperature of approx. 7–8 °C. This information is particularly valuable if the product performance of polymers at high frequencies needs to be simulated.

2.5.2.5 Free Volume

The mechanical and rheological properties of a polymer below and above the glass transition point are influenced by the temperature-dependent free volume. The presence of a free volume is what makes the molecular and chain motions possible in the first place. The relationship between the WLF equation and the free volume of a material allows the activation energies of molecular mobility to be calculated. In the WLF concept, the free volume depends on the temperature-dependent polymer chain vibrations. The activation energy of these chain vibrations can be calculated using the WLF displacement constants C_1 and C_2.

Changes in specific volume with temperature are usually explained by the Vrentas–Duda partition (Figure 2.17). Accordingly, the specific volume of an amorphous thermoplastic consists of the intrinsic volume of the molecules V_0 at absolute zero (0 K), the increase in volume with temperature V_s caused by changes in the vibration amplitudes of the polymer chain vibrations, and the free volume V_l. In the case of semicrystalline thermoplastics, the relationships are even more complex due to crystallization and the resulting shrinkage effects.

Volume fractions in polymers according to Vrentas and Duda

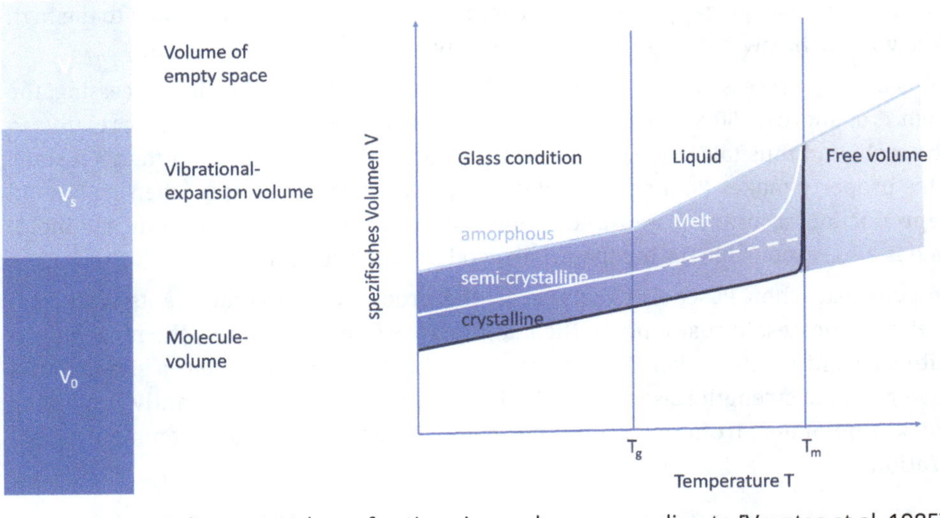

Figure 2.17 V–T diagram: Volume fractions in a polymer according to [Vrentas et al. 1985]

The polymer chain mobility depends on the chain segment mobility in a polymer. A chain segment is understood as the typical repeating unit of a chain. Polymer chain mobility includes the speed for changes in chain conformation (micro-Brownian movements: rotation, oscillations) and place change processes (macro-Brownian movements: slipping of chain segments). Since a polymer in the melt behaves approximately like a (viscoelastic) liquid, the chain segment mobility and, accordingly, the free volume must be large.

Due to the high number of available empty spaces, the change of chain segments takes place largely unhindered. On cooling, the number and size of vacancies decreases and, in parallel, segment mobility decreases. Since molecules attract each other, the decrease in free volume is associated with increasing internal friction due to increased intermolecular interactions. As a result, the viscosity (as a measure of friction between molecules) increases.

With further cooling, the number and size of the voids finally become so small that, for reasons of space, translational movements are no longer possible. The plastic falls below the glass point and "freezes", thus increasing its brittleness. The modulus of the polymer changes by several orders of magnitude in this process.

However, the free volume is not only associated with rheological phenomena, but also with chemical and mechanical properties. The size of the free volume below the glass transition point is an indication of the mechanical properties, especially the toughness of a material. A high free volume correlates with better impact strength, since introduced energy can be dissipated via rotation and vibration, if necessary. A high free volume is also a good indicator of high gas permeability.

A typical example of a material with a high free volume is polycarbonate (PC). PC exhibits an untypically high impact strength for its high T_g (approx. 150 °C). At the same time, it exhibits high gas permeability compared to its high T_g, and (related to the high free volume below T_g) a high sensitivity to solvent-induced stress cracking.

PC is usually processed at temperatures >300 °C from the melt. During processing, the high T_g of approx. 150 °C ensures that PC solidifies very quickly, shrinks little relative to the melt, and thus takes along or freezes part of the free volume from the melt state. This property makes PC a popular material in the plastics industry when high mold reproduction accuracy is required in injection molding (low shrinkage in the mold) and is also responsible for the material's good impact strength.

In contrast, when PC cools slowly, as in the production of components with high wall thicknesses, for example in the manufacture of optical lenses, the material exhibits a significantly lower free volume. As a result, the material has a significantly lower impact strength [Elser et al. 2012]. This is countered by frequently producing thick components from several thinner layers by consecutive injection molding operations.

When using PC in 3D printing via extrusion, the build temperature should be selected in such a way that good layer adhesion is achieved despite the high glass transition temperature of approx. 150 °C. This is achieved by using a high build temperature and a high extrusion temperature. The highest possible build temperature and a high extrusion temperature have proven to be advantageous for this. At the same time, care must be taken to ensure that the material is quickly quenched to a temperature below 150 °C in order to maintain the toughness of the material in the glassy state, which is typical of PC. For this purpose, in turn, the lowest possible build chamber temperatures are desirable. In the context of these conflicting requirements, build chamber temperatures of >80 °C and <130 °C and extrusion temperatures of >280 °C are recommended in practice. Due to the large contact area of the uppermost layer of the extruded polymer with the surrounding air (which applies especially to extrusion dies with a diameter ≤0.8 mm), rapid cooling to a temperature below the T_g, quite comparable to the cooling time in injection molds, is achieved even at build chamber temperatures of 90–130 °C.

Knowledge of the temperature-dependent material properties of polymers in the solid and in the melt through the use of suitable test methods, such as dynamic mechanical analysis (DMA), dynamic heat flow calorimetry (DSC), and the various capillary viscometer measurement methods (HKV, MFR, MVR; see also Section 2.5.4 and subsequent sections) thus allows fast and direct access to the selection of suitable processing conditions for the 3D printing of thermoplastic polymers.

Another decisive factor for the design of components to be printed from the melt by extrusion is the knowledge of the shrinkage of the material applied hot from the melt on its way to the cooled solid. The shrinkage has two main effects. First, it can lead to premature detachment of the part from the build platform and general warping of the printed part or of the printed component geometry. This becomes particularly critical when printing long straight component areas in a continuous build line or highly filled component areas (parts comprising high-density fill structures). Depending on the build conditions, the shrinkage can lead to uniform shrinkage in all spatial directions or to shrinkage that is selective to the build direction. Furthermore, shrinkage in highly filled component areas can lead to significant underfilling. This can result, among other things, in pronounced component porosity and/or low layer adhesion.

Since, depending on the material, processing temperature and crystallinity, a shrinkage of 5–25 % is common during the pressure-free transition from melt to solid [Frankland 2017], this behavior may need to be accommodated by over-extrusion or specific line control during extrusion. Shrinkage can be determined by matching the tabulated solid volume and the melt volume from a melt volume flow measurement (MVR, see Section 2.5.5). More precisely, the temperature-dependent volume is shown in a V-T diagram (Figure 2.18) at a given pressure.

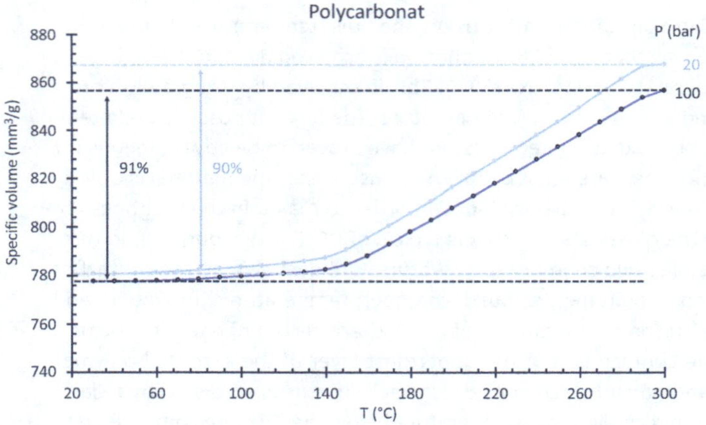

Figure 2.18 V–T diagram for a medium-viscosity PC at two different pressures [Covestro n.d.]

Typical values for the ratio of melt density to density at room temperature for standard and engineering thermoplastics at the transition from processing temperature in the melt to room temperature are given in Table 2.2, in which the density is defined as the reciprocal of the specific volume.

Table 2.2 Typical Melt and Solid Densities of Common Polymers [Beaumont 2015]

Polymer	Melt density at processing temperature (g/mL)	Solid density at 23 °C (g/mL)	Ratio between melt density and solid density (%)	Polymer type
Nylon 66	0.97	1.14	85	Semi-crystalline
PP	0.7	0.91	78	Semi-crystalline
PET	1.2	1.4	86	Semi-crystalline
PS	0.97	1.05	92	Amorphous
PMMA	1.04	1.16	90	Amorphous
PC	1.08	1.20	90	Amorphous
ABS	0.97	1.05	92	Amorphous
PLA	1.13	1.25	90	Semi-crystalline
TPU	1.13	1.20	94	Semi-crystalline

High processing temperatures and a high degree of crystallization promote a high density change and thus a high shrinkage tendency during the transition from melt to solid. High shrinkage, especially if it is asymmetrical, can lead to detachment from the component platform in 3D-printed components, as well as to internal component stress, warpage, and a local loss of layer adhesion. Accordingly, a high build temperature helps to avoid undesired direction-dependent shrinkage.

Polymers with a slow crystallization rate and a low T_g, such as thermoplastic polyurethanes (TPU), are particularly advantageous in terms of layer adhesion when processed in extrusion-based processes. What is a disadvantage in classic processing technology (for example, due to a longer cycle time in injection molding, because of slow solidification in the mold) can be an important advantage in 3D printing with regard to processing reliability.

Overall, in 3D printing of thermoplastics, both in powder-based and extrusion-based processes, preference is given to polymer types developed for processing by extrusion. These are often characterized by lower injection swelling and a higher melt stability in combination with a slower crystallization tendency. This is desired to achieve good surface qualities in extruded profiles or films, and is often achieved by selective long-chain branching or by a broad molecular weight distribution of the polymer chains.

The differences between melt and solid density (due to shrinkage) as well as the temperature-dependent properties of materials, must be taken into account either by suitable temperature and pressure conditions in the build chamber and/or in the so-called slicer when planning extrusion line laydown patterns or laser guidance in laser sintering equipment. The large differences in shrinkage, cooling, and crystallization behavior of different polymers, if not carefully planned for, can have extreme effects on the desired part quality. Professional 3D printers therefore often have fixed print profiles for selected "locked-in" materials and material formulations that take these relationships into account. Accordingly, these printers are locked to work with often cheaper but less well-characterized materials available on the open market.

2.5.3 Calorimetric Measurement Methods

While thermomechanical measurement methods such as DMA provide information about the mechanical properties of polymers over temperature and frequency, calorimetric methods such as dynamic heat flux differential calorimetry (differential scanning calorimetry, DSC), are particularly suitable for providing information on the heat flows (energies) that are required or generated during phase transitions.

Relevant phase transitions are, on the one hand, the glass points. This is a second-order phase transition in which the first derivative of the chemical potential does not change abruptly with pressure and temperature. On the other hand, there are the so-called melting points. This is a first-order phase transition in which the derivative of the chemical potential changes abruptly according to pressure and temperature. With DSC the following determinations, among others, can be carried out: phase transitions such as melting and glass transition temperatures, degree of crystallization, kinetic and enthalpic analysis of chemical reactions, specific heat capacities, decomposition points, etc.

In a DSC measurement, an encapsulated container with a sample and a second identical container without contents (reference) are subjected together to the same temperature change program in a heat bath. As a result of the heat capacity of the sample and exothermic (energy-releasing) or endothermic (energy-consuming) processes, for example during phase transitions such as melting, crystallization, or evaporation, temperature differences or different heat flows occur between the sample and the reference, which are then recorded (differential, see Figure 2.19).

In addition to physical phase transitions, DSC also records chemical reactions, which is why it is a popular method for analyzing the heat output and kinetics of chemical reactions. DSC analysis of polymers can be performed on a few micrograms of substance and takes only a few minutes to hours, depending on the problem. DSC is a cost-effective and widely used analytical method in the chemical industry, and is particularly common in the development and quality control of polymeric materials.

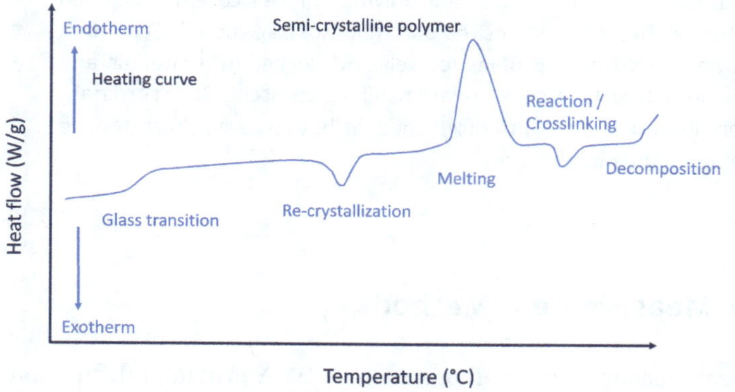

Figure 2.19 DSC measurement of a crystallizing thermoplastic up to its decomposition

In a DSC measurement of, for example, semi-crystalline thermoplastics such as PA12, in addition to the glass transition point, both the melting and recrystallization processes are recorded in one measuring cycle (heating and cooling) (Figure 2.20). In contrast to low-molecular-weight materials, polymers often do not have a sharp melting point, but a wide melting or softening range.

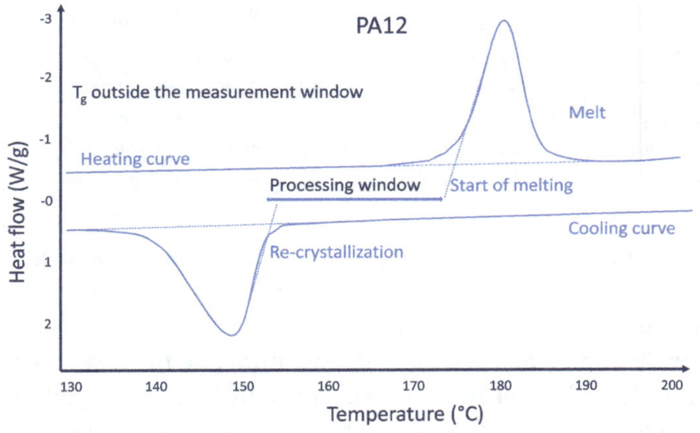

Figure 2.20 Processing window of PA12

For measurements of polymers, depending on the problem, the temperature can be cycled several times between as low as –150 °C (cooling with liquid nitrogen) and a temperature just above the melting point of the polymer (typically less than 300 °C), Two measuring cycles are common. The second measuring cycle has the advantage of a defined thermal history. The measurement usually takes place under inert gas (nitrogen or argon) to minimize chemical changes and oxidative processes during the measurement.

As with DMA, there is also the possibility of varying the heating rates in DSC. In DSC, the heating rate and cooling rate are varied in order to obtain information about melting and recrystallization kinetics and to selectively control crystal morphologies. With slow heating, lower melting points tend to be measured in the first heating cycle, since the polymer molecules have more time to react to the temperature change. Slow cooling usually results in larger and more ordered crystals, as well as higher crystallinity, and thus higher shrinkage of the thermoplastics under investigation. After slow cooling, the melting point may increase in the second heating cycle due to the higher-ordered crystal structures. The measurement signal (signal height) in the DSC depends on the physical sample history and the heating/cooling time. A higher heating/cooling rate leads to a clearer/higher measurement signal in addition to a shorter measurement time.

The temperature history of a material in the processing procedure plays a decisive role for the properties of the product, since important properties such as crystallinity, crystallite sizes, phase morphology, shrinkage, free volume, and also optical properties such as transparency and surface quality are influenced by this. This is particularly important for rapidly crystallizing semi-crystalline thermoplastics such as polyamides, polyesters, and also some polyurethanes, as well as high-T_g thermoplastics such as polycarbonate.

> DSC is an important measurement method for 3D printing, in particular powder printing. The knowledge of the energy demand and position of phase transitions helps to determine suitable process conditions such as the build chamber temperature and the minimum required and maximum permitted laser energy needed to melt and sinter a polymer. Together with the knowledge of the melt viscosity, a prediction of the layer adhesion and component density can be made.

Of particular importance for the powder printing process is the knowledge of the temperature difference between the melting and recrystallization temperatures for semi-crystalline thermoplastics (Figure 2.19). From a process point of view, this window should ideally be as wide as possible. The build temperature and the irradiated laser power must be matched to each other in such a way that only in the laser path is the uppermost layer of the powder bed selectively melted, allowing it to bond/sinter with the layer below.

Both the melt viscosity of the polymer at the given build temperature and the time until recrystallization have a strong influence on the product density and the mechanical properties of the product. If the melt viscosity of the polymer at the given laser-induced build temperature is too low, then this can cause the polymer melt to sag into deeper powder layers, which again affects the sharpness of the sintered layer image. If the viscosity of the polymer melt is too high, or the recrystallization of the melt is too rapid, then this can have a negative effect on the interlayer adhesion of the powder layers. Under certain circumstances, recrystallization that is too rapid/strong can further lead to severe distortion and thus to abortion of the construction process. In this case, the top layer of the molten polymer bulges in the build direction and can be carried away by the powder roller while adding another powder layer.

Good imaging accuracy and interlayer adhesion are always obtained when the melt viscosity is sufficiently high and the recrystallization temperature of the polymer is well below the build chamber temperature. Recrystallization should be very slow at the given build chamber temperature to avoid warping of the part during the build process due to a tension build up between layers by inhomogeneous crystallization. The material melted by the laser in the previous layer is therefore ideally not yet recrystallized and can thus combine well with the layer above. On the other hand, the laser energy should not be so high that, via heat conduction, the melted layer is bonded to a previously non-melted layer below.

A temperature gradient in the build direction is often applied to the build chamber for better control of the build behavior. The temperature decreases continuously with distance from the build surface. This allows better control of the recrystallization rate and minimizes component stresses. A temperature gradient in the build direction also improves the reuse rate of the polymer powder by minimizing stress due to temperature exposure and aging of the polymer material, especially over long build times.

A large temperature difference between melting and recrystallization temperature for the thermoplastic powders used enables a sintering process at temperatures that are as low and gentle as possible. High build temperatures favor the degradation or crosslinking of polymers by thermal and chemical side-reactions, and usually significantly reduce the reuse rate of the thermoplastic powder used.

DSC measurements also help to explain the difficulty of powder sintering of amorphous thermoplastics such as polycarbonate (PC) and polymethyl methacrylate (PMMA), which only have a glass point and no crystallite "melting point", which is useful because it can provide extra physical crosslinking points at temperatures usually significantly higher than the T_g. Here, especially with sharp glass transitions, the softening temperature and solidification temperature are almost congruent, and thus the temperature window for processing is only very small.

In contrast to recrystallization, which is kinetically controlled, the material modulus in the melt is directly related to the inherent polymer chain mobility and reacts to a temperature change without delay. In the case of crystallization, suitable chain segments must first be found, which is why the recrystallization temperature is almost always significantly below the melting temperature.

If the temperature falls below the glass transition temperature, there is an almost instantaneous jump in viscosity and modulus over several orders of magnitude. The polymer freezes because the chain mobility approaches zero. Another difference to crystallization is that, unlike recrystallization, which is exothermic, no energy is released during the T_g transition that could stabilize the transition temperature. The temperature window for successful powder printing of amorphous high-T_g thermoplastics is accordingly very narrow.

The melt viscosity of semi-crystalline polymers at the melting point is often significantly lower than the melt viscosity of amorphous polymers at the glass transition point. The semi-crystalline polymers jump at the melting point to a value of free volume that is directly related to the distance from the glass transition point ($T_m - T_g$). This jump results in a significantly higher free volume, a significantly higher chain mobility, and thus lower melt viscosity, compared to a high-T_g polymer at otherwise comparable molecular weights and melting temperatures. High-T_g polymers typically have bulky substituents and a relatively large critical chain length (see Figure 2.16), implying higher viscosities compared to crystallizing polymers with a melting point in the same range as the T_g of the high-T_g polymer. For polyamide 12 (PA12), the temperature difference between the T_g and the melting point is about 160 °C, while for polyamide 6 (PA6) it is about 210 °C. Thus, an amorphous polymer such as PC with a T_g of ca. 150 °C would not experience a viscosity decrease similar to that of a semi-crystalline polymer at the melting point until temperatures of about >300 °C.

The high viscosity jumps of high-melting thermoplastics allow for good densification of the material in powder sintering processes and enable high extrusion volume rates in extrusion processing. Amorphous thermoplastics, on the other hand, have advantages, both in sintering and extrusion processes, with regard to mapping accuracy and bridging of overhangs due to the rapid solidification processes.

Powder printing of amorphous thermoplastics often only delivers product properties and component densities that are significantly below the properties of injection-molded products of the same polymers. Precise temperature control in powder sintering processes is limited in particular by the necessary heat conduction. The temperature (heating and cooling) can largely only be set via the sides of the build chamber through the powder (as a very poor heat conductor), which is becoming increasingly difficult, especially with large build spaces and high component temperatures, as well as with high part densities in the build chamber.

Products that exhibit favorable melting and recrystallization behavior include polyamides such as PA12 and PA11, as well as some thermoplastic polyurethanes (TPU), polyester copolyamides, polyether block copolyamides (PEBA), and polyether ether ketones (PEEK).

Information from DSC analysis is also of great importance for extrusion-based processes. Similar to powder printing, layer adhesion is also influenced by the melting and recrystallization temperature. Since higher melt application temperatures can be comfortably reached in the extrusion process compared to sintering processes, additional energy can be introduced into the lower layer via the temperature of the melt strand. Therefore, extrusion-based processes are somewhat freer in material selection than powder processes. However, the following also applies here: in order to achieve good layer adhesion, both the lower and the upper layer must be melted on contact. Particularly with amorphous thermoplastics, which are difficult to process, or with rapidly crystallizing thermoplastics, a correspondingly high extrusion temperature in combination with a high build chamber temperature must be set in order to obtain good mechanical properties in the build direction (z-direction).

Thus, in extrusion-based processes, DSC measurements help to estimate whether an extruded filament can adhere to the previous build layer, or how the extrusion temperature and build chamber temperature must be selected to minimize shrinkage and warpage. Widely used polymers that are excellent for processing in extrusion-based processes include polylactic acid ester (PLA), acrylonitrile-butadiene-styrene (ABS), many thermoplastic polyurethanes (TPU), and polyesters such as polyethylene glycol terephthalate-*co*-polyester (PEBT, PETG, etc.). More difficult to process, but in high demand for their high temperature resistance and mechanics, are high-T_g, amorphous thermoplastic materials like PC, polyetherimide (PEI) and PEEK.

2.5.4 Melt Viscosity

Particularly in the case of polymers that have a viscosity that is strongly dependent on temperature and shear rate, we speak of viscosity functions in the melt. In this case, the viscosity of the melt is measured over a wide temperature and shear rate range above the melt temperature. The melt viscosity of a thermoplastic polymer is a function of the prevailing temperature, frequency, and shear rates, and determines the processing properties in the melt. Accordingly, to cover various processing and application technologies, the melt viscosity of a polymer is usually plotted for various temperatures as a double-logarithmic plot of viscosity versus the shear rate (Figure 2.21).

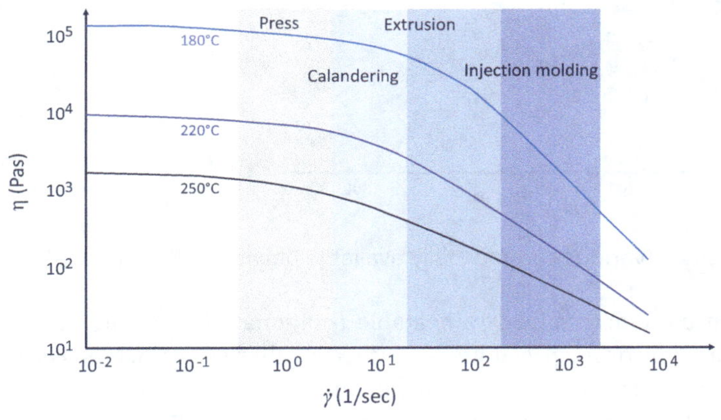

Figure 2.21 Viscosity plotted versus temperature and shear rate

Precise knowledge of the processing parameters (temperature gradients in the component, shear gradient in the extruder), and the linking of these process parameters with the measured melt viscosity function, make it possible to optimally adapt the process conditions to the polymers to be processed.

> In the range of low shear rates (0.01–0.1 s^{-1}), when the melt viscosity does not change significantly, it is also referred to as "zero viscosity". The zero viscosity is important for all processing operations where the material should flow without major external force and deformation. A prominent example of this is the powder sintering process in 3D printing. For extrusion-based 3D printing processes, on the other hand, viscosity data in the range of a shear rate of 10–100 s^{-1} are more relevant.

In general, the viscosity of melts of the same polymer class increases exponentially with increasing weight-average molecular weight of the polymer (Figure 2.16, cf. [van Gurp & Palmen 1998]). The exact relationship depends on many molecular parameters of the polymers (in addition to the shear rate and the measurement temperature) and is specific for each polymer. Of crucial importance are the monomer composition, molar mass, molecular weight distribution, degree of branching, and stereochemistry of the polymer. This results in the chain segment mobility and the critical chain length (entanglement length) of the polymer, but also defines its polarity and interactions with itself, with other polymers and with fillers, as well as the chemical and thermal bond stability of the polymer backbone [Montarnal et al. 2011].

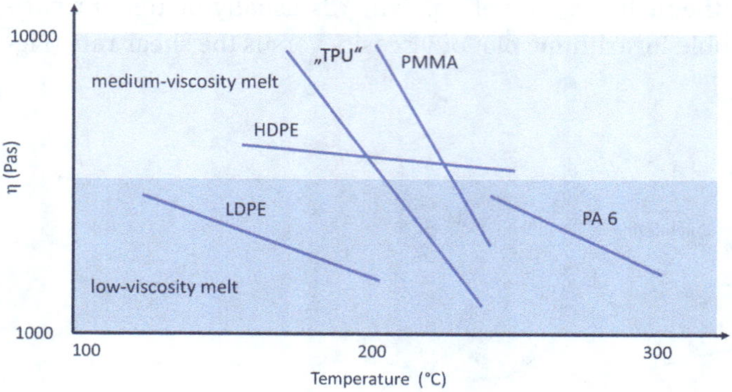

Figure 2.22 Melt viscosity of various commercially available polymers [Kaiser 2006]

The melt viscosity is usually determined in heatable rotational viscometers of the plate-cone or plate-plate type or in special high-pressure capillary viscometers. For a reliable result, a constant temperature and shear rate during the measurement is essential. The rheometer allows the estimation of the shear-dependent flow properties as a function of the feed pressure and temperature particularly well. From this, the pressure build-up or mass flow in the die can be derived for extrusion-based processes at a given extrusion temperature and die geometry. A prerequisite, however, is that the polymers are thermally stable at the measurement temperature and during the measurement, and do not crosslink or degrade.

From the melt modulus curves, if the polymer composition is known, it is possible to infer the existence of short- and long-chain branching and crosslinking of a polymer by simply plotting the complex modulus G^* against the phase angle δ in the van Gurp–Palmen plot (Figure 2.23).

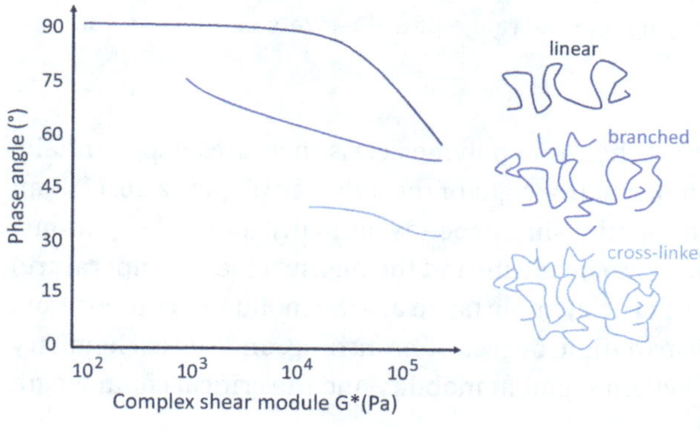

Figure 2.23 van Gurp–Palmen plot [Trinkle & Friedrich 2001]

2.5.5 Melt Volume Flow Rate

A widely used and simple method for evaluating the processing behavior of polymers in the melt is the measurement of the melt volume flow rate (MVR), formerly and still today often colloquially referred to as melt volume index (MVI). The MVR is used to characterize the flow behavior of a thermoplastic under "constant" pressure and temperature conditions at a given die geometry (Figure 2.24).

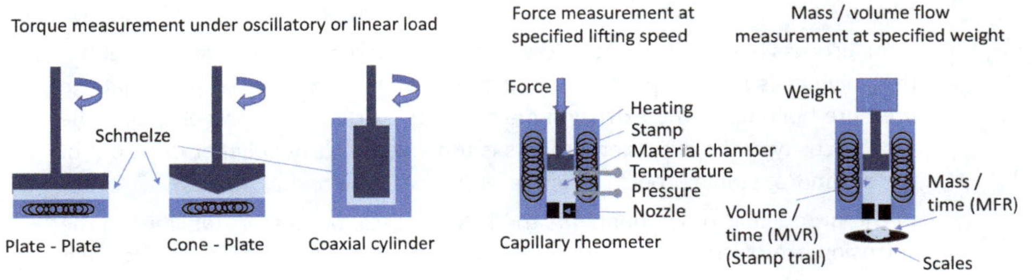

Figure 2.24 Viscosity measurement methods

The melt mass flow rate (MFR) is determined in the same way as the MVR, and differs in the measurement result only by the melt density factor. The ratio of the density of a polymer in the solid state and the density in the melt corresponds to the shrinkage of a material that is processed from the melt.

According to ISO 1133, the MVR and MFR are determined by means of a simplified capillary rheometer, where the material (granules or powder) is melted in a heatable cylinder and pressed through a defined nozzle (capillary) under a controlled pressure created by a given applied load. The exiting volume or mass of the polymer melt (the so-called extrudate) is determined as a function of time. Simply dividing the MFR by the MVR gives the melt density [Dynisco 2019].

A significant advantage of the melt volume flow rate is the simple recording of the piston stroke (with a known piston diameter) for determining the discharged melt volume. In contrast, with the melt mass flow rate, the tapped melt strands must be weighed, and as such a small additional effort is involved in the measurement. The MVR is expressed in cm^3 per 10 min and the MFR in g per 10 min. The measurement conditions include a test temperature corresponding, for example, to the typical processing temperature, and a mass of, for example, 5 kg or 10 kg exerting a uniform force on the piston of specified diameter.

The melt viscosity has become of great practical importance, especially for the design and operation of plastics processing equipment. Knowledge of melt viscosities is important in all molding processes that process melts, such as melt spinning, extrusion, injection molding, thermoforming, and also in 3D printing of thermoplastic polymers based on extrusion or powder sintering processes.

Precise knowledge of the shear- and temperature-dependent melt viscosity, as an ideally reversible change in properties, is important in AM to predict suitable processing conditions. This is especially true with respect to selecting the optimal extrusion temperature, as well as calculating the maximum possible extrusion volume per unit time in extrusion-based AM (FLM) processes. Knowledge of the shape of the flow curves at low shear and in the supercooled melt region in the powder sintering process also allow an initial prediction of the quality of the layer adhesion, and support the selection of suitable process parameters with regard to the build chamber temperature and the energy input by the laser, for example.

If the process temperature is too low in the extrusion-based process, then either the polymer is not completely melted or the melt viscosity is so high that the pressure build-up at the extrusion die prevents extrusion of the polymer. If the melt viscosity in the extrusion process is too low, then the polymer drips out of the die and/or stable product geometries are not obtained.

In the production of filaments for the FLM process, process instabilities in the extrusion are often used to produce filaments with a "shark skin" surface. For this purpose, either the temperature at the die is lowered or the volume discharge rate is increased to such an extent that a periodic break in the melt flow, the so-called melt fracture, occurs. Filaments with a structured surface can improve the grip of the filament in the conveyor unit of the printers.

In polymer sintering, the zero viscosity, i.e. the melt viscosity at a very low shear rate, is an important parameter for estimating the sintering behavior of thermoplastic polymers. Although the polymers in the sintering process are melted, the driving force for densification is provided only by the gravity and surface tension of the melt, i.e. the need of the polymer melt to minimize its surface area against air. In addition, depending on the equipment, some compaction, e.g. by means of a powder roller, takes place during the next material layer application. In any case, the shear rate is very low, and knowledge of the zero viscosity gives a good indication of whether the molten powders can sinter. If the melt viscosity in the powder process is too high, then the powder grains melted by the laser will not flow together sufficiently. If the melt viscosity is too low, then the polymer sags downward after melting, and "bellies" can appear on the lower product surfaces.

The use of fillers, which in the powder sintering application area are used in addition to their function as fillers to improve the flow behavior of the polymer powders, can at the same time exert a critical influence on the sintering behavior. This is particularly true if the fillers cover the surface of the polymer powder to such an extent that the polymer cannot flow past them to the next polymer particle. In general, the polymers in the powder sintering process tend to be of sufficiently low viscosity at the given process temperature to allow for good sintering.

The reuse rates are to a considerable extent related to the viscosity change of the polymers in the powder sintering process. If the viscosity changes (for example due to heat aging) in the direction of higher viscosities (e.g. for polyamides, post-polymerization and branching typically tends to occur), the flow behavior, component density, and layer adhesion suffer. In comparison, the polymer viscosity of polyesters tends to decrease due to polymer degradation during polyester cracking. In both cases, the build properties may have to be readjusted by adding fresh powder. For a high recycling rate, a well pre-dried build material is mandatory and a low build temperature is advantageous. Particularly good results in terms of reuse rates are obtained here for materials such as TPU and PS, which can be processed at comparatively low build temperatures of ≪150 °C.

The melt viscosity depends on temperature, shear rate, and polymer structure. Measuring the melt viscosity as a function of temperature and shear rate is a simple and widely used method for characterizing, qualifying, or selecting polymers for suitability in 3D printing processes.

2.5.6 Heat Transfer

According to the first law of thermodynamics, the heat Q is an energy that is supplied or dissipated in a system due to temperature differences without any work being done. According to the second law of thermodynamics, heat always flows in the direction of a lower thermodynamic temperature.

Equation 2.12 describes the three-dimensional distribution of heat (or variation in temperature) in a given region over time. The left-hand side of the equation represents the rate of change of heat energy in the region, while the right-hand side represents the rate at which heat is conducted through the region and any heat sources present. The heat equation is a partial differential equation used to model how a quantity such as heat diffuses through a given region. It was first developed by Joseph Fourier in 1822. The equation is derived from Fourier's law and conservation of energy. Fourier's law states that the rate of heat transfer through a material is proportional to the negative gradient in the temperature and to the area, at right angles to that gradient, through which the heat flows. The heat released or absorbed can be calculated using the specific heat capacity, mass of the substance, and the change in temperature.

$$\rho \cdot C_p \cdot u \cdot \nabla T = \nabla \cdot (\lambda \cdot \nabla T) + Q \tag{2.12}$$

ρ: Density of material

C_p: Specific heat capacity of material

u: Velocity vector of heat transport

∇: Nabla operator

T: Temperature

λ: Thermal conductivity

Q: Heat transferred

Heat is transferred either via heat conduction or heat radiation. Heat conduction is the molecular transport of energy within a solid or a fluid, which occurs due to a temperature gradient in the material. Mathematically, heat conduction is calculated according to Fourier's law with $\dot{q}_x$quantifying the heat flow density as heat energy that is transferred through an area element A:

$$\dot{q}_x = -\lambda \cdot \frac{\delta T}{\delta x} \tag{2.13}$$

$\dot{q}_x = \frac{\dot{Q}_x}{A}$: Heat flux density

λ: Thermal conductivity

$\frac{\delta T}{\delta x}$: Temperature gradient perpendicular to the surface element A

$\dot{q}_x = \frac{\dot{Q}_x}{A}$: Heat flux density

The thermal conductivity λ describes the amount of heat that flows through a unit volume per unit time when there is a temperature difference of 1 °C between two opposing surfaces. This proportionality constant depends on temperature and pressure, and in the case of isotropic materials is independent of direction.

Heat transfer by convection refers to the superposition of heat conduction with the energy transport of, for example, a flowing fluid in a tube. Since the flow velocity in the boundary layer between a solid and a fluid assumes the value zero due to the adhesion condition (see Figure 2.25), only heat conduction according to Fourier's law applies there. Within a short distance from the wall, the flow in a heated tube increases to a maximum value (core flow), so that the fluid velocity influences the absolute heat transport by keeping up a temperature difference between wall and core (until the temperature in the core reaches the wall temperature).

To determine the heat flux density in the boundary layer, the local heat transfer coefficient is defined to account for the complex dependence of temperature and velocity field.

$$\dot{q}_{\mathrm{W}} = \alpha \cdot (\vartheta_W - \vartheta_F) \tag{2.14}$$

$\dot{q}_{\mathrm{W}}$: Heat flux density occurring perpendicular to the wall

α: Heat transfer coefficient

ϑ_{W}: Wall temperature

ϑ_{F}: Fluid temperature unaffected by heat transfer

In convective heat transfer, a distinction is made between free convection and forced convection. While in free convection density differences in the fluid due to temperature differences generate the fluid flow, in forced convection the flow is caused by an external pressure difference.

With thermal radiation, part of the internal energy is converted into heat, which is emitted by electromagnetic waves and can be absorbed by neighboring bodies. According to Kirchhoff's radiation law, thermal radiation depends on the physical properties of the surface material, the geometric arrangement of the "emitting" and "target" surfaces, and the size of the radiating and receiving surfaces.

Since no matter is necessary for this radiation transport, the heat radiation also works in empty space. In gases and liquids, the emission and absorption of radiation act on the volume, but in solids, emission and absorption usually take place only in the outermost layers (a few micrometres). This is referred to as a surface effect.

The emitted heat flux $\dot{Q}_{em}$ of a solid is strongly temperature-dependent. The Stefan–Boltzmann law describes for an ideal black body (Planckian radiator) that the total radiated energy is proportional to the fourth power of the absolute temperature of the black body.

$$\dot{Q}_{em} = A \cdot \varepsilon \cdot \sigma \cdot T^4 \tag{2.15}$$

A: Surface

ε: Emissivity

σ: Stefan–Boltzmann constant ($5{,}67 \times 10^{-8}$ W/(m^2·K^4))

T: Total temperature of the body

To simulate heat transfer in a hotend for extrusion-based AM processes, it is necessary to know: (1) The initial temperature of the polymer and the hotend; (2) The heat transfer performance (contact quality) of the hotend across its contact area with the polymer; (3) The amount of heat transfer by convection; (4) The radiation losses to the outside; (5) The heat transfer coefficients of the involved materials; and (last but not least) (6) The heat capacity of the polymer.

In order to correctly describe the heat transfer at the interface between the hotend and the polymer, it is also important to know the rheological properties of the melting polymer, and the resulting velocity and pressure distribution perpendicular to the heat transfer direction as well as in the flow direction, in order to correctly represent the relationship between heat conduction and heat convection in the polymer.

For simplicity, it can be assumed that after a short time, at a given extrusion rate, a steady state of the heat transfer rate from the hot end to the polymer is established at each point along the transport direction of the filament and the resulting polymer melt.

The dependencies and laws of heat transfer on the flow velocity and the velocity gradient of the flow velocity perpendicular to the flow direction are illustrated and discussed in Section 2.5.7.

To ensure that the heat output of the hotend reaches the filament, materials with a high thermal conductivity λ should be used, such as copper (λ = 380 W/m·K), aluminum (λ = 220 W/m·K), and brass (λ = 113 W/m·K). In comparison, steel (λ = 45 W/m·K) is less suitable. However, the softer metals like brass, copper and especially aluminum have difficulties when being used with polymers highly filled with abrasives, such as glass fibers, carbon fibers, and other hard inorganic fillers, due to increased abrasion at the die. If these materials are to be processed, aluminum, for example, cannot be used despite its high thermal conductivity. A hardness comparison provides the following ranking: steel ≫ brass > copper ≫ aluminum. The combination of different metals in the hotend (heating block and nozzle) is sometimes problematic due to their different thermal expansion coefficients.

To minimize heat convection (due to poorly adjusted fans, for example) and heat radiation, hotends in the DIY sector are often encased in a replaceable "silicone sock". The silicone sock has the added advantage of reducing potential accumulations of polymer melt on the print head during the print job due to poor adhesion to the silicone.

In a powder sintering process, the heat transfer is mainly controlled by the radiation/laser power of the heat source and the radiation absorption, transmission, and reflection properties of the polymers, which are influenced by the surface geometry of the polymer powder. Heat transfer into the whole build chamber takes place largely via convection via the inert gas, but is dominated by the heat supplied via conduction through the upper layers of molten polymer by the laser/IR emitter. In this case, the initial heat transfer must take place via the existing contact surface from molten powder to the next powder grain. The generally poor heat transfer between the powder grains (small contact area, insulation by the surrounding inert gas) is one of the reasons for the still surprisingly good build precision in the laser sintering process, since molten powder adheres almost exclusively only to further molten powder.

In high-speed sintering (HSS), unlike the SLS process, an absorption-enhancing ink is selectively applied to the areas of the build surface that are to be sintered by irradiation from an IR emitter applied uniformly to the build surface. In addition to printing an absorption-enhancing ink, an aqueous ink is also printed as required on the adjacent part of the surface that is not to be melted, in order to achieve even better edge definition of the components through the cold evaporation of the water.

In both the SLS and HSS processes, heat transferred by convection as an exchange mechanism for heating/cooling over the wall surfaces of the build chamber effectively limits the usable build volume. since the complex geometry of the powder grains and the small grain-to-grain contact areas greatly reduce heat exchange by conduction over long distances. In combination with a build space "filled" with products, achieving precise temperature control in the build chamber can be extremely difficult.

The thermal conductivity of polymers is orders of magnitude lower than the thermal conductivity of metals, but about ten times that of air, which is why air and air-filled porous/foamed polymers are good insulators (polymers approx. 0.2 W/m·K; copper approx. 380 W/m·K; air approx. 0,026 W/m·K). The specific heat capacity of polymers depends essentially on the state of the polymer (glass state, (partially) crystalline state, melt). The highest heat capacity is measured at the melting point of highly crystalline polymers. Typical specific heat capacities of polymers away from first-order phase transitions (melting, evaporation) are around 2 kJ/kg·K, which is significantly higher than the specific heat capacity of most metals (copper: approx. 0.4 kJ/kg·K) and also higher than the specific heat capacity of air at normal pressure and 50 % humidity (air: approx. 1 kJ/kg·K).

Melting during processing in both extrusion and powder processes requires significantly higher amounts of heat than for heating up a solid or a melt. This energy must be applied by heat transfer. On the other hand, this energy is also released during recrystallization during the cooling process. Recrystallization in particular is a kinetically controlled process. The distance between melting point and recrystallization temperature corresponds to the process window for crystalline thermoplastics. A wide distance is good for layer adhesion in both extrusion and powder processes.

The heat released during recrystallization must be taken into account when simulating the heat balance of an application.

> Highly crystalline thermoplastics, with a recrystallization temperature just below the melting temperature, are usually difficult to process in additive processes. The same applies to amorphous thermoplastics with a narrow glass transition range. In both cases, the build chamber temperature must be precisely adapted to the process window of the polymer. This is true for both powder sintering and extrusion-based AM processes.

2.5.7 Fluid Mechanics

Tube flows, as they occur in the print head of extrusion-based AM, represent a special form of boundary layer flow. Due to the high frictional forces that occur near the wall, the fluid is strongly slowed down over the entire circumference of the tube. Starting from the layer near the wall, the velocity of the (significantly lower friction) internal flow increases to the maximum value in the center of the tube.

In general, fluid mechanics distinguishes between two characteristic flow profiles: laminar and turbulent (Figure 2.25). Laminar flows are characterized by the fact that the fluid layers do not interact, but only shift against each other, and thus no turbulence is generated. Turbulent flows, on the other hand, exhibit a three-dimensional flow field with unsteady turbulences to the main flow. The reason for turbulence is a large gradient of the flow velocity transverse to the flow direction.

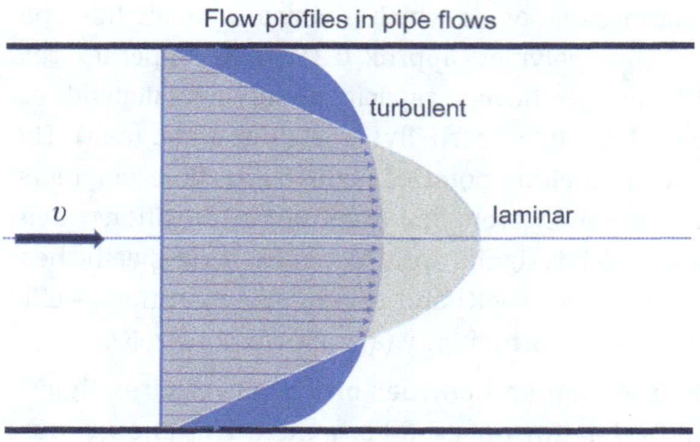

Figure 2.25
Flow profile in the pipe

To determine the flow profile, the dimensionless Reynolds number Re is used. The Reynolds number indicates the ratio of inertial to viscous forces and is as follows for pipe flows:

$$\mathrm{Re} = \frac{\rho \cdot v \cdot d_\mathrm{D}}{\eta} \tag{2.16}$$

ρ: Fluid density

v: Fluid flow velocity

d_D: Tube inner diameter

η: Dynamic viscosity

A completely laminar flow occurs below the critical Reynolds number $\mathrm{Re}_{crit} = 2300$. Only above 10,000 is a fully turbulent flow to be expected.

From Equation 2.16 it quickly becomes clear that in the case of viscoelastic polymers and the low flow velocities in extrusion-based AM processes, there is always a laminar flow profile. This also applies without restriction to common nozzle openings upwards of 0.2 mm. Only at the nozzle exit, due to the infinite diameter behind the nozzle opening (according to Equation 2.16) is turbulent flow of the polymer in the air possible. In the case of extrusion onto a build surface, however, a pressure-less flow without boundaries is not to be expected at any time, which is why laminar conditions can also be assumed during the build process with simultaneously decreasing temperature and exponentially increasing viscosity of the polymer melt.

The melt flow in the print head is a flow with simultaneous heat transfer. In order to be able to simulate the velocity, concentration, and temperature profiles of the three-dimensional flow of the polymer melt within a hotend up to the nozzle as a function of location and time, the equation for conservation of momentum (Navier–Stokes) and the equation for the conservation of energy must be solved in addition to the conservation of mass (continuity equation).

Knowledge of the velocity, concentration, and temperature profiles of the polymer melt over the pipe cross-section and length in turn provide the basis for calculating the localized temperature- and shear-dependent viscosity of the polymer melt and thus, summed over the pipe length, the expected pressure build-up to the tapered extrusion die.

The flow resistance of the polymer melt thus generates a pressure distribution within the die head as a function of the viscosity of the polymer, which is dependent on location, temperature, and shear rate, according to the boundary conditions of wall temperature, tube cross-section, tube length, and feed rate. In the simulation of typical polymers and hotend geometries, the viscosity function of the polymer melt upstream of the nozzle exhibits a strong dependence on the wall temperature and – which is surprising at first glance – only a rather weak dependence on the feed rate.

> The heat transfer capacity of a hotend is therefore the decisive factor for the performance of an extruder. In this context, it is generally advantageous if the hotend is rather short and powerful. This means heating blocks (hotends) with high heat transfer coefficients, thin tube cross sections, and high wall temperatures are the best fit to achieve high throughput rates and low residence times. High hot end temperatures are advantageous as long as the polymer in the hotend is not thermally decomposed.
>
> With conventional filament printers, linear dependencies between filament advancement and extrusion volume are found at extrusion pressures of up to 20 bar. Higher extrusion pressures are difficult to build up via common filaments with 1.5–3 mm diameter and standard extruders (Bowden/Direct Drive). At high feed pressures in the extruder, buckling effects of the filament in the filament guide can be observed. This results in a pressure loss in the hotend feed via the interaction of the filament (friction/pressure dissipation) over the walls of the guide tube in the Bowden extruder or the guided cooling section on the way to the hotend.
>
> Deviations in the planned and real volume output usually manifest themselves as under-extrusion and have a direct effect on the mechanics (layer adhesion, holes, and defects) and appearance of components.

In addition to mass shrinkage and pressure-dependent under-extrusion, the quality of the filament with regard to water content (polymer degradation, foam extrusion, and porosity) and the uniformity of the filament diameter have a major influence on the volume or mass flow of the extrusion process. Due to the manufacturing process, filaments have a fluctuating error to the specified filament diameter. The diameter tolerance of commercially available PLA filaments is usually in the range of about ± 0,05 mm. This results in a diameter deviation of up to approx. 3 % and a throughput volume deviation of up to approx. 6 % for a filament thickness of 1.75 mm.

Not only the diameter tolerance, but also the roundness tolerance of the filament has an effect on the incoming volume and mass flow. Filaments with a diameter of 1.75 mm, which are as round as possible, are the base for a precise volume and mass flow, since the slicer software calculates the volume and mass flow from the ideal geometry of the filament.

As discussed in Section 2.5.2.5, in addition to the variation in filament diameter, the difference between the polymer melt and solid volume is a decisive variable when creating the machine parameters (G-code) in the slicer. In principle, a fluctuating filament diameter over the filament length, just like the difference between melt and solids volume, is not a problem for an adaptive process control if the diameter fluctuation is known exactly.

Figure 2.26 shows the influence of filament diameter variations on the extruded strand. The variations result in an undesirable under- or over-extrusion of filament, which can be seen in the narrower or wider deposited strand. This, in turn, affects the mechanical properties of the manufactured component as well as the component surface or wall quality.

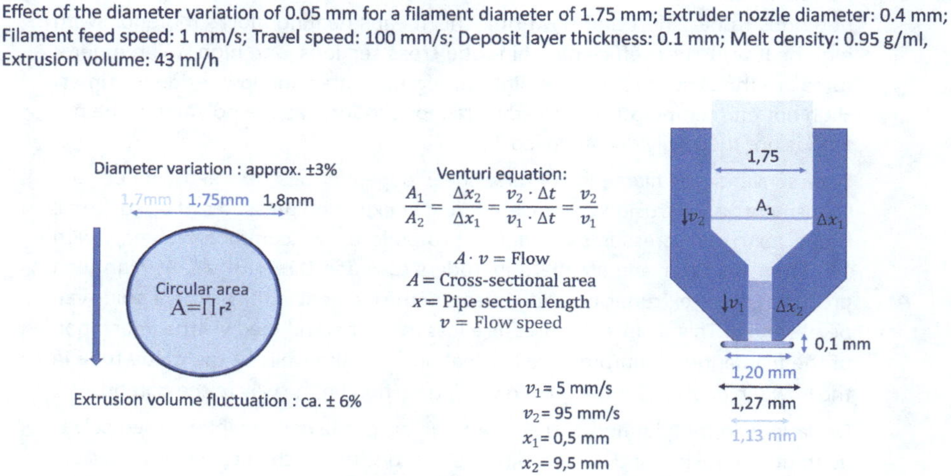

Figure 2.26 Effect of filament diameter variation

Other important parameters for the filament extrusion process are the knowledge of the rheology of a polymer via the temperature and shear rate variations. In addition to the die swelling at the nozzle, this also explains process behaviors such as tracking and stringing, and component defects due to over/under-extrusion during rapid changes in print head direction (braking/acceleration) in the build process. Die swelling or strand expansion is the swelling of a polymer strand after passing through a given die geometry [Waßner & Laun 2000]. Die swelling results from the ratio of the extrudate diameter to the die diameter (Figure 2.27). Due to the elimination of wall adhesion, the velocity profile of the flow inside the die changes from a parabolic profile to a block flow behind the die.

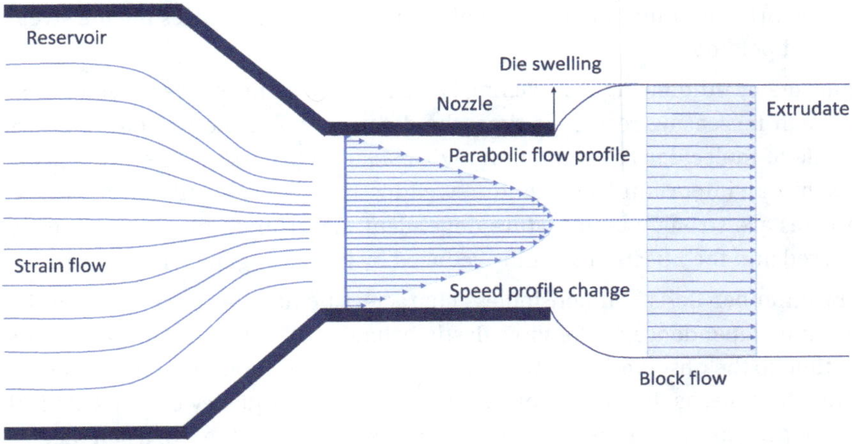

Figure 2.27 Die swelling

The size of the die swelling depends, among other things, on the shear rate. At constant temperature, the proportion of elastically stored energy in the polymer increases with increasing shear rate. The polymer clusters are stretched in the flow direction at the entrance of the die and relax back into their thermodynamically preferred shape at the outlet and downstream of the die. The proportion of relaxation of the polymer clusters before solidification of the polymer melt downstream of the die thus determines the strand expansion/die swelling. A high proportion of stored elastic energy also manifests itself in an increased pressure drop at the nozzle entrance. This information is used in the capillary rheometer to characterize polymers. High-molecular-weight, uniform, linear polymers show a higher tendency to die swelling than long-chain branched polymers or polymers with a broad molecular weight distribution.

A high dimensional stability as a combination of a low die swelling and high melt viscosity at low shear areas is also called melt stability. For extrusion and related calendering applications, long-chain branched or "broadly distributed" polymers are preferred because they exhibit lower die swelling/higher melt stability.

In addition to shear deformation, elongation deformation also occurs as a result of cross-sectional changes in an extruder. Viscoelastic fluids form flow lines in the melt area before entering the die. The ratio between the diameter in the reservoir melt area and the die opening determines the inlet angle and influences the elongation of the melt flow in the nozzle or a capillary rheometer. In order to minimize this effect, which also contributes significantly to the elastic deformation of polymer clusters and correspondingly causes a higher pressure build-up upstream of the nozzle, the inlet area from the reservoir to the nozzle is often tapered. As a result, the flow pattern is less affected and the inlet pressure drop decreases [Liang 1995]. The inlet cone can be specifically adapted for each application to optimize the flow lines, but then, in addition to the pressure buildup upstream of the nozzle, it also reduces the die swelling behavior of polymers.

Another parameter influencing die swelling is the residence time of the polymer melt in the die. With increasing residence time, the deformation of the polymer clusters decreases due to molecular re-orientation and relaxation processes. The viscoelastic behavior is thus an important factor for die swelling. The viscosity and thus the relaxation processes are strongly temperature-dependent. Accordingly, the melt viscosity can be lowered and the injection swelling reduced by increasing the temperature.

The Deborah number De is a dimensionless characteristic number of rheology. It describes the time dependence of the viscoelastic behavior of a fluid as the ratio of its relaxation time to the observation time (Figure 2.14). The observation time can be understood, for example, as the inverse of a shear rate or the frequency of a mechanical excitation – in this case the process time or the residence time in the extrusion die.

If De < 1, the material behaves similar to a viscous fluid. If De = 1, the material behaves viscoelastically. For characteristic values significantly above 1, the material behaves like an elastic solid. When entering the die, stresses are introduced into the material by strain and shear. These stresses are partially relieved by relaxation processes. Even with infinitely long nozzles, the material cannot relax completely, since new stresses are again introduced by the flow in the nozzle.

A high filler content increases the relaxation time by reducing the chain mobility. Despite the increased relaxation time, a lower swelling behavior occurs compared to polymer blends with lower filler content. This is due to the proportional substitution of the potentially viscoelastic polymer by non-deformable filler particles. The effect of volume replacement thus tends to be greater than the relaxation time extension. In particular, the use of fillers with low specific surface area reduces the swelling behavior.

> Due to the use of short nozzles and the associated low residence time, relaxation of the chains takes place only to a small extent in extrusion-based additive processes. The low extrusion rates compared to classical extrusion processes and the lower melt pressure as well, as the simple geometry of the transition from the melt reservoir (filament diameter) to the extrusion die, however, cause only moderate die swelling in the extruded polymer strands.
>
> Since the extrudate is extruded directly behind the die onto a substrate with a layer thickness less than the die diameter, the condition of extrusion into a free, unconfined volume is also only partially given. Die swelling therefore manifests itself mainly in a change in extrudate width and layer height, with the extrudate width tending to be slightly reduced and the extrudate height slightly increased. Since the layer thickness is also determined by the subsequent layer, this effect largely balances itself out. Only when the set layer height corresponds to the die diameter, and when extruding larger volumes from pellet extruders with extrusion screws, does the die swell become relevant again as an effect on the component geometry and the component surface.

Melt or flow instabilities are often divided into the surface defect "sharkskin" and the volume defect "melt fracture". (Figure 2.28). Sharkskin refers to a flow instability in the extrusion direction, which manifests itself in the form of a rough, dull surface of the extruded profile [Rauwendaal et al. 2014]. Sharkskin occurs at the die exit. The rearrangement of the velocity profile from a steep parabolic profile at the die exit to a block flow outside the die is considered to be a cause for the occurrence of sharkskin.

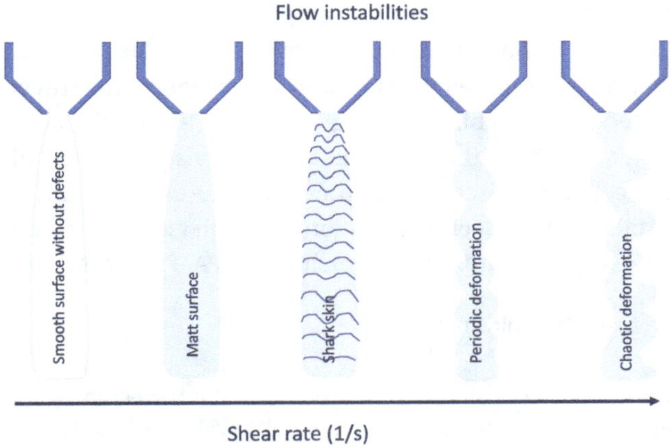

Figure 2.28
Flow instabilities/melt defects

Polymer processing aids (PPAs) can be used to avoid sharkskin. The PPAs form thin layers on the inner die surface. These low surface energy layers allow the polymer to glide over the walls with only a minimum of adhesion, reducing the occurrence of sharkskin, and also reducing the pressure drop across the die so that higher throughputs can be achieved.

> However, the excessive use of PPAcan significantly affect/endanger the layer adhesion in components when 3D printing thermoplastics, and must therefore be carefully examined. Alternatively, permanent surface coatings of the hotend and die with fluoropolymers, such as PTFE, are used. Analogous to PPA, the low surface energy leads to increased sliding of the thermoplastic melts on the nozzle walls, whereby the velocity differences in the parabolic profile become significantly less pronounced.

The melt fracture represents a deformation of the polymer strand. In comparison to sharkskin, this defect is not only found on the surface but, depending on the degree, in the entire extrudate. It is thus a volume defect that negatively affects the strength of the extrudate and thus of the component produced [(Noriega & Rauwendaal 2010]. At high throughputs, the extrudate deforms periodically at the beginning due to unsteady flow processes and randomly or chaotically in the later course (when the throughput is further increased). The critical wall shear stress is assumed to be between 0.1 and 0.4 MPa, regardless of the material used. Above this shear stress, flow instabilities occur.

The extrusion speed in AM processes is usually too low for the development of flow instabilities/melt defects. With higher output volumes, this issue becomes more important, as discussed above. The problem of melt fracture in particular can then exert a considerable influence on mechanical component properties, in addition to other issues like layer adhesion (melt/build temperature too low) and "foam extrusion" (insufficient drying or melt temperature too high, and thus decomposition and foaming of the polymer).

Another problem with a rheological origin is the so-called stringing. Here, the extrudate flow does not stop reliably during an "empty run". An empty run is the repositioning of the print head without the material being extruded (travel move). The material is then pulled out of the die into threads between the start and stop points of the travel move. This phenomenon is caused by a combination of gravity flow, residual elastic stress in the direction of flow, low viscosity of the melt, and a specific viscosity jump of the material after leaving the die. A high viscosity jump outside the extruder die is particularly observed for materials with a fast crystallization or a high glass transition temperature that require a high extrusion temperature, such as PC, PA, PET, or ABS.

Various measures can be taken to avoid stringing:

- The traversing speed between the start and stop points of the travel move should be increased and empty runs should be generally avoided or made as short as possible. Due to the higher traversing speed, less material can flow, a thread breaks more easily, and is not tensioned between the start and stop points.
- During or before the empty run, the thread should be retracted. For each material and each printer setup, there is an optimal retraction length and retraction speed in relation to the filament geometry, the hotend length, the material viscosity (hotend temperature), and the nozzle diameter. Again, a short powerful hotend is advantageous to avoid filament pullback issues.
- The fan speed at the extruder nozzle should be increased. This increases the viscosity so far downstream of the extruder nozzle that dripping of the material is prevented.
- Using wiping mode: Here, the nozzle is "emptied" over an existing layer before it makes an empty run.
- Coasting: The extrusion is stopped shortly before the end of the printing line. However, this function should preferably only be used for internal structures (infill), as otherwise gaps may occur in the part.

2.5.8 Thermal Decomposition Temperature

Every polymer has a decomposition temperature at which the polymer is in kinetic equilibrium with its building blocks (monomers) (ceiling temperature [Hanselmann 2006]) or is decomposed by external influences, such as hydrolysis with water present or oxidation by atmospheric oxygen. The decomposition temperature is usually determined by means of thermal analysis methods, in which the weight degradation of the polymer is recorded under controlled conditions and the degradation products can be collected and analyzed. The best-known method is thermogravimetric analysis (TGA, Figure 2.29). Process parameters of a TGA measurement are the temperature increase (K/min) and the set gas flow (L/h). Air as well as oxygen or inert gases (nitrogen, argon) can be used as gas. From the TGA, information is obtained about all reactions of the polymer at a given temperature and time that are associated with a change in mass (predominantly mass loss). Degradation reactions that occur without mass change are not registered in the TGA.

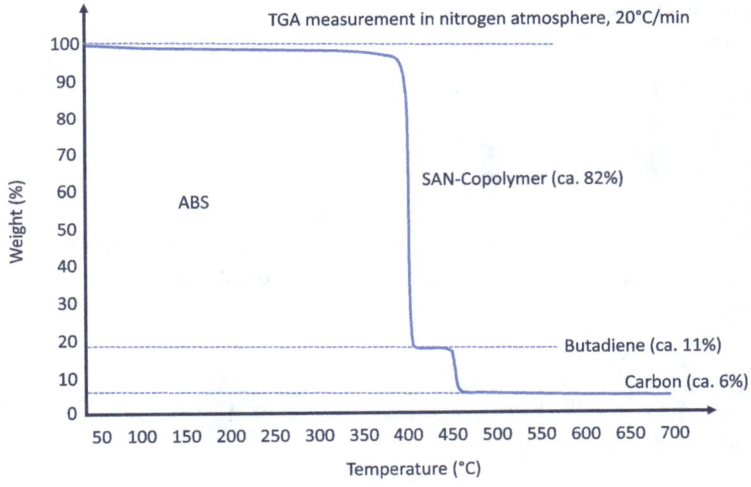

Figure 2.29 TGA of ABS [Sichina 2007]

> As a method, TGA is very sensitive and, in addition to thermal degradation, it also indicates the formation of stable residual materials, e.g. coking, as a residue. Such coking and deposits are particularly common with highly filled thermoplastic formulations in extrusion-based processes (MEX), if material is allowed to remain in the nozzle or at the hot end for too long in an air environment. It is precisely these residual deposits that are often critical for the stability of the printing process.
>
> The TGA can also run targeted temperature ramps and be maintained at a desired temperature. By analyzing the thermal degradation products at a given process temperature, it is thus possible to infer the toxicological potential of processing a material in extrusion and sintering processes at realistic temperatures and also in the event of temperature deviations. If the critical temperatures are known, appropriate measures (temperature control, cleaning cycles, ventilation, and air filtration) can be taken to ensure that no hazards arise from the additive build process.

2.5.9 Hardness Measurement

In the technical field, the modulus of a material is often also expressed as hardness in the units Shore A0, Shore A, or Shore D. The measurement of Shore hardness is a simple measuring method in which a standardized "needle" (depending on the test method A0–D), which is coupled with a spring, is pressed into a material test specimen of a given thickness to a given depth. The resulting indentation force is indicated as Shore A0/A for gels, rubber materials, and thermoplastic elastomers (TPE), and as Shore D hardness for hard elastomers, thermoplastics, and duromers (Figure 2.30), whereby the hardness ranges clearly overlap (Figure 2.31).

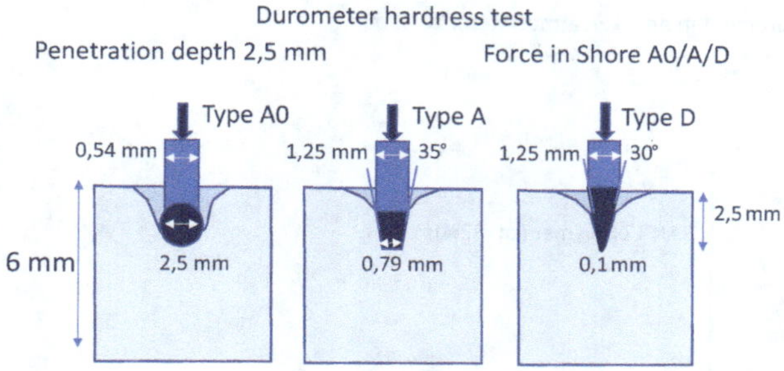

Figure 2.30 Hardness measurement in Shore A0/A/D

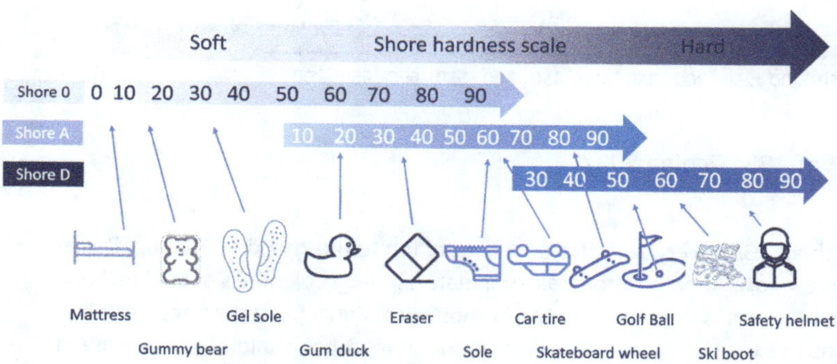

Figure 2.31 Examples of the hardness of typical products

> In AM, too, input materials are frequently specified according to material hardness. However, the material property of hardness can only have an orienting character for the component properties here. In order for the property of hardness to be meaningfully transferred to the properties of 3D-printed products, these must represent the specifications of a typical hardness test specimen or even just the properties of the materials used. In reality, only a few 3D-printed products meet the specifications for wall thickness or material density.

In the case of products produced by the powder sintering process, the density of the product and the porosity of the walls have a significant influence on the real hardness. Local variations of the product density and the surface quality (porosity) produce considerable variations when compared with injection-molded products.

Products with closed surfaces, which are additively manufactured using extrusion-based processes, benefit in particular from the reinforcing internal structures, which cannot be achieved using conventional processes. Here, too, it is not really possible to predict the properties of such products on the basis of the hardness of the materials used according to Shore A or D, since the contribution of the internal structures also varies greatly depending on the location. Product simulations based on the material modulus are more useful in this context, where porosity and structural design of the product can be better taken into account by clever parametrization.

2.5.10 Tensile Test

The tensile test belongs to the group of destructive material tests. In the tensile test, essential mechanical properties and characteristic values of a material are determined. These characteristic values form the basis for the calculation and dimensioning of statically stressed components and structures and are required for characterizing the results of the processing behavior of materials. The characteristic values determined are also used in quality control for assessing the uniformity of polymer production and in material selection for comparing properties of materials in virgin form with their properties after heat, cold, and media storage.

The characteristic values that can be obtained from a tensile test are as follows:

- Tensile stress σ (in MPa): Force F related to the initial cross-sectional A of the specimen
- Change in length ε (in %): Strain, change in gage length Δl referred to initial gage length
- Tensile modulus E (in MPa): Slope of the curve in the stress–strain diagram
- Yield point (in %): Stress and strain at the curve point where the slope is zero
- Fracture point (in %): Stress and strain at the moment of specimen breakage.

The simultaneously registered quantities of the tensile test are the stress σ (MPa) and the change in length ε (%). By plotting the stress on the vertical axis and the resulting strain on the horizontal axis, the stress–strain diagram of the tested material is obtained (Figure 2.32).

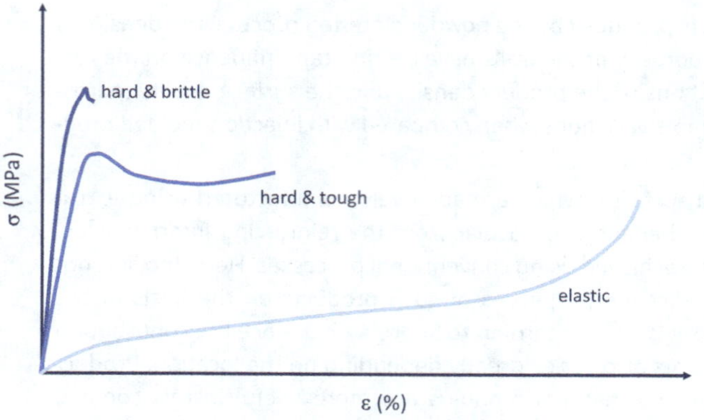

Figure 2.32 Typical stress–strain diagrams of polymer materials

2.5.10.1 Heat Tinting in Tensile Tests on Plastics

When a material is deformed outside the energy-elastic range, work is done and consequently heat is generated. If, during the test, the material is in a "temperature-sensitive" area, for example near a phase transition (glass/melt transition), then the deformation behavior is already influenced by small local heat effects or temperature changes in the specimen.

The plastic elongation that occurs after exceeding the yield point often starts from a localized necking point (Figure 2.33). The material is stretched locally and pulls out of the unstretched material. The resulting flow zones migrate over the entire test specimen up to the shoulders of the test rod. This significantly increases the molecular orientation in the stretched area.

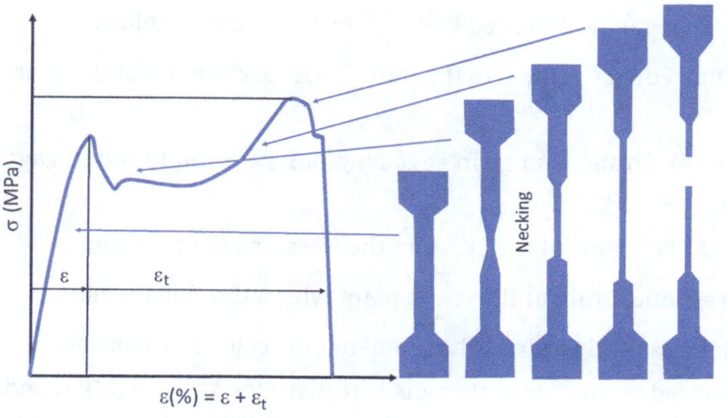

Figure 2.33 Stress–strain diagram of a ductile material with yield strength

This effect of the orientation of polymer chains is readily detectable from density measurements in the "unstretched" and stretched deformation regions (stretched is denser).

At the same time, the local deformation speed is lower in the already stretched specimen areas.

With the emergence of the necking zone, the deformation behavior of a material localizes and a hotspot is formed in this area. As the deformation increases, two flow fronts often develop, which move at different velocities towards the shoulders of the specimen and consequently exhibit different temperature rises. This process is velocity-dependent and can produce temperature differences at the surface of up to 10 °C. For many plastics, this temperature increase at standard testing conditions is already sufficient to noticeably influence the deformation process and test results.

The first stress maximum in the tensile test is called yield strength and the corresponding stress is called yield stress. Up to the yield point, one can conditionally expect an energy-elastic behavior of the material. After the yield point, the material is irreversibly deformed and the stress apparently decreases. However, the onset of cold flow actually leads to a reduction of the specimen cross-section (necking). The true tensile stress, on the other hand, almost always increases monotonically with strain. When the chains are stretched, thus aligned and parallel to the direction of force, the tensile stress often increases again significantly with further elongation due to crystallization effects, until fracture occurs due to the applied stress at any defect in the specimen. The quality of the test specimens thus has a direct influence on the absolute values of the fracture stress and fracture strain, while the initial range of the stress–strain curves is typically very reproducible.

Both ISO 527-1 and ASTM D638 specify test methods for tensile testing of plastics. Both standards are technically equivalent, but do not provide fully comparable results, since the specimen shapes, test speeds, and methods of determining results differ in some respects.

In the standardized tensile test, test results are presented in relation to a defined pull-off speed on a defined specimen (as described above). In the practical use of a product or a more complex geometric body, however, the stresses that occur can lie in a very wide range of deformation speeds. Due to the viscoelastic property of the polymers, significant variations in the mechanical properties usually result under changed strain rates compared to measurements on standardized specimens. For this reason, the characteristic values determined in the tensile test are only of limited use for product design. However, they do provide a very reliable basis for comparing and characterizing similar materials and material quality variations.

> In AM, the tensile test is often used as a comparative parameter to illustrate material properties, but also print quality. From the previous explanation of the massive dependencies of the results of the tensile test on even the smallest defects and material irregularities, however, it is clear that instead of tensile test measurements, it is better to carry out product tests in the main load direction.

Furthermore, the direction of construction plays an essential role for the product properties in all AM processes. Standardization and specification of printing process conditions seems to be of little use compared to injection molding processes, since the design for AM of components follows different logical aspects than the design for conventional manufacturing processes. Despite this, tensile tests of 3D-printed test specimens provide good indications for developing a correct product design and suitable process conditions and printing directions for this purpose.

2.5.10.2 Aging Tests

The tensile test provides a good basis for showing the change in the mechanical properties of a polymer after aging, heat or media storage or weathering. For this purpose, the characteristic values of the tensile test are determined in the freshly extruded or printed state and after defined aging or weathering times.

2.5.10.3 Temperature-dependent Mechanical Testing (Heat Resistance)

Heat resistance is a measure of the thermal endurance of plastic materials. It describes the ability of a test specimen to retain its shape up to a specified temperature under a specified load or to deform only within specified limits at a given test temperature.

To describe the high temperature capabilities (heat resistance) of a polymer, various measurements are possible each providing a specific limiting temperature that can be used for comparison with other materials.

The Vicat softening temperature (VST) is the temperature at which a standard indenter penetrates 1 mm into the surface of a plastic test specimen under a given constant load when the temperature is increased at a uniform rate.

The heat deflection temperature (HDT), or heat distortion temperature, is a measure of a polymer's resistance to alteration under a given load at an elevated temperature. This material test determines the temperature at which a specified deflection (0.25 %) occurs in a specimen positioned on bearings at either end under a defined, concentric load when the temperature is increased at a uniform rate.

The two common loads used in heat deflection testing are:

- 46 MPA for softer or lower melting grades of plastic like PE, TPE, or PLA
- 8 MPA for more durable grades of plastic like PEEK, PA, or PC.

Also, the tensile tests can be performed at different temperatures. The measurement of mechanical properties of plastics at different temperatures helps to characterize the temperature dependence of the material behavior. Knowledge of the temperature dependence is particularly important with regard to the change in properties such as brittleness at low temperatures (for example, bumpers in automotive applications) and softening under load at high temperatures (for example, engine compartment applications).

2.5.10.4 Other Destructive Test Methods

In addition to the tensile test, there are a number of other important destructive mechanical test methods that test specific properties of materials. Particularly worthy of mention here are, in analogy to the tensile test, the three-point bending test, the compression test, the dynamic-mechanical fatigue test, various abrasion test methods, the notched bar impact strength, and various heat resistance tests.

2.5.10.5 Destructive Testing of Additively Manufactured Components

If additively manufactured components and test specimens made of thermoplastics are tested according to the above-mentioned test methods, it is noticeable that most products manufactured in this way have significantly lower material properties than classically manufactured products. There are several reasons for this, which are discussed below.

Reduced layer adhesion: A product produced by extrusion or injection molding ideally has no distinguishable layers, but shows mainly uniform material behavior across the entire component. All additive processes based on thermoplastic materials are layer-build processes. The number of layers increases linearly with the imaging accuracy of the component compared to the template. If the interlayer adhesion is lower than the strength of a homogeneous material, the weakest layer in the direction of the load in each case determines the strength of a component. In extrusion-based processes, the effect of interlayer adhesion can be partially compensated by matching the build direction with the preferred load direction, i.e., the preferred load direction ideally corresponds to the XY position of the layers, i.e., is orthogonal to the build direction (Z) (because XY > Z). In powder printing, the parameterization of the laser power can be optimized accordingly so that the component properties correlate only slightly with the build direction (XY = Z).

Number of flaws: As the number of layers increases, so does the number of potential flaws in a component. The simplest method of quantifying the number of defects is to compare the density of an additively manufactured component with the density of an "ideal" component with the same volume. It is known from tensile testing that tensile strength and elongation at break in particular correlate directly with the quality of the test specimen. Any defect can lead to premature failure of the component, since it represents the starting point of a potential crack. If the tear energy is less than the tensile stress, the component will tear. The tear propagation energy further decreases with the number of flaws, since cracks can show a jump in crack growth with a high number of flaws. Many additively manufactured components exhibit typical flaws in the XY and Z directions. The absolute number and especially the local accumulation of flaws as well as the location and type of flaws have a great influence on the properties of an additively manufactured component. Intelligent slicer programs have an equally significant influence on product properties as the orientation of the component in the build volume (on the build platform). These are programs that translate the part geometry into traverse movements of the printer/laser as well as locally defined extrusion volumes/energy output.

Another option for improving additively manufactured products is offered by downstream treatment methods that use surface smoothing, material infusion in defective areas, and post-sintering in suitable inert media, which can lead to a densification of the component and a reduction in defects, especially at the critical surface, as well as to a general improvement in layer adhesion.

2.6 Thermoplastics

Thermoplastics are plastics that can be "thermoplastically" deformed within a certain temperature range. This process is reversible, i.e. it can be repeated as often as desired by cooling and reheating to the molten state, as long as the so-called thermal decomposition of the material does not set in due to overheating. In this respect, thermoplastics differ from thermosets (duromers) and elastomers, which cannot be melted without the material decomposing.

Organic thermoplastics are composed of linear or sparsely branched carbon-containing chain molecules that interact with each other only through physical bonds. A distinction is made between amorphous thermoplastics and semi-crystalline thermoplastics. Amorphous thermoplastics soften above the glass transition temperature, semi-crystalline thermoplastics have both a glass transition temperature and a crystalline phase and therefore soften only above the melting point of the crystalline phase. Semi-crystalline thermoplastics thus have at least two transition temperatures (glass transition temperature and melting temperature) at which the mechanical properties change significantly. If two or more incompatible thermoplastics are mixed together, the product is called a polymer blend. A phase morphology typical of each blend is produced [Sole & Solc 1982]. The blend of compatible/blendable polymers is called a polymer alloy [Meyer 1987].

Thermoplastics can be welded under the influence of heat and pressure [Induux 2021]. The materials to be welded are heated above their glass transition temperature or melting temperature and brought into a flowable state. Since this is done at different temperatures for different plastics, only plastics with similar melting indices can be easily welded together, in addition to plastics of the same type, e.g. PMMA with PVC. In principle, the welding temperature must be higher than the melting point of the higher-melting polymer. Since thermoplastics are non-crosslinked polymers, they can often also be dissolved with suitable solvents and thus chemically welded or bonded [Industrieverband Klebstoffe 2021].

Weldability of thermoplastics in AM

Thermal weldability is the basis of AM processes with thermoplastics. In both powder-based processes and extrusion-based processes, the component is heated layer by layer above the softening temperature (melting temperature) and thermally welded. To ensure good seam adhesion between the layers, not only must the melt temperature and melt viscosity of the polymers to be welded match, but the polymers must also be sufficiently compatible with one another so that mixing of the polymers at the interfaces is chemically preferred to segregation. If this is not the case, brittle joints will result, supported only by contributions from surface tension and weak polymer interactions. Since the compatibility of polymers basically decreases with polymer chain length (for thermodynamic reasons), there are not many classes of polymers in which two different materials produce stable bond lines with each other.

When combining polymeric materials in products, sufficient compatibility is made possible in part by using surface modifying/bonding substances such as adhesives, phase mediators, or by chemical reaction and crosslinking. Multimaterial components in 3D printing of thermoplastics are currently limited to materials where these conditions of compatibility are met. A good example of this is thermoplastic polyurethanes (TPU), where the necessary material compatibility can be achieved through co-crystallization and hydrogen bonding due to the structural similarity of soft and hard TPU phases. In 3D printing, multi-material components offer the opportunity to incorporate another property dimension into the component design, in addition to the obvious dimensions of material property and component geometry of "single-material 3D-printed products". These would include variations in both mechanical and chemical properties such as modulus, damping, softening, swelling, chemical/weather resistance, and abrasion.

A controlled "poor" adhesion of different polymer types on or to each other is exploited in the selection of suitable support materials in the FLM process. Here, the support materials should be easy to separate from the desired component material after completion of the printing process, either mechanically or chemically, and rarely also thermally. A support material that can be easily separated mechanically has the advantage of fast processing, but still must have sufficient adhesion in the build process to enable filigree structures (e.g. combination of TPU and PLA as a support material). An alternative chemical separation of the support material requires a strongly varying resistance to suitable release media (e.g. alkaline bases for the ABS/PVOH system). Thermal separation, on the other hand, is less common and is mainly used for materials where the support phase can be burned out at the end of the build process, for example in the production of metal bodies from highly filled plastic filaments [Fraunhofer IFAM 2021] or in the SLA process for the production of post-sintered Teflon bodies [3M 2021].

The partial solubility of thermoplastics is exploited in common post-processing procedures in extrusion- or powder-based 3D printing to dissolve and smooth the process-related uneven surfaces in a so-called solvent (vapor) bath [Rajan et al. 2019]. Furthermore, a solvent can also be the transport medium for a dye or pigment, making colored products available from previously "colorless" raw materials. The introduction of dyes/pigments via the thermoplastic itself is not always advantageous, especially in the powder process, as this can cause the laser or IR absorption to be changed in an uncontrolled manner. An excessively high absorption or scattering of the introduced laser and/or IR radiation can lead to unwanted melting and caking in the powder. Subsequent coloring of the products is often the more cost- and production-efficient option here [3D HUBS 2021].

2.6.1 Classification of Thermoplastics

Thermoplastics are known for their plastic and partially flexible properties. In order to be thermoplastically deformable, thermoplastics must not have any chemical crosslinking points. They therefore consist of long, mainly linear molecular chains of atoms (mainly carbon), forming polymers. The rigid and solid form of thermoplastics is caused by the close packing of strongly interacting molecular chains. However, if the polymers are heated, then the chains become more mobile and beyond the glass transition point or melting point can slide past each other. The material begins to flow and becomes malleable. As the temperature increases further, the viscosity continues to decrease until the material can be easily formed into any desired shape. Typical molding processes include injection molding, extrusion, and blow molding. Most thermoplastic materials are also available as semi-finished products in sheet or rod form, from which they can be further processed via application of temperature and pressure using thermoforming processes such as film thermoforming. Since a thermoplastic can also be heated and remelted several times, it is often easy to recycle.

Within the group of thermoplastics, there are major differences in terms of properties and possible applications. Important distinguishing criteria for the technical use of thermoplastics, in addition to the mechanical properties, are the heat deflection temperature (HDT) and the continuous service temperature (CST). In the example shown in Figure 2.34, selected typical representatives of amorphous and semi-crystalline thermoplastics are listed and sorted according to their respective maximum continuous service temperature. The thermoplastics are divided into high-temperature plastics (also called high-performance plastics), engineering plastics, and standard plastics.

Since the temperature stability often correlates with mechanical properties, chemical and physical resistance, as well as manufacturing costs and sales revenues, the plastic pyramid with its vertical stratification is also a suitable form of representation for the cost and property position of the plastics shown. In a representation of the production volume, however, the base of the pyramid would have to be much broader in order to correctly represent the share of standard plastics of around 80% (and growing!) of the absolute production volume of thermoplastics.

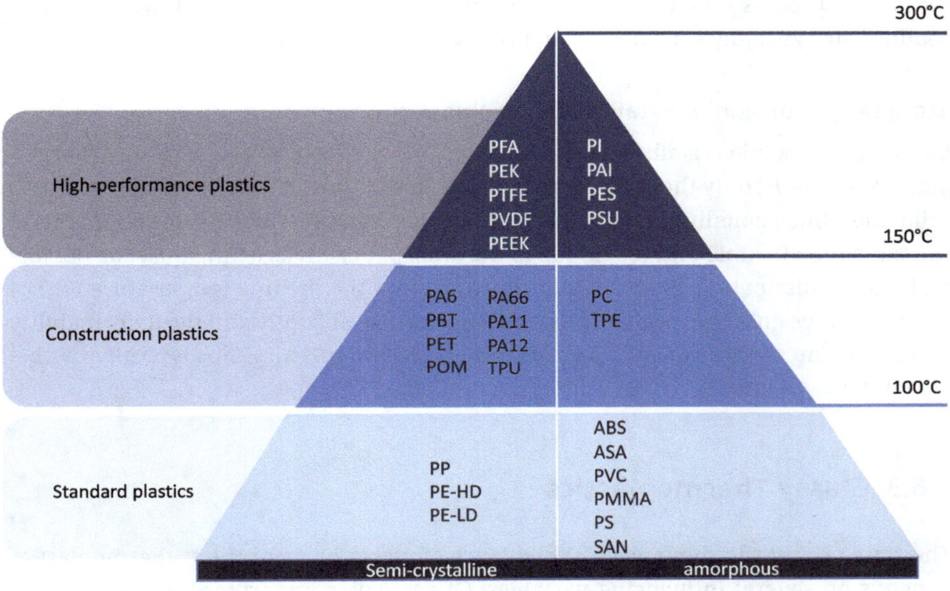

Figure 2.34 Pyramid of properties of thermoplastics

2.6.2 Semi-crystalline Thermoplastics

Semi-crystalline thermoplastics show both chemically and geometrically regular regions, i.e. there are areas where crystallites form. Crystallites are regions with parallel bundles of molecular segments or folds of molecular chains that have a melting point significantly above the glass transition temperature, i.e. the softening point of the amorphous phase of the polymer. Some of the chain molecules stretch across the crystalline and amorphous regions and sometimes may even belong to several crystallites at the same time. These boundary crossers ensure good bonding of the amorphous and crystalline phases and are the basis for the good mechanical performance of semi-crystalline thermoplastic polymers.

Semi-crystalline thermoplastics often have a whitish inherent color. This is due to light refraction at the dense arrangement of molecules in the usually quite large crystal structures with their higher refractive index. If the crystallites have a size significantly below the wavelength of visible light, semi-crystalline thermoplastics can also be translucent and even transparent in thinner layers.

The degree of crystallization greatly influences the properties of the semi-crystalline thermoplastic materials in the unfilled state. The tendency of the plastic to crystallize can be significantly influenced in all common processing methods via the applied temperatures during the process, but also via additives such as nucleating agents. For reasons of process efficiency (i.e. cycle time in injection molding), however, a high cooling rate is usually specified when processing thermoplastics.

State ranges of semi-crystalline thermoplastics

Below the T_g, semi-crystalline thermoplastics are similarly brittle as glassy thermoplastics. At the T_g, only the amorphous regions thaw. Above the T_g, the material is still solid and dimensionally rigid due to the molecular cohesion in the crystallites, but at the same time flexible and tough due to the amorphous areas acting quasi as flexible and energy dissipating "joints". Only shortly before the melting temperature (T_m) of the crystallites does the mechanical strength decrease abruptly and the deformability increases. The service range of semi-crystalline thermoplastics thus typically lies between the T_g and the T_m.

2.6.3 Glassy Thermoplastics

The state of order in the molecular structure of glassy or amorphous thermoplastics depends on several influencing variables. Of particular importance is the chemical structure of the chain molecules (macromolecular structure). Large and bulky side chains on the macromolecules prevent an ordered spatial structure and at the same time reduce chain mobility, which increases the T_g and decreases crystallinity. The modulus and tensile strength of a thermoplastic below the glass transition temperature are often high. Only when the T_g is reached does the modulus decrease sharply due to the abrupt onset of molecular mobility.

The elongation at break of amorphous thermoplastics is typically low in the glassy state, with a few exceptions (for example, the engineering thermoplastic polycarbonate and the high-performance thermoplastics polyimide and polyaramide). The material is comparably brittle, however less brittle than glass. The elongation at break and toughness increases rapidly, passing through a maximum, shortly before and during the T_g and decreases again after the T_g until the melt stage is finally reached.

The product class of amorphous thermoplastic elastomers is an exception here. They exhibit the typical behavior of thermoplastic elastomers of high elasticity and extensibility. The amorphous thermoplastic elastomers each comprise at least one low-T_g phase ($T_g \ll$ service temperature) and one high-T_g phase ($T_g \gg$ service temperature). Often, these phases are arranged in blocks with the order ABA in a polymer chain, where A describes the high-T_g phase and B the low-T_g phase. The polymeric phases are formed from different monomers in each case, most of which have been successively polymerized by means of living block copolymerization, but are part of a single polymer chain. Typical representatives are the block copolymers of styrene and butadiene (polystyrene-butadiene-styrene block copolymers: SBS) and variants in which the butadiene has been subsequently hydrogenated to polyethene (SEBS).

2.6.4 Thermoplastic Vulcanizates

Thermoplastic vulcanizates (TPVs) have replaced the classic crosslinked and elastic rubber materials in various applications since they were introduced to the market at the end of the last century. TPVs, which belong to the thermoplastic elastomers (TPE), combine to some extent the positive characteristic properties of thermoplastics and elastomers.

As a rule, these systems consist on the one hand of a softer crosslinked elastic phase (mostly EPDM) and on the other hand a thermoplastic semi-crystalline hard phase (mostly PE or PP). The soft phase determines the elasticity and the properties of the system to tensile and impact loads, while the hard phase, in turn, determines the resistance to higher temperatures and enables the thermoplastic processing of the material.

Thermoplastic vulcanizates based on EPDM/PP are obtained by intensive mixing of the hard thermoplastic phase (PP) and the soft synthetic elastomer (rubber) phase (EPDM). The phases are mixed in the melt until a stable mixture is formed, which is subsequently fixed by chemical crosslinking of the EPDM. The elasticity, UV and ozone resistance of EPDM are thus ideally combined with the simple thermoplastic processability of PP through, for example, injection molding and extrusion.

Areas of use of thermoplastic vulcanizates

By combining the elastomers and thermoplastic polymers used, positive properties can be generated. These include the increased resistance of PP and PE plastics to external influences and the low residual deformation of the typical crosslinked EPDM elastomers. For this reason, TPVs are finding application in many areas previously reserved for rubber materials:

- Automotive industry: Lining of pipelines, gaskets, trunk and window seals, vibration dampers, and elements in the starter system
- Architecture and construction: Ceilings and floors, decorations, and window profiles
- Electricity, electronics: Cable insulation, computer housings, and telecommunication equipment
- Others: Medical and food industry, elastomeric items, and small parts.

Despite this combination of advantageous properties, TPVs are still a niche application in AM and are not very widespread, mainly due to the difficulty of processing PE and PP in additive processes with regard to platform and layer adhesion [Leapfrog 2021; Statista 2015].

2.6.5 Thermoplastic Polymer Blends

Polymer blends belong to the group of polymeric multiphase systems. A trend that has been increasingly observed in the plastics sector in recent years is the production of blends of different polymers, so-called polymer blends. Polymer blends are the mixtures of two (or more) thermoplastics, which distinguishes them from other polymer blends, such as TPV (an elastomer-reinforced thermoplastic).

Compatible mixtures generally exhibit an additive overlapping of their properties, whereby the "worse" properties often prevail and determine the usability of the material. In incompatible mixtures, the possible dependencies are incomparably more complex, and ideally positive properties of the mixture components can be combined or even new better properties can be obtained. Usually, all properties depend more or less on the concentration in the mixture. Concentration ranges are sought in which the properties of the components are optimally combined in one material, according to the application requirements.

Ideally, multiphase polymer blends make it possible to combine the properties of the blend partners in an advantageous way or to create completely new property patterns that cannot be predicted from the characteristics of the individual components. In some cases, known property patterns can be finely differentiated and achieved more cost-effectively and more easily in terms of processing technology. The characteristic feature of all incompatible polymer blends is their multiphase structure, which is characterized by the fact that areas that differ in their typical parameters (such as chemical composition, density, relaxation behavior, mechanical properties, etc.) alternate and are separated from each other by more or less sharp phase boundaries.

The properties of incompatible polymer mixtures depend to a large extent on the processing conditions. The flow and solidification behavior, as well as the interactions of the polymer components at the phase boundary, have a significant influence on the morphology resulting from the production process. The number, size, and shape of the phases present, as well as their mutual interactions, control the macroscopic properties of the products thus obtained. Specifically, the phase interactions in the boundary layer are of high importance.

Known typical polymer blends are, for example, polyacrylonitrile-butadiene-styrene (ABS), polycarbonate/polyacrylonitrile-butadiene-styrene (PC/ABS) and polycarbonate/polybutylene terephthalate (PC/PBT), which are also used in 3D printing as typical raw materials in extrusion-based processes.

2.6.6 High-performance Plastics

High-performance thermoplastics are of particular interest for specialty applications in industrial manufacturing. This interest stems from the unique properties these polymers exhibit under extreme operating environments. Their resistance to wear stress, stability at high temperatures, inertness under chemical attack, and high specific strength make them important alternatives for applications where metals have traditionally been used. High-performance polymers in general, specifically as carbon or glass-fiber reinforced components, play a major role in applications such as transportation, where functional weight is closely coupled with energy consumption. The ability to design high-strength and high-stiffness systems with low weight is seen as key to achieving good energy/fuel economy.

The classification of polymers in the pyramid has changed over time. While aliphatic polyamides (PA6 and PA66) and polyethylene terephthalate (PET) were considered extraordinarily high-performance plastics when they were first introduced, they are now considered ordinary engineering plastics, with annual production volumes in the range of several Mt each.

In some cases, high-temperature thermoplastics allow continuous service temperatures of between between 150 °C and 260 °C, although this is still well below the possible service temperature of predominantly metallic and ceramic materials.

Improving the mechanical properties as well as the temperature and flame resistance has always been an important goal in the research into new plastic materials. Since the early 1960s, the development of high-performance plastics has been driven by corresponding needs, mainly in the aerospace and nuclear industry. Among the best-known high-performance plastics are the following:

- Polyaryls, where aromatic rings are linked via oxygen or sulfur atoms or via CO or SO_2 functions. This group includes polyphenylene sulfides (PPS), polyarylether sulfones (PAS), and polyarylether ketones (PAEK)
- Aromatic polyesters (polyarylates) and aromatic polyamides (polyaramides), such as poly-*m*-phenylene-isophthalamide (PMI)
- Heterocyclic aromatic polymers, such as polyimides (PI), polybenzimidazoles (PBI), and polyarylether imide (PEI).

Polyimides are heterocyclic polymers that are heat-stable up to 550 °C. They have very good chemical resistance and are suitable as materials in mechanical engineering (valves, bearings) and in electrical engineering for coils, as cable insulation, and substrates for printed circuit boards.

Polyarylether imides have a lower softening point (approx. 360 °C) and can be processed using standard processes in the plastics industry like injection molding and extrusion.

Aromatic polyamides based on PMI are suitable for fiber spinning processes from solution (Figure 2.35). The resulting fiber has a very high tensile strength and good temperature stability. Only at about 400 °C does the fiber start to deform and carbonize. Such fibers are i.e. used as protective clothing against heat or as interior linings in aircraft. Also PBIs can be used for protective suits and for fire-resistant fabrics in aircraft.

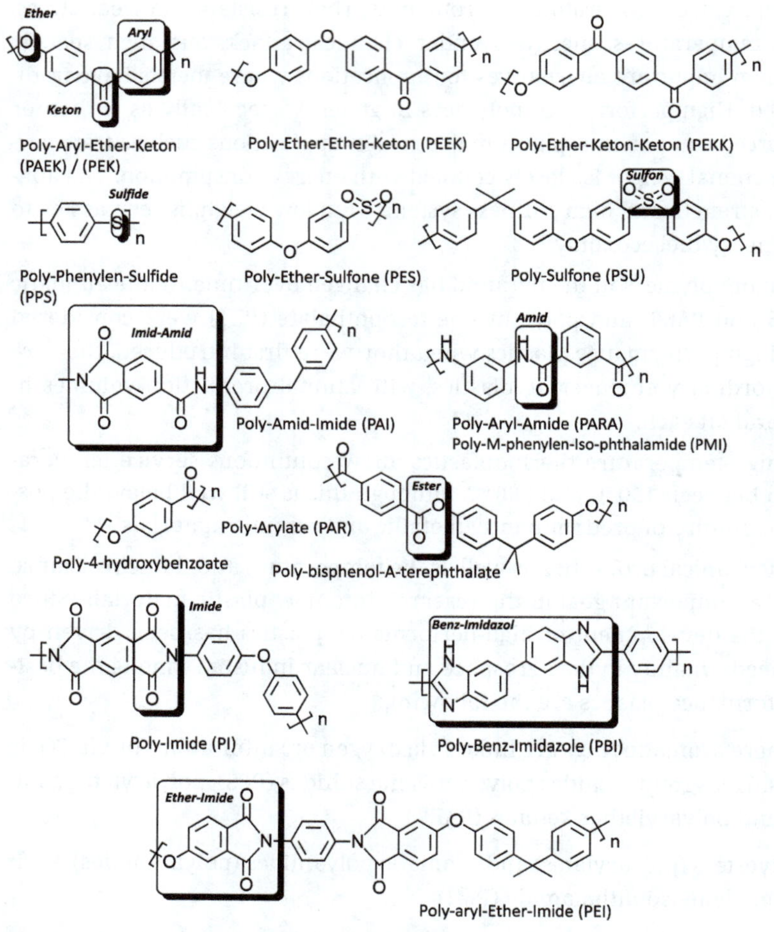

Figure 2.35 Aromatic high-performance thermoplastics

Some aromatic high-performance plastics belong to the class of liquid-crystalline polymers (LCPs), which are used, for example, in LCD screens due to their specific optical properties.

A polymer material such as polyether ether ketone (PEEK), which belongs to the class of polyarylether ketones (PAEK), has proven to be highly competitive with engineering metal materials such as titanium, steel, and aluminum alloys due to its biological inertness, chemical resistance, specific strength, and wear resistance.

Typical applications are found, for example, in thermo-mechanically loaded gear systems. PEEK gears operate with significantly less lubricant compared to metal gears. This advantage leads to longer intervals between gear maintenance and thus cost reduction, as gear parts need to be replaced less frequently.

Fluoropolymers such as polytetrafluoroethylene (PTFE, also known as Teflon) and polyperfluoroalkoxyalkane (PFA) with a temperature resistance of up to 260 °C are only in some cases counted amongst high-performance plastics, despite their high temperature and chemical resistance (Figure 2.36).

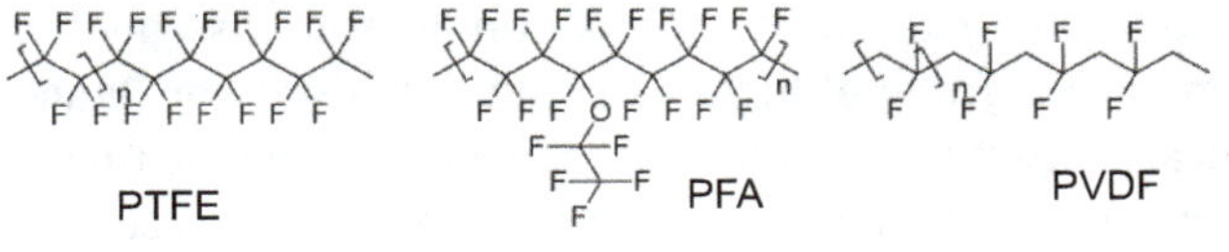

Figure 2.36 Fluorinated high-performance thermoplastics

2.6.6.1 Manufacturing Processes for High-performance Thermoplastics

Synthesis routes for polyphenylene sulfide (PPS), polyarylether sulfone (PES), and polyarylsulfone (PSU) were not developed until the 1960s, and each entered the market only in the early 1970s. Industrial processes for the production of polyarylether ketones (PAEK), including polyarylether ether ketones (PEEK) and polyarylether imides (PEI), via polycondensation were developed even later, in the 1970s.

Today, the majority of high-performance plastics are produced by polycondensation processes. In manufacturing processes by polycondensation methods, the high purity of the starting materials is decisive. In addition, the stereochemistry plays an important role in achieving the desired properties. The development of new high-performance plastics is closely linked to the development and industrial availability of the underlying monomers, which are usually difficult to access.

Both the mechanical and thermal properties of high-performance plastics and many standard engineering plastics depend on their molecular structure (structure–property relationship), as illustrated below using the example of polyarylamides (sometimes shortened to polyaramide) and polystyrene.

2.6.6.2 Stereochemistry of Polymers

Stereochemistry is a branch of chemistry [EPFL 2021]. It deals with molecules that have the same chemical bonds and composition but a different arrangement of atoms, which determines the three-dimensional structure of a molecule (stereochemical isomerism). In polymers, too, the stereochemistry of the building blocks or their spatial linkage has a significant influence on their properties. In most standard and engineering materials, the stereochemistry is statistical, i.e. random, due to the manufacturing processes. Particularly in the case of high-performance polymers, however, stereochemistry plays a significant role, since it is often only the desired stereochemical alignment of molecules that produces the intended properties.

> The positions of each two substituents on an aromatic carbon ring (benzene ring) – which is represented in the basic structure of most high-performance polymers – can be *ortho* (*o*-, next to), *meta* (*m*-, between) and *para* (*p*-, opposite). Depending on the orientation of the substituents and their regularity, polymers with significantly different properties are obtained.

Among the aromatic polyamides, the so-called polyaramides, a distinction is made, for example, between *m*-aramides and *p*-aramides. The *m*-aramides are assigned to the high-performance fiber materials group of high-temperature-resistant or flame-resistant fibers with mechanical properties in the range of conventional textile fibers. They are characterized by exceptional chemical resistance and high heat resistance. Compared to that, the *p*-aramide fibers belong only to the group of high-strength synthetic fibers with somewhat increased temperature resistance. The aromatic polyamides are technically produced by polycondensation reaction from the diamine and the acid chloride of the dicarboxylic acid or directly from the dicarboxylic acid. The products have a high melting point, which is often close to the decomposition temperature. They are therefore typically polymerized in suitable solvents and also processed directly out of the solvent (for example, spun from sulfuric acid).

> The high melting point and poor direct processability of polyaramides is a major reason why they are only used indirectly in additive processes as a fiber reinforcement in the form of prefabricated fiber materials. As fibers, they can be used, for example, in a modified FLM process together with molten filament, or in a paste extrusion (PAM) followed by UV curing [Mason & Gardiner 2020]. The value of these processes lies in the fact that with the aid of the additive process, the fibers can be partially oriented in the direction of the expected main load on a part, thus achieving particularly high strength and stiffness per weight, which can compete with the properties of metallic materials.

Figure 2.37 shows aromatic polyamides (polyaramides), where the starting compounds are *p*-substituted aromatic diamines and *p*-substituted aromatic diacids or *m*-substituted aromatic diamines and *m*-substituted aromatic diacids.

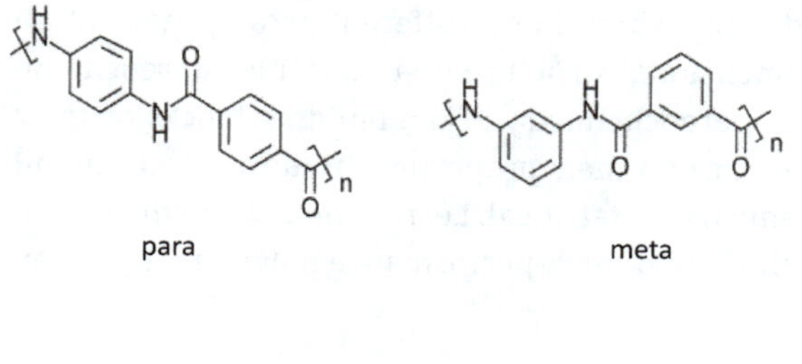

Figure 2.37
p- and *m*-polyaramides

Polystyrene provides another good example illustrating the influence of the stereochemistry on the properties of polymers, even in the range of standard polymers (Figure 2.38):

- Atactic styrene: amorphous, T_g = 100 °C
- Isotactic styrene: slowly crystallizing T_g = 100 °C, T_m = 240 °C
- Syndiotactic styrene: rapidly crystallizing T_g = 100 °C, T_m = 270 °C.

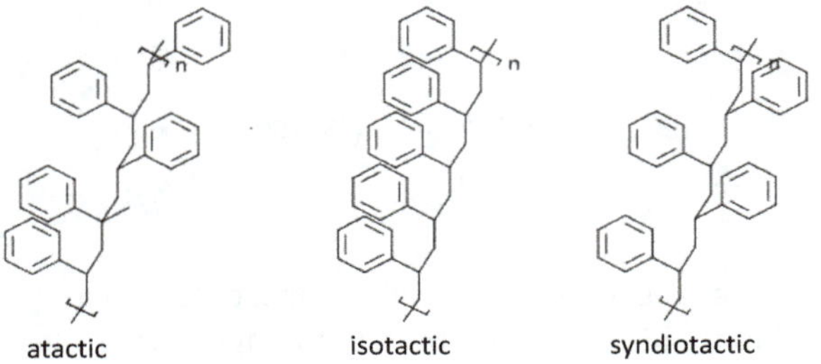

Figure 2.38 Polystyrene

If the arrangement is atactic (random), the resulting polystyrene polymer has a T_g of 100 °C, is transparent and relatively brittle. The polystyrene obtained in this way is a mass product in the technical sector with a wide range of applications. If the arrangement of the phenyl group is syndiotactic (alternating), a rapidly crystallizing polymer with a melting point of 270 °C is obtained. Syndiotactic polystyrene is a high-performance engineering material that is easy to process. Isotactic polystyrene, with all benzene groups on the same side, has a melting point of about 240 °C. The polymer crystallizes so slowly that technical utilization is not economically interesting.

Depending on the stereochemical arrangement (tacticity) of the regularly recurring side chains (substituents) in a polymer, significantly different properties are obtained and thus different applications are made possible. The most prominent examples are the phenyl group in polystyrene (Figure 2.38) and the methyl group in polypropylene. In an atactic arrangement, the substituents are randomly distributed. In an isotactic arrangement, the substituents are all on one side, while in a syndiotactic arrangement, the substituents are distributed alternately on both sides of the chain.

2.6.6.3 Mechanical Properties

High-performance plastics meet higher requirements than standard plastics and engineering plastics. They have better mechanical properties, higher chemical resistance and/or higher heat resistance. Accordingly, thermoplastic processability is often more difficult, and special processing methods, such as sintering or solvent processing, are necessary. Table 2.3 compares the properties of selected high-performance plastics, engineering plastics and standard plastics [Wikipedia 2021].

Table 2.3 Comparison of Properties of Selected Plastics

Polymer	Density (g/cm^3)	Melting temperature (°C)	HDT/A (°C) (at 1.8 MPa)	Modulus of elasticity (MPa)	Yield stress (MPa)	CTI/A (—)
PE-LD	0.915–0.935	105–118	<23	200–400	8–10	600
ABS	1.03–1.07	235	95–105	2200–3000	45–65	600
PA6.6 (dry)	1.13–1.15	255–260	70–100	2700–3300	75–100	600
PEEK	1.30	343	152	3500	100	150

In most cases, high-performance plastics specialize in a single property (e.g. heat resistance), depending on the application. This contrasts them with engineering plastics, which are suitable for a wide range of processing technologies and applications. Table 2.3 shows some comparative properties of the standard plastics low-density polyethylene (PE-LD) and polyacrylonitrile-butadiene-styrene (ABS), the engineering plastic polyamide 6.6 (PA6.6), and the high-performance plastic polyether ether ketone (PEEK) (unreinforced in each case).

It can be seen that towards the high-performance plastics, the density, the thermal properties as well as the melting temperature and heat deflection temperature (HDT) [Grellmann & Seidler 2015], the mechanical properties (Young's modulus and yield stress), and the electrical properties comparative tracking index (CTI) [Hellerich et al. 2004] all increase.

2.6.6.4 Thermal Stability

The thermal stability is a central property of high-performance plastics. Based on the properties of standard plastics, certain mechanical and thermal improvements can already be achieved by adding reinforcing materials (e.g. glass and carbon fibers), adding stabilizers, and increasing the degree of polymerization.

An even higher service temperature can be achieved by completely eliminating aliphatic building blocks and closely linking aromatic groups through functional groups such as ether, sulfone, or imide groups. As a result, service temperatures of 200 °C can be achieved in the case of polyether sulfone (PES) and up to 260 °C for polyamide imide (PAI).

The increase in temperature stability through the incorporation of aromatic groups is related to the fact that the temperature stability of a polymer is determined by its resistance to thermal degradation and oxidation. Thermal degradation occurs primarily through random chain scission. Depolymerization and cleavage of low molecular weight compounds play only a minor role in high-performance polymers.

> The thermal-oxidative degradation of a polymer already starts at lower temperatures than the purely thermal degradation. Both types of degradation proceed via a radical mechanism. Aromatics offer good protection against both types of degradation, since free-radicals are delocalized and thus stabilized by the π-electron system of aromatics, and the aromatic binding energy is particularly high here. The thermal stability increases strongly in this way.

All high-performance plastics contain aromatic structures. Aromatic structures combine the two most important characteristics for resistance to high temperatures: On the one hand, they are resistant to oxidation, since the aromatic C–H bond (with 435 kJ/mol) is significantly more stable than the aliphatic C–H bond (with 350–400 kJ/mol), making the radical formation that occurs during thermal decomposition or fire more difficult. On the other hand, the chain stiffness of aromatic polymers is greater than that of aliphatic polymers, which increases the T_g (and the T_m in the case of semi-crystalline polymers). Thus, via the low free volume and high density up to high temperatures, both solubility to media and accessibility of oxidizing gases are reduced, which in turn increases the chemical and oxidative stability.

In practice, the highest temperature resistances are generally achieved with fluoropolymers. In fluoropolymers, fluorine atoms replace the usual hydrogen atoms of hydrocarbons. PTFE has the largest market share here, accounting for 65–70 % of fluorine-containing polymers [Acmite 2012]. However, fluorine-containing polymers are not suitable as construction materials due to their poor mechanical properties. They exhibit comparatively low strength and stiffness, and high creep under load. They are therefore not always categorized as high-performance plastics.

2.6.6.5 Crystallinity

Like all polymers, high-performance plastics can be divided into amorphous and semi-crystalline polymers (Figure 2.34). This helps to understand some basic property differences. Crystalline polymers (especially those reinforced with fillers) can also be used above their glass transition temperature (T_g). This is because, in addition to the T_g, semi-crystalline polymers also have a crystallite melting temperature (T_m), which is usually considerably higher than the T_g. For example, PEEK has a T_g of 143 °C, but can still be used permanently up to 250 °C due to a melting point of >300 °C (continuous service temperature, CST)). Another advantage of semi-crystalline polymers with a particularly high T_g is their high resistance to chemically aggressive substances. For example, PEEK has a high resistance to aqueous acids, aqueous alkalis, and organic solvents. Due to its high T_g and high T_m, the polymer is very poorly soluble and offers only a small surface area for attack.

2.6.6.6 Solution Properties

High-performance plastics are only very slightly soluble in organic and aqueous media due to their generally high T_g. Some high-performance polymers, such as polyether ether ketone (PEEK), are only soluble in strong acids such as concentrated sulfuric acid and nitric acid, whereby attack and chemical modification of the aromatics by the strong acids can occur (sulfonation, nitration).

2.6.6.7 Applications

High-performance plastics are expensive by comparison: the price per kilogram can be between US $5 (for PA4.6) and US $100 (for PEEK). The average value is somewhat less than US $15/kg, depending on which products are included in the high-performance plastics. High-performance plastics are thus around three to twenty times more expensive than engineering plastics. No significant price drop is to be expected here in the future either, as it can be assumed that the investment costs for production facilities, the complex development, and the high distribution costs will remain high. Their production volume is very low, totaling about 1 Mt as of 2023 [Fortune Business Insights 2023]. Their market share is well below <1 % of the production volume of all plastics in total. However, this is combined with a market value of approximately $21 billion. High performance polymers continue to have the largest projected growth of all polymer classes at 7–9 %.

Among the high-performance polymers, fluoropolymers have a ca. 45 % market share (main representative: PTFE), sulfur-containing aromatic polymers have a ca. 20 % market share (PPS), aromatic polyarylethers and polyether ketones have a ca. 10 % market share (PEEK) and liquid-crystal polymers (LCP) such as polyaramides have a ca. 6 % market share. The electrical and electronics industry uses ca. 41 % of high-performance plastics, and the automotive industry uses ca. 24 %, making them the largest users. All other industries (including the chemical industry) have a share of ca. 23 % [Wikipedia 2021].

2.6.6.8 Additive Manufacturing with High-performance Thermoplastics

High-performance thermoplastics may be the most promising material candidates for early adoption of AM, particularly in the areas of prototyping, small batch production and individualization, lightweight construction, as well as in diverse areas of application in medicine.

For this reason, a good understanding of existing AM processes and the resulting requirements for the processability of high-performance thermoplastics is important. This is the only way to predict the properties of additively manufactured components made from high-performance thermoplastics for use in critical environments. Furthermore, this indicates the development fields for the development of new polymer modifications and processing techniques in order to make even more high-performance polymers usable in AM.

Depending on the processing method chosen, various high-performance polymers are already well suited for 3D printing. For material extrusion-based processes (MEX) including fused layer modeling (FLM), fused filament fabrication (FFF), and pellet extrusion modelling (PEM), they are PEEK, carbon PA, PEI, PVDF and POM-c. For powder-based processes like selective laser sintering (SLS), they are PEEK, PEKK with encapsulated carbon, and PPS with and without fibers.

Studies show the success of high-performance polymers in AM especially for the service sector, where the highest priority is to offer end-users customized products and functional products in limited series sizes and in a fast and cost-effective manner. Already, the manufacturing costs of optimized additively manufactured components can rival the material costs of CNC-milled parts, depending on the degree of filling, especially for expensive high-performance polymers. Here, AM increases efficiency, reduces costs, and positively influences the availability of high-performance components and medical devices.

Polyether ether ketone (PEEK) is an ultra-high-performance polymer that sits at the top of the performance pyramid. PEEK and its relatives the polyarylether ketones (PAEK), including polyether ketone ketone (PEKK), are known for their ultra-high performance in aerospace, automotive and other industries where strength and high-temperature resistance are required. PEEK is a semi-crystalline material with very fast crystallization kinetics that make it well suited for many conventional processing methods like profile extrusion and injection molding. Like most semi-crystalline materials, it undergoes a significant dimensional change (shrinkage) during crystallization. In conventional processing techniques, this dimensional change is countered by process technology. However, in AM processes, this shrinkage can lead to warping and failed prints. The rapidly formed crystalline domains are almost completely resistant to further chain diffusion and entanglement, drastically reducing layer-to-layer adhesion. To achieve adequate part quality and print reliability, many 3D printer manufacturers have taken extreme measures (enabling very high build space heating, single layer heating etc.). These process requirements currently still limit the overall size and final mechanical properties of parts printed from PEEK.

More and more PEKK grades are available on the market, offering most of the excellent properties of PEEK with a wider and more flexible processing window. PEKK and PEEK are similar in structure, but have some significant differences. PEKK has an additional ketone group that replaces an ether group in the backbone. The presence of more ketone groups in PEKK produces a stiffer polymer backbone compared to PEEK and thus increases the T_g of the material, the temperature at which the polymer first begins to soften. The ketone groups can be in either the *para*-position (opposite) or *meta*-position (between) relative to the next functional group. The *para*-position produces a "straight" polymer structure, while the *meta*-position produces a more kinked polymer structure. By controlling the ratio of straight to kinked units in the polymer backbone, it is possible to precisely control the melting temperature, crystallinity, and crystallization rate of PEKK. This control allows PEKK to be produced and processed in a way that is difficult or impossible with many similar polymers such as PEEK. Dedicated PEKK grades are much easier to print with extrusion-based processes. This comes at the cost of a slightly lower crystallinity, overall temperature resistance, and mechanics when compared directly to unmodified PEKK or PEEK; however, due to a higher T_g this is balanced by a slightly higher compressive strength, improved barrier properties, and improved wear and friction properties.

The controlled adaptation of high-performance plastics to the processing realities of AM are significantly expanding the applications of high-performance plastics by offering manufacturers increased print reliability. Fewer failed prints and the reduced need for fine-tuning result in significant reductions in manufacturing time and cost.

In order to print polyetherimide (PEI)-based filaments, an extrusion head (hotend) is required which reaches a temperature of at least 350–370 °C. In return, available PEI filaments offer short-term heat resistance of up to 220 °C and continuous heat resistance of up to 170 °C. PEI is also very durable in printed form and resistant to many chemicals and liquids, such as automotive fluids, hydrocarbons, alcohols, and aqueous solutions. PEI is further characterized by a high creep resistance and dimensional stability, so that printed parts have a long service life. PEI is inherently flame retardant and performs very well in FST (flame, smoke, and toxicity) tests. PEI thermoplastics are widely used for tooling, whether to construct molds, fixtures, or various fasteners. PEI has a high strength-to-weight ratio and (thanks to its mechanical properties) can partially replace metals. Compared to PEEK, PEI is more cost-effective in 3D printing. It is a material certified for a number of aerospace applications and meets highly standardized requirements. However, printed PEI has lower impact and temperature resistance than PEEK.

The 3D printing of polytetrafluoroethylene (PTFE) is a special process. PTFE is not suitable for extrusion and powder-based printing processes, but can be processed in a similar way to metal powder via a resin printing process. In this process, the PTFE is first brought into a mold in a resin matrix via an SLA process. The resin is then pyrolyzed and the PTFE is sintered [3M 2021]. As a special technology, this process is offered as a service by a major PTFE supplier. The end result is an individually manufactured component with properties and performance comparable to conventionally processed and CNC-machined PTFE.

Polyvinylidene fluoride (PVDF) can be processed as filament on extrusion-based standard printers. PVDF is characterized by very good thermal resistance of up to 150 °C. PVDF is also highly resistant to a wide range of chemicals, such as acids and bases. All liquids commonly used in the automotive sector (oils, gasoline, and brake fluid, for example) also pose no problem.

PVDF is self-lubricating and thus to a degree abrasion-resistant. PVDF is UV-resistant and has a good flame resistance (V0). Typical printing conditions are an extruder temperature of 245–265 °C and a print bed temperature of approx. 90–110 °C. Suitable adhesion promoters should be used to obtain good print bed adhesion. The recommended printing speed is rather slow at about 25–40 mm/s and the part fan should be set to low power to keep the last layer interface surface hot. At a temperature above 290 °C, the material can decompose, releasing toxic fumes. Areas of application for PVDF are mainly the chemical process industry, the oil and gas industry, and highly stressed equipment parts such as fittings and valves, especially as spare parts and functional parts where high chemical, temperature, and UV resistance is required.

2.6.7 Engineering Plastics

Engineering thermoplastics are distinguished from standard plastics and high-performance plastics in terms of properties and price. Although the classification is not clear-cut, engineering plastics often have better mechanical properties than standard plastics, such as greater impact strength and a higher modulus of elasticity at temperatures above 80 °C. In keeping with their name, they are used mainly in engineering and construction applications where load-bearing or mechanically highly stressed components are required. Engineering thermoplastics are a subgroup of engineering plastics that includes duromers (crosslinked, non-meltable plastics) and thermoplastics (meltable plastics). Widely used engineering thermoplastics are: polyamides (PA), polybutylene terephthalate (PBT), polyethylene terephthalate (PET), polycarbonate (PC), and polyoxymethylene (POM).

2.6.7.1 Polyamide (PA)

The name polyamide is derived from the typical structure of a carboxylic acid amide. In biochemistry, the same structure is found in polypeptides and proteins. Peptides are organic compounds consisting of several amino acids. If there are more than one hundred amino acids linked together, they are called proteins. All natural proteins consist of twenty amino acids found in nature (natural amino acids). They all have in common the presence of at least one carboxylic acid and one amino group, which are linked together to form carboxylic acid amides. The many interactions of the building blocks of proteins with each other, in particular the hydrogen bond interactions, condition the folding and organization of proteins, for example, into highly efficient biocatalysts such as enzymes.

Engineering polyamides are among the oldest plastics in use and are accordingly used in a wide range of applications. With a production volume of approx. 10 Mt and a share of approx. 2.5 % of the total volume of plastics produced, polyamides are still among the most important engineering plastics and, due to their close links with the automotive, textile, electronics, and construction markets, continue to grow at >4 %/year, faster than global economic output (as GDP, gross domestic product) [Mordor 2021].

2.6.7.1.1 Production

Aliphatic polyamides, i.e. polyamides without aromatic structures (Figure 2.35) in the polymer backbone, are prepared by the reaction of carboxylic acids with amides (Figure 2.39). In the first case, the starting materials are lactams or ω-aminocarboxylic acids. Lactams are cyclic carboxylic acid amides in which the polymerization reaction is a ring-opening polyaddition. ω-Aminocarboxylic acids are carboxylic acids with an amino group on the C atom furthest from the carboxylic acid function, in which the polymerization reaction is a polycondensation reaction while undergoing the elimination of water.

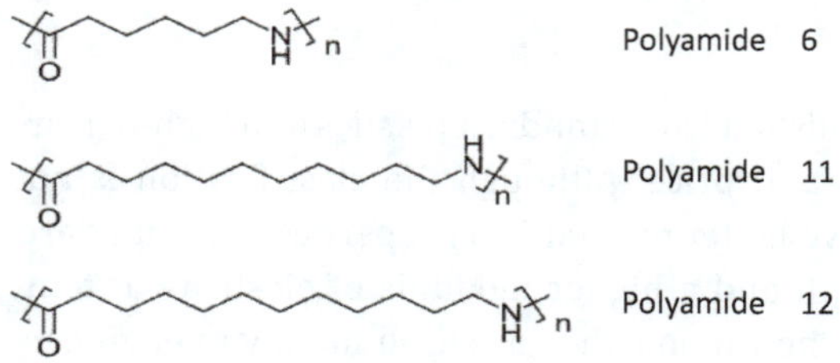

Figure 2.39 Polyamides based on the reaction of cyclic amides or ω-aminocarboxylic acids.

In the production of polyamides by polycondensation reaction from at least two starting materials, a combination of diamines and dicarboxylic acids is used (Figure 2.40).

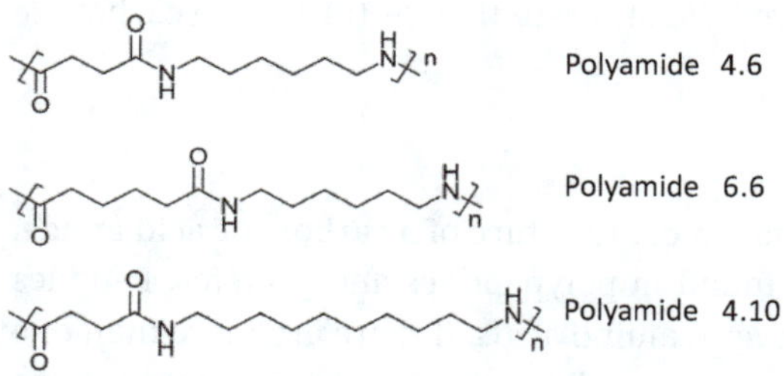

Figure 2.40 Polyamides based on the reaction of diamines and dicarboxylic acids

Both processes produce linear polyamides. They have a strong polarity and are mostly highly crystalline. The CO and NH groups protruding from the polymer chains cause a strong attraction of polymer molecules to each other by hydrogen bonds to other polymer chains or even chain segments of the same polymer chain. This shifts both the T_g and the T_m to significantly higher values (Figure 2.41).

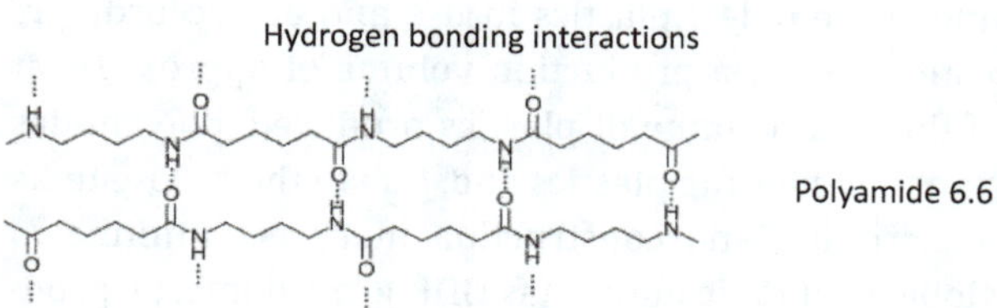

Figure 2.41 Formation of hydrogen bonds between two polyamide chains, in each case between the carbonyl group C=O and the amide group N–H

The ability to form hydrogen bonds and the high polarity are also the reason for the high water absorption of polyamides. As a result of the water absorption, the T_g drops sharply from temperatures above room temperature and can then also be below 0 °C. The water acts as an effective plasticizer and the polyamide changes its service properties at room temperature from hard-brittle to hard-tough.

When using polyamides for the production of dimensionally accurate components, water absorption in equilibrium must be taken into account. To avoid bubbles and thermal degradation during the processing at high temperatures, the polyamides are pre-dried to a residual water content of ≪0.5 % before use. After fabrication, the part is conditioned in a standard humidity climate at 23 °C and 50 % humidity to allow the polymer to swell back to its equilibrium state. Thus, not only the shrinkage (shrinkage) due to crystallization from the melt, but also swelling due to moisture absorption after processing, must be taken into account in the part design.

For AM, the crystallization and moisture behavior of polyamides causes a higher design and manufacturing complexity than in injection molding, since the shrinkage cannot be compensated – at least partially – by high holding pressures in the mold. In AM, therefore, polyamides with a comparatively low proportion of hydrogen bonds, such as PA11 and PA12, are mainly used. Alternatively, polyamides can be copolymerized, blended, and formulated with polyesters or polyethers, for example, in order to reduce the effect of hydrogen bonding (see also polyester-*co*-polyamides/polyether-*co*-polyamides).

The lower number of hydrogen bonds results in a lower melting temperature and lower mechanical strength of PA11 and PA12 compared to the significantly more commonly used polyamides PA6 and PA6.6. However, the advantages clearly outweigh the disadvantages in AM, especially since high melting temperatures and rapid crystallization as well as high component shrinkage cause significant processing disadvantages for both the extrusion-based processes and the powder-based processes. In particular, the necessary high extrusion temperatures of >300 °C and the high build-chamber temperatures of >200 °C should be mentioned here. The long build times in AM processes force working under inert gas (nitrogen or argon) and require very good pre-drying of the materials. Despite this, significantly poorer reuse rates are found due to thermal ageing when processing PA6 using the powder process than when processing PA12.

2.6.7.1.2 Features

Common aliphatic polyamides have a T_g <100 °C and are therefore often reinforced with glass fibers. In addition to increasing strength and modulus of elasticity, this can also significantly increase the continuous service temperature to temperatures even above 100 °C. Depending on the structure of the polyamides, the typical operating limits of aliphatic polyamides range from about –30 °C (PA11 and PA12: –70 °C) to above +100 °C. For short times peak temperatures of around 140–180 °C and above are possible for higher-melting polyamides like PA66 (PA4.6: up to 280 °C).

Compared to other thermoplastics, polyamides absorb a lot of water depending on the type, up to about 3 % at 23 °C and 50 % humidity (other thermoplastics absorb typically well below 1 %). This is associated with a volume increase of up to 0.3 % per 1 % water absorption, which must be taken into account in mold design.

Polyamides are polar and resistant to dilute alkalis, aliphatic and aromatic hydrocarbons, and fuels as well as alcohols, esters, ketones, fats, and oils. They are unstable in strong acids and bases and in chlorinated hydrocarbons, as hydrolysis and cleavage of the polyamide group may occur at elevated temperatures. However, polyamides are significantly susceptible to hydrolysis than polyesters. The use of polyamides for electrical applications is limited by the water absorption.

Table 2.4 List of Important Representatives of Polyamides and Their Typical Properties [Elser et al. 2012]

	PA6	PA6.6	PA4.6	PA6.10	PA11	PA12
Melting point (°C)	220	255	295	215	180	175
Glass transition temperature (°C) (dry)	40–45	50–60	75	50	40–45	40–45
Density (g/cm^3) (semi-crystalline)	1.12–1.15	1.13–1.16	1.18	1.07–1.09	1.03–1.05	1.01–1.04
Moisture absorption (%) (23 °C, 50 % RH)	2.8	2.3	3.2	0.4	0.8	0.7
Moisture absorption (%) (23 °C, 100 % RH)	9.5	8.5	13	3.3	1.9	1.5
Tensile modulus (MPa) (dry)	3000	3000	2800	2400	1500	1600
Tensile modulus (MPa) (moist)	1200	1500	1200	1200	1100	1200

Polyamides, especially PA12, are not very susceptible to stress cracking. When used outdoors, polyamides should be stabilized against the influence of UV radiation. PA12 has particularly good weathering resistance when appropriately stabilized.

Depending on their structure, polyamides are mostly opaque (non-transparent) due to their partial crystallinity. Based on amorphous polyamides with a high T_g or by adding crystallization nuclei to obtain crystallites with a size below the wavelength of visible light, it is in principle possible to produce crystal-clear transparent films and even optical-quality injection-molded parts from polyamides.

Polyamides, like many other materials produced by polycondensation, must be dried prior to processing by injection molding or extrusion in order to minimize degradation and imperfections (hydrolysis/bubble formation) during processing at high temperatures. However, they reach their final properties and dimensions only after reabsorption of water to equilibrium moisture content, which requires some time and storage under controlled conditions.

2.6.7.1.3 Applications

Polyamides are processed in large quantities into fibers, but are also among the most important engineering thermoplastics. They are tough materials with high strength and stiffness, excellent impact strength and good abrasion and wear resistance. Polyamides have displaced many metal parts in vehicle construction due to their high mechanical strength. Examples of applications include lamp housings, rigid fuel lines, fuel tanks, and accelerator and clutch pedals. Glass fiber-reinforced polyamides have a significantly higher heat deflection temperature than unreinforced grades and, due to their good oil resistance, are used for many applications in the automotive engine compartment. Due to the weight savings and the possibility of rationally manufacturing highly stressable parts with complicated geometries, glass fiber-reinforced polyamides are also used in hybrid design (combination of different materials) with steel and aluminum for structural components such as bumpers.

Polyamides are also used as housing material for electronic components and power tools, since they can be equipped with halogen-free flame retardants, and some polyamide grades (e.g. PA4.6) can withstand temperatures of over 260 °C for short periods without significant deformation, as occurs when components are soldered. Other applications include filter and pump housings, special seals and drive belts, and plain bearings. In precision engineering, polyamides are used to manufacture gears, rollers, screws, and nuts. Polyamides also serve as a material for filter fabrics, e.g. in dialysis filters.

2.6.7.1.4 Bio-based Polyamides

Depending on whether one or more monomers for a polyamide have been produced from renewable raw materials, the polyamide is considered to be partially or completely bio-based. The monomeric components (diamine, dicarboxylic acid, α,ω-dicarboxylic acid amines, and cyclic lactams) for bio-based polyamides do not differ structurally from the monomers of petrochemical polyamides, so the polymers themselves do not differ structurally either. The use of bio-based raw materials in polyamides is largely driven by the availability and cost-effectiveness of these raw materials.

2.6.7.1.5 Recycling

In principle, polyamides can be recycled both chemically and mechanically. Since the products are usually contaminated by use and are gathered in a compounded form with glass fibers and pigments, recycling them into an equivalent application has not been economically attractive in most cases to date. Exceptions are the hydrolytic degradation of PA6 fibers from carpets and fishing nets to the starting monomer (caprolactam), which can then be polymerized to new high-quality PA6 [Recycling Portal 2016].

2.6.7.1.6 Polyamides in Additive Manufacturing

In AM, modified polyamides, especially PA11 and PA12 powders, play a prominent role as materials in powder-based processes such as laser sintering (SLS) and high-speed sintering (HSS). Especially in the case of PA11 and PA12, the low water absorption (<1 %) and low melting points of PA11 and PA12 are exploited (<190 °C). Other interesting building blocks for use – especially in sintering processes – are polyester-*co*-polyamides and polyether-block-*co*-polyamides (PEBA), which are characterized by a comparatively lower melting temperature, low hardness, and particularly high elasticity. They are therefore used in applications where elastic, rubber-like properties are desired (see also the use of other similar thermoplastic elastomers/block-copolymers such as TPU).

The processing of the more common and higher-melting polyamides PA6 and PA6.6, which also have a higher water absorption, is difficult to achieve in the SLS process, since the polymers tend to discolor, oxidize, or hydrolytically decompose after the required residence time at the high sintering temperatures. Furthermore, the high crystalline content and the resulting high shrinkage are a problem for AM processes, since the shrinkage cannot be easily compensated for by increasing the injection pressure in molding processes or by adjusting the die design in extrusion processes.

In extrusion-based processes, too, the high water absorption and the high processing temperature combined with the high shrinkage are are a problem when processing polyamides. They are therefore hardly used in the hobby sector. In the professional sector, these disadvantages can be partially compensated by high build chamber temperatures and thoroughly pre-dried filaments. By copolymerization with polyesters (PES) or by mixing diamines and dicarboxylic acids in the polymerization process, polyamides and co-polyamides that are more processable for AM can be obtained.

2.6.7.2 Polyethylene Terephthalate (PET)

Polyethylene terephthalate (PET) is one of the most important thermoplastics, accounting for around 15 % of the total amount of plastics [Statista 2015]. Originally used only for fibers, about a quarter of the volume produced is now used for packaging and semi-finished products. The large-scale production of fibers based on PET began as early as 1953, and the 1970s in particular saw a strong expansion in the production of these fibers worldwide. Fibers made from PET now occupy the leading position among synthetic fibers. The main producing country today is the People's Republic of China. PET fibers are partly produced from recycled PET bottles. However, the majority of European PET production is now recycled again for use in the beverage packaging industry.

2.6.7.2.1 Production

The monomers from which PET is produced are terephthalic acid (1,4-benzenedicarboxylic acid) and ethylene glycol (1,2-dihydroxyethane). Figure 2.42 shows the reaction scheme of PET production by esterification of terephthalic acid with ethylene glycol with elimination of water (a polycondensation reaction). For most applications, PET in semi-crystalline form is preferred. However, it crystallizes spontaneously only very slowly, which is why nucleating agents are added for rapid crystallization.

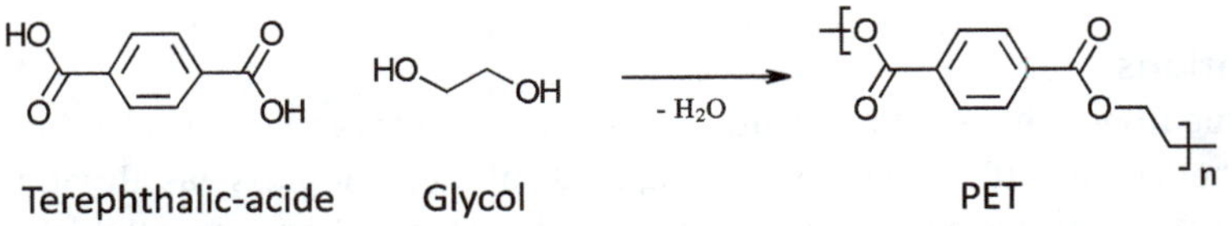

Figure 2.42 Reaction equation for the production of PET

2.6.7.2.2 Features

PET is a polar, linear and symmetrical polymer that exhibits strong dipole–dipole interactions between the polymer chains. These are the prerequisites for semi-crystalline domains and its use in textile fibers. These domains, in combination with a high T_g of about 80 °C, provide a high fracture strength and dimensional stability, up to a temperature of 80 °C. However, the impact strength of PET compared to polyamides is low, while the sliding and wear behavior is good. The melting point of PET is between 235 °C and 260 °C, depending on the degree of crystallization and polymerization. The density of amorphous PET (A-PET) is 1,33–1,35 g/cm^3 and of semi-crystalline C-PET 1,45–1,58 g/cm^3, with the degree of crystallization rarely exceeding 70 %.

The properties of PET depend strongly on its crystallinity. PET is typically a semi-crystalline polymer, but due to its very slow crystallization rate, it can also exist in an amorphous-transparent state, depending on the respective type and processing conditions. Semi-crystalline (C-PET) has a high strength and stiffness, but only low impact strength. The tendency to deformation due to creep is low. C-PET also has good abrasion resistance. The application limits are approximately between –20 and 120 °C (continuous temperature). In the short term, temperatures of up to 200 °C are possible. However, the modulus of elasticity of uncompounded drops significantly above the T_g of about 80 °C.

A-PET is transparent and surprisingly has a slightly better impact strength, but lower hardness and stiffness than C-PET. The application limits of –40 to 60 °C are well below those of C-PET. At temperatures above 80 °C, crystallization begins. A-PET is transparent up to a wall thickness of about 5 mm.

PET has only medium electrical insulating properties. It is polar and resistant to dilute acids as well as aliphatic and aromatic hydrocarbons and oils, fats, and alcohols, but not to halogenated hydrocarbons and ketones. Since PET belongs chemically to the group of polyesters, it is unstable in continuous use in the presence of hot water and alkalis. The comparatively low permeability to oxygen and carbon dioxide allows PET to be used in established packaging materials instead of glass bottles and metal cans, i.e. also for carbonated beverages. To further increase the barrier properties, such containers are usually plasma-coated on the inside with a thin silicate layer. The (co) polymerization of PET in the presence of bulky molecules such as isophthalic acid or 1,4-cyclohexanedimethylol disturbs the chain symmetry and further reduces the tendency to crystallize, but tends to increase the T_g. In this way, transparent parts can also be produced from modified C-PET.

2.6.7.2.3 Applications

In addition to textile fibers, the main applications for PET are blow-molded packaging goods, i.e. mainly beverage bottles. Here, packaging made of metal or glass has already been displaced to a considerable extent. PET beverage bottles have different wall thicknesses; thin-walled bottles are single-use bottles, thick-walled bottles are returnable bottles.

As a textile fiber (polyester), PET is used due to various useful properties. It is wrinkle-resistant, tear-resistant, weather-resistant, and absorbs very little water. The material is used in continuous filaments, staple fibers, crimped fibers, felts, or spun yarns for apparel fabrics, home textiles, and industrial purposes. They are also used in blends with other fibers such as cotton, wool, silk, and man-made fibers in the form of woven and knitted fabrics to make outerwear and underwear, coats, ties, upholstery covers, and bedding. The puffy variants are suitable as filling material for jackets, sleeping bags, pillows, and quilts. Because of its weather resistance, PET is also used in curtains, awnings, sailcloth, tents, and flags. The range of applications for PET as a recycled fiber is also growing in carpets and floor coverings.

Technical applications include conveyor belts, drive belts, tire cord, seat belts, fire hoses, ropes, filter materials, sewing threads, and yarns. However, PET is also used in appliance manufacturing and precision engineering as well as household appliances, photographic films, magnetic tapes, sausage casings, and food packaging. Injection molding is used to produce heavily stressed molded technical parts such as bearings, gears, cam discs, plugs, screws, and nuts.

The near-cousin material polybutylene terephthalate (PBT) has similar properties and applications to PET, but its melting point (230 °C) is lower than PET, and it is therefore easier to process.

2.6.7.2.4 Recycling

Non-refillable PET bottles are collected in Germany, partly recycled and partly thermally recovered. In processing plants, PET from beverage bottles is chopped into small pieces, known as “flakes”, cleaned of foreign matter, sorted by color, and reprocessed into granules. Some of this material can be added to virgin material and used to produce new preforms in an injection molding process for beverage bottles.

Chemical recycling processes also open up possibilities for using 100 % recycled PET. Two processes have been developed for this purpose: In the first process, the previously mechanically shredded material is chemically dissolved, filtered, and precipitated again; in the second process, the material is hydrolyzed (split into its components) with the aid of heat and a catalyst and subsequently converted back into pure PET [Ügdüler et al. 2020].

Recycled PET is also used for the production of polyester textile fibers. The world market leader in this field has so far been China. In 2006 alone, almost 4 Mt of PET waste were exported to China, equivalent to around 100 billion plastic bottles. The market value of sorted PET is between 400 and 500 €/t. Although technically feasible, the economics of recycling depend heavily on the cost of collecting and reprocessing PET. China stopped importing plastic waste at the beginning of 2018.

Since around 2010, a PET variant has been on the market in which the ethylene glycol component (approx. 30 % of the product) is sourced from renewable raw materials. In 2017, around 1,09 Mt of this were produced, making it the largest share among all (partially) bio-based plastics. In mid-2015, a PET bottle made entirely from renewable raw materials was presented for the first time at the World Expo in Milan [Business Wire 2015].

2.6.7.2.5 PET in Additive Manufacturing

Polyethylene terephthalate glycol (PETG) is frequently used in extrusion-based AM process. This is a co-polymer of polyethylene terephthalate (PET). It is a hard-tough material with good impact resistance, but has a slightly softer surface than PET, making it susceptible to wear. The material benefits from the excellent thermal properties of PET. There are several variations (co-polymers) of this material on the market (Figure 2.43).

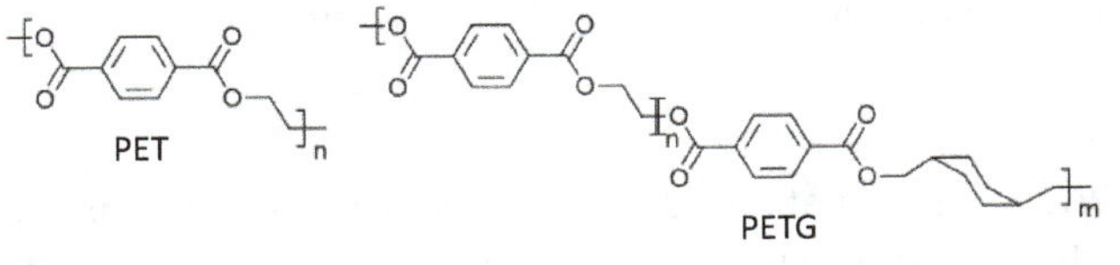

Figure 2.43 Structures of PET and PETG

The PETG variants preferred for 3D printing in the FLM process are characterized by particularly high transparency and low viscosity. Other advantages include a glossy and smooth surface. PETG is also mostly odorless when printed. This is offset by some disadvantages, such as poor bridging properties and the tendency to pull threads (stringing), as well as – depending on the printing conditions – a tendency to warp on the build platform, which may lead to a loss of adhesion.

PETG can be printed between 210 °C and 250 °C, depending on the quality. The print bed temperature is ideally set between 70 °C and 90 °C, and the printing speed is approx. 50 mm/s. PETG is now widely used as a filament for use in DIY FLM 3D printers and is also popular in a number of different professional applications, thanks to its material properties and its comparable ease of printing.

When working on new concepts, designers and engineers can take advantage of the transparency, moisture resistance, and chemical resistance of PETG to create functional prototypes and small numbers of such as containers for liquids, signage, graphic displays, and housings for electrical devices, etc. The most obvious uses of PETG in 3D printing are the design and prototyping of packaging for consumer products that will later be mass-produced from PET.

PETG, compared to other common printing materials like PLA, is also suitable for more demanding environments where heat and chemical resistance are required in addition to mechanical resistance. As a result, manufacturing engineers can use PETG to produce durable custom parts such as manufacturing tools and aids, test components or robotic end-effectors. Specifically compounded PETG (including e.g. glass or carbon fibers) can be further used to produce end-use parts, including snap-fits and living hinges, guards, or custom machine parts.

2.6.7.3 Polybutylene Terephthalate (PBT)

Polybutylene terephthalate (PBT) is very similar to PET in chemical structure and also in properties (Figure 2.44). During the polycondensation reaction, 1,4-butanediol is used instead of ethylene glycol. PBT is also a semi-crystalline thermoplastic, but it crystallizes faster and is therefore more suitable for injection molding than PET. PBT is mainly processed by injection molding at melt temperatures of 230 °C to 270 °C.

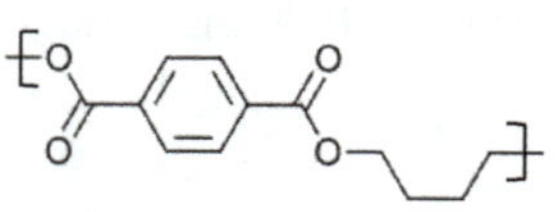

PBT

Figure 2.44
Structure of polybutylene terephthalate (PBT)

2.6.7.3.1 Features

The strength and stiffness of PBT is somewhat lower than that of PET, while toughness at low temperatures is somewhat better. The sliding and wear behavior is very good. The glass transition of the amorphous phase is at 60 °C. The maximum short-term use temperature is correspondingly lower at 180–200 °C, but the long-term use temperatures of 100–120 °C are comparable to PET. PBT is a good electrical insulator, with a low water absorption and low temperature and frequency dependence of the insulation properties. Its permeability to CO_2 is much greater than PET, and resistance to chemicals, weathering, and fire behavior are comparable. Resistance to hot water is better, and PBT grades are also approved for contact with food.

PBT grades are often modified. A thermoplastic elastomer modification (PBT-TPEE) to increase flow properties and improve flame retardancy and impact strength provides a polyamide substitute. PBT can be highly filled with glass fibers, for example, to increase stiffness, strength, and wear resistance, or reduce the coefficient of sliding friction. The material also behaves advantageously in blends with other semi-crystalline and amorphous thermoplastics with high T_g (PBT-PET, PC-PBT, PBT-ASA).

2.6.7.3.2 Applications

PBT is used to manufacture products similar to those made from PET: plain bearings, roller bearings, valve parts, screws, plug connectors, pump housings, and wheels, as well as parts for household appliances such as coffee makers, egg stoves, toasters, hair dryers, vacuum cleaners, and cooking appliances. Processed into fibers, it is used, among other things, in toothbrushes, in the sheathing of optical fibers, and as a fiber material for technical and textile applications.

2.6.7.3.3 Recycling

PBT can be recycled in the same way as PET. However, the prevalence of PBT in predominantly higher-value products and smaller components and the preferential blending with other plastics make it difficult to collect as a single-grade material. The availability of a single-grade material still forms the basis for simple material or chemical recycling. As a result, PBT and its blends and mixtures are mainly thermally recycled.

2.6.7.3.4 PBT in Additive Manufacturing

In 3D printing, PBT is currently not much used as a pure filament. The higher crystallization tendency makes it more difficult to use in the extrusion process due to the higher shrinkage and poorer layer adhesion. For processing, it requires build chamber temperatures of approx. 120–140 °C and extrusion temperatures of >240 °C. As a blend with polycarbonate (PC), PBT is used in filaments in order to improve the flow properties and lower the extrusion temperature of the PC.

PC-PBT filament combines the good chemical resistance of PBT with the strength and toughness of PC. In addition to mechanical strength, heat resistance, and excellent chemical resistance, the material also offers high print quality. The polymer blend has good resistance to hydrocarbon-based chemicals and can still be used at temperatures as low as –20 to –30 °C. PC-PBT retains its good toughness and ductile fracture behavior even at low temperatures, and is therefore particularly suitable for demanding technical applications, for example in the automotive sector. However, even with the PC-PBT blend, a build chamber temperature of >90 °C and bonding aids for the build platform are recommended, as the required high extrusion temperatures of >280 °C can lead to thermally induced distortion in the product and thus also to detachment from the build platform. After production, products are preferably annealed for a further time at approx. 90 °C in order to relieve internal internal stresses.

Recent developments also allow the use of modified PBT variants in powder-based processes such as laser sintering (SLS). For this purpose, the melting point and recrystallization properties, as well as the viscosity of the PBT, are specifically modified in such a way that a larger processing window is created [Boissonneault 2019]. In the laser sintering process, printed PBT parts exhibit high ductility compared to other sintered polyamides and polyesters, similar to that of conventional injection-molded parts. In conjunction with the good mechanical properties, PBT powder exhibits particularly low water absorption, comparable to PA12, largely eliminating the need for costly drying prior to processing. Expected areas of use for the new PBT powders are industrial applications in the automotive and aerospace industries.

2.6.7.4 Polyoxymethylene (POM)

Polyoxymethylene (POM) is a semi-crystalline, largely linear thermoplastic produced by chain polymerization or chain copolymerization. It is one of the typical engineering thermoplastics with good mechanical properties and high dimensional stability, as well as excellent sliding and wear behavior. The production volume of POM in 2016 was 1.2 Mt [Plastics Insight 2021].

Due to the presence of a repeated acetal (-CHR–O-) group in its chemical structure, polyoxymethylenes are also called polyacetals. A distinction is made between POM homopolymers and copolymers with incorporated -O–$(CH_2)_2$- or -O–$(CH_2)_4$- units (Figure 2.45).

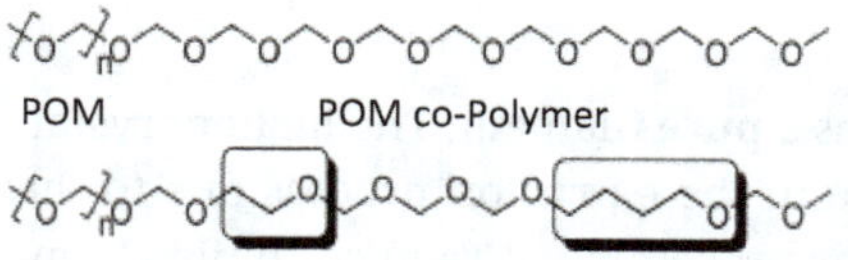

Figure 2.45
Polyoxymethylene (POM) and co-polymers

2.6.7.4.1 Manufacturing Process

Polyoxymethylene (POM) can be obtained by various manufacturing processes. Raw materials are mainly formaldehyde ($O{=}CH_2$) or formaldehyde trimer (trioxane, -(O–CH_2-$)_3$). It is important for the production of POM that the end group in the polymer is terminated with an ester to avoid depolymerization in an acidic environment, or at elevated temperature with the release of formaldehyde. By copolymerization with cyclic ethers, such as ethylene oxide and tetrahydrofuran, the properties of POM can be varied to a certain extent.

2.6.7.4.2 Features

POM belongs to the group of engineering thermoplastics and has medium to high impact strength and hardness, and excellent sliding and abrasion behavior. The high elasticity of recovery finds many applications, particularly in the field of snap-fit connections.

The crystallinity of POM-H (homopolymer) reaches up to 80 % and is thus somewhat higher than that of POM-C (copolymer) with up to 75 %. The melting point of POM-H is

175 °C and of POM-C 164–172 °C. POM-H has slightly better heat resistance, while POM-C is slightly more resistant to alkalis and hot water.

Pure POM has a low resistance to weathering. Excessive processing temperatures or heating above 220 °C lead to thermal decomposition, whereby formaldehyde may be released (pungent and irritating odor).

Without special surface treatment, POM can only be bonded by adhesives to a limited extent. The adhesion of selected adhesives can be improved by special pretreatment (e.g. plasma treatment) of the surface.

The operating limits of POM are around –40 to 85 or 100 °C continuous service temperature (up to 120 °C for short periods). Its heat resistance can be further increased by adding glass fibers. The thermal expansion of POM is in the medium range. Due to its semi-crystalline structure, parts made of POM are opaque and white, but can be opaquely colored. POM exhibits good electrical insulating properties.

Due to their high crystallinity, POM thermoplastics exhibit good resistance to numerous chemicals. They are resistant to dilute acids (pH > 4) as well as alkalis, aliphatic, aromatic, and halogenated hydrocarbons and oils, as well as alcohols. They are not resistant to concentrated acids, hydrofluoric acid, or oxidizing agents. POM shows low moisture absorption and exhibits high gas tightness. Without the addition of UV stabilizers (including carbon black), weathering resistance is low. POM burns as a result of the specific degradation mechanism, namely the splitting off of gaseous formaldehyde when exposed to sufficient temperature, but achieves class HD in the UL 94 fire test. POM is available variants approved by the FDA (US Food and Drug Administration).

2.6.7.4.3 Processing

POM is mainly processed by injection molding, especially also by “micro-precision injection molding”, so that it is suitable for the production of dimensionally accurate small parts with low tolerances. Other processing options include extrusion, blow molding, and welding (except HF welding). POM is also excellent for chip-removal processing. Due to its high degree of crystallization, POM exhibits high shrinkage. Despite the high shrinkage, there is only a small risk of warpage, since the shrinkage differences along and across the flow direction are small in POM.

2.6.7.4.4 Applications

The main application of POM is in mechanical engineering and the automotive industry, including gears, miniature gears, switch gears, ball bearing cages, parts of tank caps, fuel pumps, and carburettors, loudspeaker grilles, snap and spring elements, parts and housings for water pumps, valves, and fittings, as well as screws, nuts, bearings for clockwork mechanisms, etc. Other applications are found in electrical engineering, including precision parts for telephone sets, radio, television, sound reproduction, copiers, conductive molded parts (with appropriate filling), and in the packaging industry for spray cans, gas lighter tanks, and gas ampoules. In the construction and furniture industries, POM's high mechanical resistance and self-lubricating effects are exploited for use in fittings, hinges, lock parts, door and window handles, and coupling parts for garden hoses. Other areas include mechanically highly stressed small parts such as hooks, zippers, ski binding parts, push-on toothbrushes, and toys.

2.6.7.4.5 Recycling

It is almost impossible to recycle acetal resins like POM as they would have to be absolutely type-clean. They are also not too stable against (multiple) processing steps. However, if they are depolymerized in strong acidic medium, the acetal resins decompose into trioxane, the original raw material for the synthesis of new POM. This can be directly fed back into POM polymerization. Due to the comparatively low prevalence of POM, this process has only been carried out in experimental plants to date.

Solvolytic depolymerization: Many polymers are formed by condensation reactions with removal of water or alcohols or also by addition reactions to "pre-dehydrated" groups (anhydrides, isocyanates, ring esters, ring ethers etc.). Today, this is true for many thermoplastic materials such as amide, ester, acetal, terephthalate, and urethane resins. Reversing this reaction, in principle, opens up the possibility of attacking them again hydrolytically in an acidic or basic environment with water, alcohols, or amines at this junction and breaking them down into monomeric components. These processes are collectively referred to as solvolytic depolymerization, since at least part of the polymer must be dissolved in the reaction mixture or reaction vessel during depolymerization.

Thermal depolymerization: An alternative to thermal recycling in which the energy of the plastics can be recovered in the combustion process, is pyrolytic recycling. This involves the decomposition of hydrocarbons in the absence of air or oxygen (reducing conditions). Depending on the temperature (500–850 °C), various products are obtained under reducing conditions – from hydrocarbons to waxes or oils to gasolines. Beyond these products, pyrolytic recycling has the additional advantage over incineration that the amount of flue gases is greatly reduced, with hazardous substances remaining in the coke-like residue.

2.6.7.4.6 POM in Additive Manufacturing

Polyoxymethylene (sometimes called polyacetal) is becoming increasingly popular in the field of 3D printing, especially in extrusion-based processes. POM-C has good thermal and chemical resistance and is easy to extrude. Applications for POM in 3D printing, similar to injection molding applications, are mainly automotive, medical, packaging, and small parts with low volume tolerances.

Polyoxymethylene is particularly suitable for 3D printing due to its wide range of applications. A component printed with POM remains stable and does not deform even when exposed to temperatures ranging from –40 to 140 °C. The absorption of moisture is very low. It therefore does not need to be dried before printing and has good electrical insulation properties. The high impact and abrasion resistance and good sliding properties are also maintained in 3D printed components, as high adhesion between the melt layers is observed. The insensitivity of POM to polar chemicals also allows disinfection of components in medical use without damaging the part.

When printing POM, a heatable printing plate should be used at a temperature of 120 °C, since initial adhesion to the printing plate is low due to the poor bondability of the material with other materials. An extrusion temperature of 190–210 °C should be selected, being aware that this temperature is already close to the decomposition temperature of POM. Constant temperature management during the printing process is important to achieve a high-quality result and to prevent the material from decomposing in the print head. The component fan should only run at a low level during the build process, and the filling level should not exceed 25 % to reduce unwanted deformation, shrinkage, or warping. Accordingly, the printing process should only take place in well-ventilated rooms.

In order to glue objects printed with POM filament, a few preparatory steps are required. POM has a low surface energy, therefore, the surface must be pre-treated before bonding is possible; for example, by means of surface oxidation and/or plasma treatment. POM filament can be ground, polished, milled, as well as drilled. POM should be stored in a cool and dry place. Especially as a pure, unstabilized material, it should not be exposed to direct UV radiation. Started filament rolls should be re-packed in dyed plastic bags for this purpose.

2.6.7.5 Polycarbonate (PC)

Polycarbonates (PC) are formally speaking polyesters of carbonic acid. In 2020, the global PC consumption was around 5 Mt [Grand View Research 2016]. In addition to pure PC, there is also a large market for PC blends with polyesters (PBT, PET) and ABS, for example. The high growth of PC volumes is mainly due to increasing applications in the automotive industry and the boom in the IT sector.

2.6.7.5.1 Production

Polycarbonates can be produced by polycondensation of carbonic acid dichloride (phosgene) with diols or by transesterification of the phenol ester of carbonic acid. Production by the phosgene route is done by an interfacial polycondensation reaction (Figure 2.46). The aqueous phase consists of a sodium hydroxide solution, in which bisphenol A dissolves as a sodium salt. The gaseous phosgene is introduced into the separate organic phase, consisting of an inert solvent. The polymerization starts immediately at the phase boundary of the water and organic phase, producing hydrogen chloride (HCl) and a thermoplastic polycarbonate polymer. HCl immediately reacts with excess sodium hydroxide (NaOH) in the water phase to form sodium chloride (NaCl). NaCl in turn serves as a raw material in chlor-alkali electrolysis for the production of chlorine and caustic soda. Chlorine is reacted with carbon monoxide (CO) in a further step to produce phosgene ($COCl_2$). The bottom line is that chlorine and sodium are fed back into the PC manufacturing process in a closed process loop using "only" electrical energy.

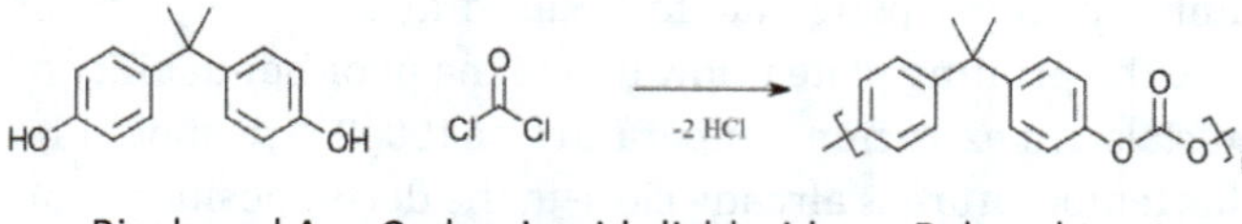

Figure 2.46 Reaction of bisphenol A and carbonic acid dichloride to polycarbonate

Alternatively, the preparation can be carried out via transesterification with diphenyl carbonate. In this melt condensation, the reaction takes place under inert gas at low pressure. The reaction product phenol is subsequently removed from the reaction mass by low-pressure distillation at temperatures of up to 300 °C.

2.6.7.5.2 Features

Polycarbonates generally have a crystallite content of less than 5 % and are therefore considered to be amorphous with a glass transition temperature of around 148 °C. They are characterized by high strength, impact resistance, stiffness, and hardness. In addition, polycarbonates are good insulators against electrical voltage.

With its high optical transparency combined with a surprisingly low density of approx. 1.2 g/cm^3 and a high Young's modulus of 2400 MPa, a tensile strength of approx. 65 MPa, an elongation at break of up to 120 % and a heat deflection temperature (HDT A) of approx. 125 °C, PC is one of the most important engineering materials for a wide range of demanding applications in the automotive, optical, and electronics sectors, where good surface quality, hardness, and toughness are essential.

Polycarbonates are resistant to water, many mineral acids, and aqueous solutions of neutral salts and oxidizing agents. Some non-polar organic solvents such as hydrocarbons and many oils and greases also do not attack them. On the other hand, they are not resistant to some chlorinated hydrocarbons such as dichloromethane, and are also attacked by alkaline aqueous solutions, amines, and ammonia, as well as some organic solvents like acetone.

Polycarbonates are flammable, but the flame extinguishes after removal of the ignition source. PC meets the requirements of fire class B2 according to DIN 4102. In layer thicknesses of 1–6 mm, it is classified in fire class B1 ("flame retardant") in the case of interior applications. It also meets the fire behavior requirements for automotive glazing according to approval guidelines such as TA29 (national), ECE43, or ANSI Z26.1 (USA).

The water-clear plastic is characterized by glass-like light transmission levels (88 % at a thickness of three millimeters according to DIN 5036-1) and a high refractive indices

(1.59 according to ISO 489-A), making it useful for many optical applications with a high demand for toughness and low weight (e.g. ophthalmic glasses for children/sports, and automotive flood lights). However, PC is sensitive to UV light in the wavelength range around 340 nm. Irradiation with light of this wavelength (among other factors when used outdoors) causes the material to become brittle and yellow over time, unless a protective coating or compounding with a suitable stabilizer package is used.

Copolymerization with other monomers can change the properties of PC to a certain extent. Copolymerization with BPTMC (a bisphenol A derivative) has proved to be particularly advantageous (Figure 2.47). In this way, the T_g could be increased by almost 50 °C without significantly changing the optical properties.

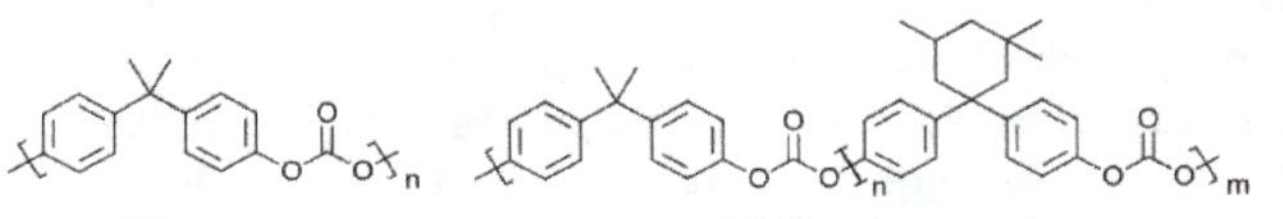

Figure 2.47 Polycarbonate and co-polycarbonates with BPTMC

2.6.7.5.3 Processing

Polycarbonates can be processed by all the usual methods for thermoplastics. In injection molding, high injection pressure is required due to the high viscosity of the melt. The processing temperatures are between 280 and 320 °C for molding, and between 240 and 280 °C for extrusion. Before processing, however, the residual moisture must be brought to below 0,01 % by drying (4–24 hours at 80 °C). The processing shrinkage of PC is very low at about 0.6 to 0.8 %. PC exhibits virtually no post-shrinkage, which explains its particular popularity for injection molding components with high surface quality and high dimensional accuracy. It can be easily glued with solvents and reactive resin adhesives, and is also ultrasonic and high-frequency weldable.

2.6.7.5.4 Applications

The largest PC consumers in 2012, with a combined share of 35 %, were the manufacturers of electrical goods, information/communication technology, and household appliances. In second place came the automotive industry (17 %). The construction industry, which mainly uses PC extrusion applications such as transparent sheets, recorded a share of 15 %. The consumer goods industry and medical technology together accounted for 17 %. Still surprisingly high, although currently declining at annual rates of around 5 %, is PC consumption for optical data storage such as CDs, DVDs, and Blu-Ray discs (16 %) [Abts 2020]. The world's largest single PC market is China, which was responsible for 35 % of global consumption in 2012, driven mainly by a rapidly growing IT sector and continuing growth in the automotive industry, especially for vehicles in the mid-range and luxury segment.

2.6.7.5.5 Recycling

Various processes are available for the recycling of PC:

- Mechanical recycling, whereby single-variety PC can be proportionally melted down and recycled by mixing with fresh material [Covestro 2019]
- Solvolytic recycling, in which PC can be dissolved from PC blends and mixtures, extracted, selectively precipitated, and recycled in a cleaned state
- Hydrolytic recycling, whereby PC can be hydrolytically separated into its components and returned to polymerization.

In the case of PC, efforts are being made to recycle in all of the above areas.

> The higher the quality of a material, the more a recycling process may cost in order to be economically viable. On the other hand, the production volume must be sufficiently large to make sorting worthwhile, and the product application must result in recyclates that are as pure as possible. For many standard polymers, the price of the raw materials is so low that more costly recycling processes, in addition to thermal recycling and pyrolysis, are often not worthwhile. In the case of high-performance polymers, production volumes are often so low that collection is not worthwhile.

2.6.7.5.6 PC in Additive Manufacturing

PC is known in the 3D printing market for its durability and transparency, and is often used for professional applications. In 3D printing, PC enables the production of complex heat-resistant components via extrusion, but it is not as easy to print as PLA or PETG, as it requires relatively high extrusion and build chamber temperatures. PC is highly valued by the AM industry for its strength and transparency. It has a much lower density than glass, which makes it particularly interesting for the design of optical parts, protective screens, or decorative objects such as lampshades. The PC filament can withstand temperatures from –150 to 140 °C, which significantly expands the application space of AM parts.

PC must be kept in a dry environment or pre-dried before use in 3D printing. PC tends to swell at the extrusion die, which can lead to extrusion problems. Due to the high application temperature, PC tends to cause adhesion problems on the build platform and in the layers. The build platform should have a temperature of at least 100 °C and the build chamber one of 100–140 °C depending on the PC type/blend. The extrusion temperature is high (between 260 and 310 °C), and production in a heated enclosed build chamber is strongly recommended to minimize temperature fluctuations. Pure PC is therefore suitable for hobbyists only to a limited extent.

In addition to pure PC, it is also available in various blends that are much easier to process, such as PC-PBT and PC-ABS, which combine the strength of PC, for example, with the good extrudability of PBT and the flexibility of ABS. PC can also be reinforced with glass fibers or carbon fibers to further increase the strength of the material, but then again requires suitable printers with nozzles reinforced against abrasion and high extrusion temperatures.

In the powder-based printing process, PC can be used as a powder with high imaging accuracy. In order to obtain good layer adhesion and component density, downstream tempering of the component in an inert medium at temperatures slightly above the T_g of PC is required here [Goulding 2021].

2.6.7.6 Polylactic Acid (PLA)

Polylactic acid (PLA), also called polylactide or polylactic acid ester, is a thermoplastic polyester produced from sugar via a multi-step synthesis. PLA plastics are biocompatible. PLA does not appear in the common representation in the polymer thermoplastics pyramid, because it is still a small product with a production volume well below 0.5 Mt. PLA's properties, as well as its price, place it between engineering materials and standard thermoplastics. However, apart from starch blends, PLA are currently the material group with the greatest volume potential within bioplastics. The basic material for PLA is lactic acid, which occurs in two stereoisomeric variants, D-lactide and L-lactide. Accordingly, there are also two PLA variants: poly-D-lactide (PDLA) and poly-L-lactide (PLLA). Depending on how the building blocks are incorporated in the PLA chain, the morphology of the material also differs, from amorphous to crystalline, with a corresponding influence on the thermal and mechanical properties.

2.6.7.6.1 Features

PLA is transparent, crystalline and rigid (Young's modulus up to 5000 MPa) with a T_g of 45–65 °C and a melting point – depending on composition – of 150–160 °C. PLA can be processed by conventional thermoplastic methods and has a high resistance to oils and greases, but becomes brittle when exposed to UV radiation. The bioplastic is successfully used in a temperature range from –10 °C to 40 °C. The mechanical properties of pure PLA at room temperature are similar to those of PET. In particular, its transparency and low migration values predestine PLA for use in the food packaging sector. However, compared to PET, PLA has much higher CO_2, oxygen, and moisture permeability, is more brittle, and has a significantly lower maximum use temperature. The price for PLA of about €2/kg is currently higher than that for PET. However, with increasing production volumes, production costs are expected to decrease in the coming years. PLA is biodegradable according to EN 13432, but this requires environmental conditions that are usually only found in industrial composting facilities. At room temperature and normal environmental conditions, PLA is stable. Furthermore, biodegradability can be controlled by its chemical composition as well as the use of suitable copolymers.

2.6.7.6.2 Production

Polylactides are accessible mainly by the ionic chain polymerization of a cyclic dilactate ester, a ring-shaped compound of two lactic acid molecules. The lactide itself can be produced by fermentation of molasses or by fermentation of glucose. Figure 2.48 shows the reaction scheme of the preparation of polylactic acid from the cyclic dilactate ester. The preparation is carried out by ring-opening polymerization in bulk at temperatures between 140 and 180 °C in the presence of catalytic amounts of a tin catalyst. Plastics with a high molecular weight and strength can be obtained this way.

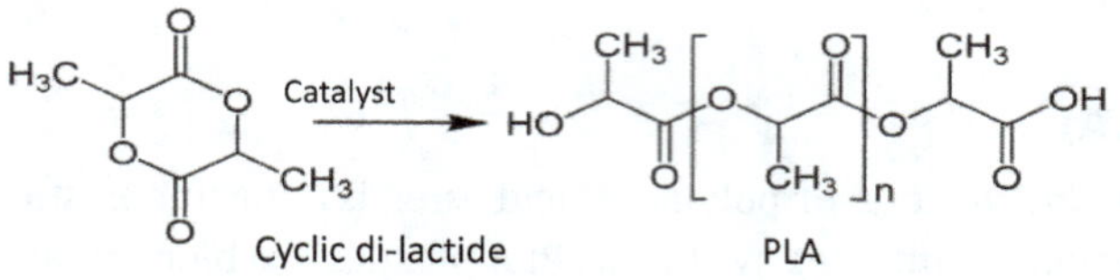

Figure 2.48 Reaction of a cyclic dilactate ester to PLA

2.6.7.6.3 Applications

Due to its physical and mechanical properties, PLA is a good substitute for petroleum-based thermoplastics. PLA and PLA blends are used as pellets in various grades for the plastics processing industry, for the production of molded parts, cans, cups, or for the production of fiber materials, as well as packaging films, mulch films, and biodegradable brackets and clips in agriculture.

Packaging, films, molded parts: The strong PLA growth in recent years is largely based on the use of PLA blends in the packaging of short-lived goods as an alternative to conventional packaging plastics such as PE and PP. Suitable PLA blends are used for this purpose. Here, particular emphasis is placed on biodegradability.

Injection molded items: Writing instruments and other office utensils can be made from injection-moldable PLA blends. Here, the mechanical properties are similar to polypropylene or even ABS, depending on the PLA content. Good flow behavior of the melt is very important for such complex geometries.

Medical technology: One of the most lucrative areas of application for PLA is medical applications. PLA is available for numerous applications due to its degradability and biocompatibility, for example as a suture material, for implants such as bone nails and screws, but also as plates or stents, as a scaffold material for tissue engineering, and others. What most of these applications have in common is that the PLA decomposes in the body after a time specified by the compound, and does not need to be removed again.

Composites: In addition to applications in packaging and medical technology, PLA also has potential as a matrix material for composites. By combining PLA with natural fibers, it is possible to produce biodegradable composites from renewable raw materials that can be an alternative to conventional fiber-reinforced or filled plastics.

In addition to the comparatively high costs, the low temperature resistance of PLA is a particular problem in industrial applications. Since the plastic softens at temperatures as low as 50–60 °C, it has so far only been suitable for applications in the low temperature range.

2.6.7.6.4 PLA in Additive Manufacturing

PLA is, with ABS, one of the most commonly used materials in 3D printers that use material extrusion-based processes (MEX). PLA is also particularly valued here for its ease of printing and its biocompatibility. When PLA is printed at high temperatures, it produces a slightly sweet odor. PLA's low moisture absorption, combined with its comparably good UV resistance, makes for easier storage conditions prior to printing compared to other materials. Its poor flammability is advantageous in processing and in a number of potential applications, for example, as electronic housings. In 3D printing PLA has some good properties, such as high surface hardness, a high Young's modulus, and a high tensile strength. However, products made from PLA show a brittleness and only moderate impact strength compared to other engineering polymers. The PLA in 3D printing filament is often not pure PLA, but a so-called PLA blend, with its basic structure enriched with additives to obtain certain desired properties. As a basis for 3D-printed products, PLA is only moderately resistant to temperature and weathering. PLA is therefore only of limited use for more thermally demanding applications and objects.

Processing: Unlike ABS, PLA exhibits little to no warping during printing. This makes 3D printing easy even for beginners. It is possible to print PLA even with the most rudimentary 3D printers. A low melt temperature accommodates the extruder and hotend, whereas the low T_g of approx. 50–65 °C in most cases allows a heated build chamber or bed to be dispensed with. Having said this, the T_g is sufficiently high to make PLA a suitable material to build unsupported overhangs. It solidifies quickly behind the die with sufficient fan cooling, but at the same time the extrusion temperature (190–220 °C) is high enough to re-melt the last extruded layer sufficiently to produce good inter-layer adhesion.

Post-processing: Mechanical post-processing procedures are mainly determined by the impact strength value and brittleness of the printed material. Unfortunately, this is not very high for PLA. Milling, drilling, and similar machining methods are only possible to a limited extent. Particular care should be taken with fragile parts, as they can break easily. Gluing, painting, sanding, and other surface treatments, on the other hand, can be carried out easily and without hesitation in order to modify the 3D print as desired. Pure PLA is resistant to acetone, but can be postprocessed (smoothed) with dichloromethane. Any use of solvents should only be done with appropriate safety measures applied (protective wear like googles, gloves and lab coats, as well as very good ventilation).

Applications: PLA is the most widely used build material in 3D printing for models and non-functional prototypes. PLA can be dyed well and is commercially available in a variety of blends and variants. When PLA is not used as a pure material, its biocompatibility, biodegradability, and harmlessness, for example as food packaging, are no longer necessarily given. PLA filament is used, among other things, for medical applications (for example, the production of true-to-life models of organs used for surgery planning), the production of transport containers and shopping bags, hygiene products, toys and other dimensionally stable figures, housings, and operating elements. PLA can also be used for the production of functional prototypes and spare parts, provided these do not have to be heat-resistant.

2.6.7.7 Thermoplastic Elastomers (TPEs)

The main representatives of thermoplastic elastomers are phase-separated block copolymers of styrene and butadiene, as well as styrene and hydrogenated butadiene-polystyrene block-copolymers (SBCs), thermoplastic olefins (TPO), thermoplastic vulcanizates of polypropylene/polyethylene and lightly crosslinked EPDM (TPV), see Figure 2.5), thermoplastic polyurethane elastomers (TPU), thermoplastic polyester elastomers (TPE-E) or copolyester elastomers (COPE), and polyether-block-copolyamides (PEBA).

The production volume of TPEs was approximately 4.7 Mt in 2017 [Ceresana 2018]. By far the most widely used class of TPE is the styrenic-block-copolymer (SBC) group, which accounts for around half of the consumption volume worldwide. Their use ranges from blending in adhesives to asphalt modification in road construction. Thermoplastic polyolefins (TPOs) make up the second largest TPE market, with the automotive industry being the main area of application. Another important type of TPE is thermoplastic polyurethanes (TPU), which are used, for example, for dashboards and housing shells and, above all, in applications subject to high dynamic or mechanical stresses. Comparatively, TPEE and PEBA are specialties, with applications at low temperatures and also in dynamically demanding applications (Figure 2.49).

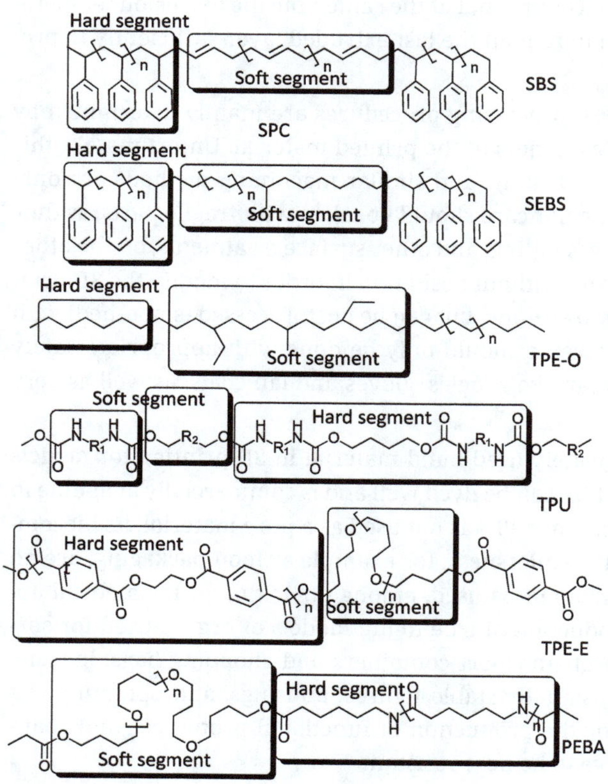

Figure 2.49 Common thermoplastic elastomers

2.6.7.7.1 Manufacturing Process

The diversity of the class of TPEs is also reflected in the number of manufacturing processes. SBCs are manufactured by a living anionic or radical chain polymerization process. TPO manufacturing processes are based on modified processes as used for the production of polyolefins (PE and PP). They are mostly chain polymerization processes in the presence of homogeneous or heterogeneous catalysts. TPVs are typical representatives of a reactive blending process of two different, non-compatible (non-miscible) polymers, resulting in complex phase structures of thermoplastic (PE and PP) and partially crosslinked phases (EPDM). TPE-E, COPE, and PEBA are produced via polycondensation reactions from dicarboxylic acids and diols/diamines with elimination of water. TPU polymers are produced via polyaddition reactions of diisocyanates with diols/diamines in a reactive extrusion process.

Prices and properties increase from TPO to TPV and from SBC to TPE-E, TPU, and PEBA. Of all the TPEs, the TPU product class has by far the widest range of properties. Thanks to the large number of possible combinations of diisocyanates, diols, and diamines, these products can cover the entire range from soft highly elastic materials to hard thermoplastics. This ranges from weather-resistant materials to materials approved for food contact and for use in medical devices. Another advantage of TPU is that they can be produced economically in small quantities using a simple reactive extrusion process. In this way, tailor-made polymers for special applications become accessible using an economical and efficient process. Many TPU polymers are also compatible with each other, are miscible, and adhere well to each other.

2.6.7.7.2 Features

TPEs combine the advantages of rubber and plastic. At room temperature, they are elastic, pliable and flexible like vulcanized elastomers, but can be processed in the melt like conventional standard thermoplastics, and recycled accordingly.

While the recycling of most TPEs involves a loss of properties, in the case of TPU materials, the recycled material can be chemically refreshed by returning it to the reactive extrusion process.

The ease of processing, as well as the variability of properties, allow TPEs to be used in many areas. The main areas of application are sealing profiles and "rubber mats" in vehicles and the construction industry, other industrial products, footwear, sporting goods, medical devices, and other consumer goods where a soft, elastic feel is required, such as in household appliances, tool handles, toothbrushes, and toys, among others.

2.6.7.7.3 TPEs in Additive Manufacturing

In 3D printing, TPU polymers in particular are playing an increasingly important role. They are very well suited for use in both powder-based and extrusion-based systems. The great variability of this product class, combined with the possibility of tailoring products in terms of both material and processing properties, makes them the "Swiss army knife" of 3D printing.

In the powder-based processes, it is possible, via the targeted variation of the recrystallization temperature, to carry out the build process at such low build temperatures (<100 °C) that almost complete recycling of the unbuilt powder material is possible. The compatibility of different TPU also allows powders of different hardnesses to be mixed in order to set specific intermediate hardnesses.

In extrusion-based processes, the same behavior allows TPU components to be printed at room temperature and extrusion temperatures of 180–240 °C, making production equipment extremely cost-effective and enabling printing in a new generation of direct extruders even for entry-level equipment and hobbyists. Again, TPU polymers are very suitable for combining materials with different hardnesses, usually obtaining excellent layer adhesion even between polymers of different hardness. For the printing of soft TPU filaments, direct drive extruders or pellet extruders are particularly suitable.

Accordingly, applications for TPU in 3D printing range from complex or customized sealing profiles, drive belts and wearables, to orthopedic inserts, hard hats, and even body-fitted seats for automobiles, as well as mattress applications.

TPE-E COPE and PEBA materials are also used in powder-based processes. Here, too, the recrystallization temperature can be specifically modified by suitable combination of the polymers so that good process properties in powder printing are possible. The products obtained from the TPE-E and COPE are mostly hard and tough, and are therefore popular in orthopedic applications or in shoe soles.

The processing conditions of the TPEs depend on the melting points or softening points. For TPE-E and PEBA, extrusion temperatures of 200–260 °C and build chamber temperatures of 60–90 °C are recommended. PEBA is also available as a powder for sintering processes, where the build temperature should be 100–140 °C, depending on the type. The SBC-based polymers can be processed at extrusion temperatures of 200–240 °C, but show only moderate layer adhesion due to the T_g of the hard-block segment polystyrene. Accordingly the build chamber temperatures should be 60–90 °C. TPE-O exhibit the build platform adhesion and interlayer adhesion problems typical of polyolefins and have yet to find wider use in 3D printing. Particularly soft TPEs are advantageously processed on 3D printers equipped with pellet extruders.

2.6.8 Standard Plastics

Standard plastics, such as polyethylene (PE), polypropylene (PP), polyvinyl chloride (PVC), and polystyrene (PS) are produced and used in very large quantities compared to other plastics. In total, the above products combine more than 70 % and thus more than 250 Mt of the produced tonnage of about 400 Mt of thermoplastics in 2020. Together with some other thermoplastics, they are called standard or bulk plastics. Due to their many advantageous performance properties and low prices, they are used in numerous applications in everyday life. New product developments also allow their use in more demanding technical applications. The largest standard thermoplastics are the four mentioned above, but in addition, the following polymers are also considered standard thermoplastics: polystyrene-acrylonitrile (SAN), polyacrylonitrile-butadiene-styrene (ABS), and polymethyl methacrylate (PMMA).

2.6.8.1 Polyethylene (PE)

PE is one of the most widely used plastics, accounting for about 20 % of total thermoplastic polymers [Fortune Business Insights 2019]. PE is a semi-crystalline and non-polar thermoplastic. By choosing the polymerization conditions, the molecular weight (chain length), molecular weight distribution, and degree of branching can be adjusted almost at will, which is used to optimize the properties of the material to a wide extent.

2.6.8.1.1 Manufacturing Process

Due to the different densities, PE is divided into four main types, although the abbreviations are not always used consistently:

- High-density polyethylene: PE-HD or HDPE
- Medium-density polyethylene: PE-MD or MDPE
- Low-density polyethylene: PE-LD or LDPE
- Linear low-density polyethylene, PE-LLD or LLDPE.

Other important product variants are ultra-high-molecular-weight polyethylene (PE-UHMW or UHMWPE) and very-low-density polyethylene (PE-VLD or VLDPE).

There are a number of major processes for the polymerization of PE which, depending on the choice of catalysts and conditions, can produce different types. Important processes are, for example, the high-pressure process, the solution and suspension processes, gas-phase polymerization, and catalytic polymerization according to Ziegler. PE can be polymerized radically and coordinatively. Particularly high-molecular and linear types are preferably polymerized via coordinative processes.

2.6.8.1.2 Molecular Structures of PE Types

The different material behavior of the various PER types can be explained by their molecular structure. Molecular weight and crystallinity play the most important role, with crystallinity in turn being influenced by the molar mass and the degree of branching. The less the polymer chains are branched and the lower the molar mass, the higher is the crystalline content in polyethylene. The crystalline fraction is between 35 % (PE-LD/PE-LLD) and 80 % (PE-HD). Within crystallites, PE has a density of approx. 1.0 g/cm^3, and in amorphous areas 0.86 g/cm^3. There is an almost linear relationship between the density and the crystalline fraction. Schematically, the degree of branching of the individual PE types can be represented as shown in Figure 2.50, in which the number and length of the branching points can be seen. The branching points are divided into short-chain branching (just a few ethylene monomers attached locally to the main chain) and long-chain branching.

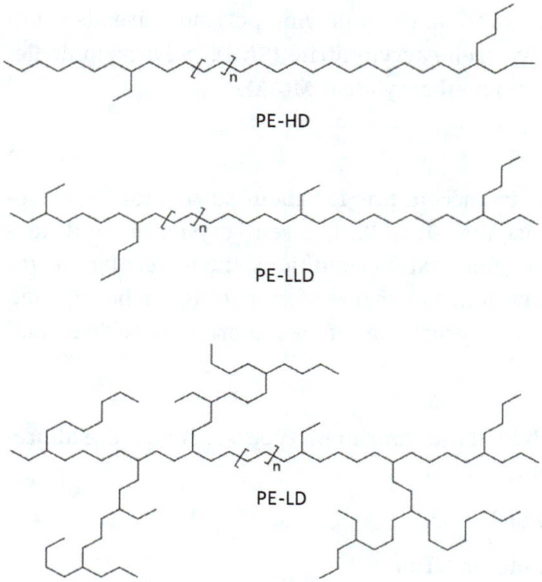

Figure 2.50
Structures of different PE types

2.6.8.1.3 Features

By choosing the production process, production conditions, and the proportion and type of co-monomers as well as catalysts, polymers with desired properties can be produced. In addition to hard or soft plastics, waxes, fats, and even oils are also possible from PE.

In this context, properties that require only a small movement of the chain segments relative to each other are more dependent on crystallinity and less on molecular weight. These properties include melting and softening point, elastic and flexural modulus, as well as yield strength, and surface hardness (Table 2.5). Properties that require more extensive movement of the chain segments, on the other hand, are more dependent on molecular weight. Tensile strength, embrittlement point at low temperatures, and tensile strength should be mentioned here.

Table 2.5 Properties of the Most Important Representatives of the Polyethylene Polymer Class [CAMPUS 2021]

Property	PE-LD	PE-HD	PE-LLD
Yield stress (N/mm^2)	10	26	10 (50 % elongation)
Yield strength (%)	16	11	–
Density (g/cm^3)	0.918	0.953	0.925
Melting point (°C)	106	130	116
Modulus of elasticity at 23 °C (N/mm^2)	120	1200	150

The general mechanical properties as a function of crystallinity and chain length are shown in Figure 2.51. It can be seen that substances with high crystallinity tend to be hard and brittle and those with low crystallinity tend to be soft and tough. High crystallinity favors high density, stiffness, hardness, abrasion resistance, service temperature, and chemical resistance. Low crystallinity or a high amorphous polymer content improves strength, toughness, notch insensitivity, and stress cracking resistance.

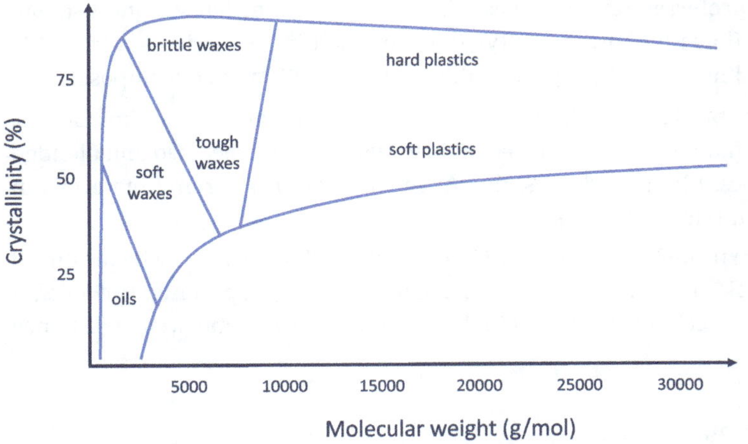

Figure 2.51 State forms of PE [Richards 2007]

As its molecular weight decreases, PE eventually loses its plastic character and first becomes waxy, then greasy, and finally oily. This is also used, for example, to produce PE wax from PE by thermal cracking. In Figure 2.51 the crystallinity of PE is plotted against the molecular weight. The areas mark the resulting properties.

2.6.8.1.4 Chemical Properties

PE is a non-polar, saturated, high-molecular weight hydrocarbon. It therefore resembles kerosene waxes in its chemical behavior (hydrophobic, non-polar, tending to crystallize). The individual macromolecules are not chemically bonded, and due to their symmetrical molecular structure, they tend to crystallize. Overall, PE is therefore semi-crystalline. Higher crystallinity increases density as well as mechanical and chemical stability.

PE has a high resistance to acids, alkalis, and many other polar chemicals, as the saturated PE chain offers only a few points of attack. PE hardly absorbs any water. Its gas permeability to polar gases, such as water vapor, is lower than that of most plastics; however, it is highly permeable to oxygen, carbon dioxide, and aromatic substances. PE can become brittle when exposed to sunlight, so carbon black is usually used as a UV stabilizer. PE burns without residue with a dripping bright flame and continues to burn even after the flame that caused the fire has been removed. Without suitable pretreatment, PE is difficult or impossible to print or bond.

2.6.8.1.5 Processing

PE can be processed by all known primary molding processes and forming methods. These include injection molding, extrusion, blow molding, compression molding, thermoforming, rotational molding, compression molding, casting, and sintering. Injection molding is preferred for PE grades with a narrow molecular weight distribution and high melt flow rate (low viscosity). Injection molding is carried out from 170 up to 300 °C depending on the PE type, but above 310 °C all PE types decompose.

Extrusion is used to produce PE pipes at 160–240 °C, for example, while films, sheets, and monofilament filaments are produced at up to 300 °C. PE is processed and blended with other polymers, fillers, additives, dyes/pigments, and reinforcing materials in specially designed mixing extruders.

PE-UHMW has an extremely high viscosity in the melt, so that it can only be produced by compression molding, ram extrusion, gel extrusion, and gel spinning. However, it is characterized by excellent strength, toughness, and abrasion and wear resistance in applications.

Finishing and joining

The joining of PE moldings is almost invariably carried out by means of heat. Bonding or printing on PE is made more difficult by the non-polar and hydrophobic surface. The surface must therefore first be pretreated with laser, plasma (high-pressure plasma e.g. "corona discharge" or low-pressure plasma), strong acids (e.g. chromosulfuric acid), or a polyolefin primer. PE-HD has excellent machinability but poor formability. Conversely, PE-LD can be formed well but has poor machining properties.

2.6.8.1.6 Applications

PE-LD and PE-LLD are mainly used in film production. Typical products for PE-LD are cling films, carrier bags, agricultural films, milk carton coatings, garbage bags, and shrink films. An important application is as a sealing medium in laminated films. To a lesser extent, PE-LD and PE-LLD are also used in the manufacture of cable sheathing, as dielectric in coaxial cables, and for pipes and hollow bodies. PE-LLD is mainly used in stretch film, films for industrial packaging, thin-wall containers, and for high-performance bag films.

The most important area of application for PE-HD is hollow bodies produced by the blow molding process, for example bottles for household cleaning agents, but also large-volume containers with a capacity of up to 1000 L. Another area of application that is growing strongly, especially in emerging countries, is cable lines and pipes, for example for gas and drinking water supply. Here, PE pipes are replacing pipes made of concrete or PVC.

Ultra-high molecular weight (UHMW) PE is used for its high wear resistance for pump parts, gears, sliding bushings, implants, and the surfaces of endoprostheses, where particularly smooth running with the lowest possible friction and abrasion is required. Fibers made of PE-UHMW are among the strongest known man-made fibers, and can thus be processed into webbing slings and reeving cords.

2.6.8.1.7 Networking

Crosslinked PE (PE-X, also XLPE) can be produced from thermoplastic PE (PE-LD, PE-LLD or PE-HD) by various processes. Crosslinking allows low-temperature impact strength, abrasion resistance, and stress cracking resistance to be significantly increased, whereas the hardness and stiffness of the material are somewhat reduced. PE-X no longer melts (analogous to elastomers) and can withstand higher thermal stresses (up to 120 °C over longer periods, up to 250 °C for short periods without applied mechanical stress).

PE-X is used as insulation material, as an electrical insulator of medium and high voltage cables, as pipes for gas and water installations, and for molded parts in electrical engineering, apparatus engineering, and automotive engineering. PE-X has very poor welding properties, with welding only possible by means of electrofusion.

The specialty grades and copolymers listed below have gained importance as technical materials based on PE. Chlorinated PE (PE-C) is used to increase the flame retardancy of polyolefins or to increase the impact strength of polyvinyl chloride (PVC) and chlorosulfonated polyethylene (CSM). Rubber-elastic PE-C is used after crosslinking for cable insulation, coated fabrics, roofing films, tubular blankets, and inflatable boats.

Ethylene vinyl acetate copolymers (EVACs) are used with up to 10 % vinyl acetate for greenhouse films, pipes, and bellows with high resilience. Copolymers with up to 30 % vinyl acetate are used for thermoplastic cable sheathing and up to about 40 % vinyl acetate are used in adhesives. With more than 40 % vinyl acetate and crosslinked as elastomers (EVM) with a very high filler content, PE-EVA copolymers are mainly used for cable insulation (FRNC cables).

Ethylene-vinyl alcohol copolymers (EVOH) with 53–68 % vinyl alcohol have minimal gas permeability and are used as so-called barrier films.

PE compounds with about 6 % starch are used for the manufacture of biodegradable products (carrier bags, trash bags). Only the starch is degraded, but not the surrounding PE matrix. However, the sensitivity to UV light already present in PE can be further increased by incorporating special UV-sensitive molecular groups, so that photochemical degradation is possible.

Cycloolefin copolymers (COC) are produced, among other things, by copolymerizing ethylene with cyclic olefins such as norbonene. Semi-crystalline grades exhibit high heat distortion temperature and good chemical resistance.

2.6.8.1.8 Recycling

The starting product of PE, ethylene is mainly obtained from crude oil or natural gas. However, ethylene can also be produced from non-fossil plant carbon sources. In Brazil, for example, a sugarcane-based PE is marketed ("Green PE"), which does not differ chemically or in terms of its processing properties from conventional PE. PE used for retail packaging is already collected and separated from household waste in many developed countries.

Using near-infrared technology, the individual types of plastic can be separated in the sorting plants. Today, a grade purity of up to 98 % is achieved, which enables high-quality mechanical recycling. Standard PE is up to 100 % recyclable. With the aid of various mechanical processes, the used plastic packaging can either be remelted directly into new products or processed into re-granulate. This granular recycled plastic is a cost-effective alternative to virgin material, and a high-quality starting material for the plastics processing industry. The product range for recycled PE is diverse: films, garbage bags, canisters and drums, garbage cans, drinking water pipes, landfill seals, or cable insulation.

2.6.8.1.9 PE in Additive Manufacturing

When using PE in extrusion-based additive processes, firm anchoring of the resulting component on the carrier plate of the printer is absolutely essential for precise application of the melt at the correct position. Non-polar materials such as PE or PP generally exhibit only low surface adhesion and are therefore difficult to bond to the build plate.

Formulated PE grades (for example glass-fiber-reinforced grades) are also commercially available for 3D printing. PE is rather difficult to process and requires a nozzle temperature of approx. 240–280 °C as well as a build plate temperature of 100–120 °C. Due to the high thermal expansion coefficient of approx. 100–200 × $10^{-6}\,K^{-1}$ and the fast crystallization rate, PE tends to exhibit strong warping/shrinkage. The poor adhesion to the printing platform further exacerbates the warping. The difficult printing profile combined with the rather moderate properties in comparison makes the processing of PE in extrusion-based processes, despite its low material price, attractive only in special cases where high chemical resistance to acids and bases and high water resistance are required. Accordingly, there are only a few suppliers of PE filaments. However, the possibility of producing PE based on sugar cane allows the use of bio-based PE in 3D printing.

Despite some efforts, PE as a raw material has so far found little use in the field of AM using powder-based sintering processes. PE as a material must have a high molecular weight in order to be processed into products with good mechanical properties. As in extrusion-based processes, the required high molecular weight necessitates a very tightly controlled build temperature in the range of 100–130 °C and a comparatively high energy input is required to enable sufficient flow and adhesion of the powder grains. Both of these factors have so far made the use of PE in classic sintering processes difficult, slow, and therefore expensive.

Copolymers of ethylene, such as polyethylene-co-vinyl acetate (EVAC) and the saponified variant EVOH, have been little studied in 3D printing, but should be of interest for various food and packaging applications. A fully saponified variant of pure polyvinyl acetate, known as polyvinyl alcohol (PVA), is used in 3D printing alongside other starch-based polymers as a support material with very good water solubility. The filament adheres well to printing materials such as polylactide (PLA), polyamide (PA), polyvinyl butyral (PVB), and polyurethane (PUR), and can be printed at similar build and extrusion temperatures. The dissolved material can be disposed of with normal wastewater.

2.6.8.2 Polypropylene (PP)

PP is a thermoplastic produced by chain polymerization of propene. PP belongs to the group of polyolefins and is semi-crystalline and non-polar. Its properties are similar to PE, but PP is somewhat harder and more heat resistant. With approx. 74 Mt in 2016, PP is one of the most frequently used standard plastics alongside PE [P&S Intelligence 2021].

2.6.8.2.1 Production

PP is produced by catalytic polymerization of propene using various industrial processes. These include the suspension (slurry) process, the bulk process, and gas-phase polymerization.

2.6.8.2.2 Features

PP is similar to PE in many properties. The additionally present methyl group improves the mechanical properties and the thermal resistance, while the chemical resistance decreases. As with PE, the properties of PP are strongly dependent on the molecular weight distribution, crystallinity, and especially tacticity (Figure 2.52). In technical use PP with a high degree of isotacticity of about 85–95 % is mainly used. The melting point is about 160 °C and the T_g is about –10 to 0 °C.

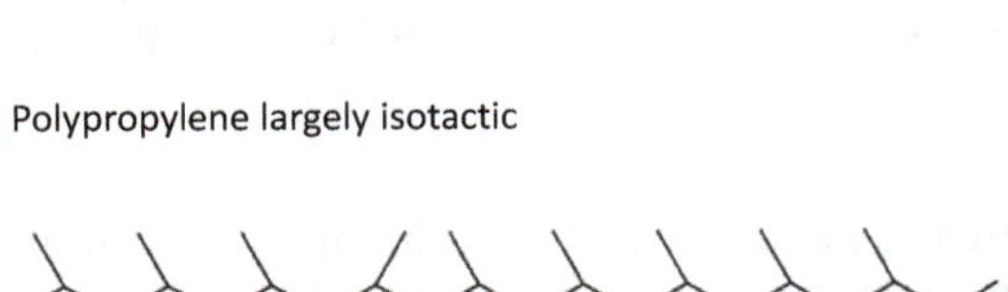

Figure 2.52 Polypropylene

The methyl side-chain can be incorporated isotactically, syndiotactically, or atactically. This arrangement affects the physical properties. The isotactic structure, which can be produced with the aid of Ziegler–Natta catalysts, for example, results in a partially crystalline structure of the PP. This can be attributed to the methyl radical always being located on one side of the molecular chain, which forces the macromolecule into a helical shape. The atactic variant of PP, on the other hand, is less crystalline and contains higher amorphous proportions. Isotactic PP has a high degree of crystallinity, ranging from 30–60 %, and is preferred for use in industrial products. Syndiotactic PP is somewhat less crystalline and has fewer commercial uses.

The density of PP is between 0.895 and 0.92 g/cm^3. This makes PP the standard plastic with the lowest density. Unlike PE, crystalline and amorphous areas differ only slightly in density. The modulus of elasticity of PP is between 1300 and 1800 N/mm^2. PP often has very good fatigue resistance.

2.6.8.2.3 Thermal Properties

The melting and continuous service temperatures of PP are higher than those of PE, as are numerous mechanical properties (stiffness, hardness, and strength). PP homopolymers can be used permanently between 0 and 100 °C. Below 0 °C, PP becomes brittle. This temperature range can be increased, for example, by copolymerization. The thermal expansion of PP is given by the coefficient of linear expansion, α, which at 100–180 × 10^{-6} K^{-1}, is similar to that of PE.

PP can be filled with mineral fillers, such as talc, chalk, or glass fibers. This again significantly extends the range of mechanical properties (stiffness, service temperatures, etc.).

2.6.8.2.4 Chemical Properties

PP is resistant to greases and almost all organic solvents at room temperature, except for strong oxidizing agents. Non-oxidizing acids and alkalis can be stored in PP containers. PP is odorless, skin-compatible, and suitable for food and pharmaceutical applications, as it is physiologically harmless and biologically inert.

2.6.8.2.5 Processing

PP is suitable for injection molding, extrusion, blow molding, hot forming, welding, deep drawing, and machining. Injection molding of PP takes place at processing temperatures of up to 260 °C. Due to its low surface energy, PP is difficult to glue or print. Relevant quantities of PP are drawn into fibers each year. This is used to produce yarns, nonwovens, and fabrics for a wide range of technical and textile applications.

2.6.8.2.6 Applications

Important applications of PP are in particular PP foam (EPP) as cushioning, insulation, and packaging material, unstretched PP film (CPP), cast PP as versatile packaging material, stretched PP film (oriented PP, OPP, and biaxially oriented PP, BOPP) as packaging materials with very low water vapor permeation, improved mechanical strength, and good transparency.

For price reasons, PP is increasingly displacing engineering thermoplastics such as ABS and PAin a wide range of applications in the temperature range 0–100 °C. Typical applications continue to be found in mechanical engineering and vehicle construction for the interior trim of passenger cars, for crash absorber elements in vehicle construction, in children's seats and in bicycle helmets, as well as in electrical engineering for transformer housings, wire and cable sheathing, and insulating films. In the construction industry, PP is used for valves, fittings, and piping, and in ventilation and air-conditioning technology in corrosive environments and for conveying corrosive gases, mostly in the form of PP-S (S = flame retardant).

PP fibers can be added to reinforced and prestressed concrete to improve the fire protection properties. The fiber, which melts and burns when heated, leaves pores that allow expansion and escape of water vapor. PP fibers are further processed into home textiles, carpets, sports textiles, packaging materials, hygiene products, medical products, buoyant ropes, and geotextiles, among others. In the food industry, packaging technology and in the household, PP is used in many ways, for example as bottle caps, inner parts for dishwashers, boil-proof films, reusable containers, thermal transport boxes or heat-retaining containers (EPP), packaging parts, drinking straws, cling film, and more. In general medical surgery, nets made of PP are used to close hernias, such as inguinal hernias.

2.6.8.2.7 Recycling

PP – just like PE – is a fossil-based plastic in conventional production. In the case of PP, the switch from crude oil to efficient natural gas as a raw material base has been very consistently pursued in recent years, so that PP is now one of the most energy- and raw material-efficient (low carbon footprint) standard plastics per unit mass of new plastic produced. PP is generally well suited for recycling. Thermal recycling is also non-critical, producing only water and carbon, releasing a large amount of energy and thus saving fossil raw materials ("cut-resistant crude oil").

With the help of the various material processes, the used plastic packaging can either be directly remelted into new products or processed into re-granulate. This recycled plastic is increasingly becoming a technically and economically convincing alternative to virgin material and a high-quality starting material for the plastics processing industry [AGVU 2021]. In contrast to HDPE, LDPE, and PET, which have been traded worldwide for some time, PP recyclates have only been available for sale in significant quantities for a few years.

2.6.8.2.8 PP in Additive Manufacturing

When using PP in extrusion-based additive processes, a firm anchoring of the resulting component on the build plate of the printer is absolutely necessary for the precise application of the melt at the correct position. Non-polar materials such as PE or PP generally exhibit only low surface adhesion and are therefore difficult to adhere to the build plate. In addition to surface adhesion, adhesion to the previous layer is also a problem due to rapid recrystallization, which is why PE and PP typically require high extrusion temperatures and high build temperatures to ensure good layer adhesion, in addition to adhesion to the build platform.

While PP is widely used in injection molding, it is less commonly used in AM. As it is usually a semi-crystalline material, it is important to control the temperature throughout the process, as PP has a very narrow melting point range and a high recrystallization rate.

When processed from the melt, PP shows pronounced material shrinkage, which can lead to significant warping. PP should therefore preferably be printed in a well-controlled heating chamber. In addition, it is a material that does not adhere to all surfaces. For this purpose, various manufacturers have developed appropriate aids, such as adhesive foils that can be placed directly on the printing platform to improve the adhesion of the printed part to the platform [Flynt 2020].

PP can also be advantageously blended with fillers such as calcium carbonate or glass fibers for use in 3D printing, which improve its mechanical strength and fire behavior. Furthermore, PP can be coated for 3D printing or blended with other materials that are easier to process, such as TPE.

Recent developments in PP filaments, for example through copolymerization of PP with crystallization-disturbing co-monomers or polar co-monomers, address the problems mentioned and give rise to hopes for simpler processing and better product properties in the future when PP is processed in extrusion-based processes.

In the last ten years, a number of companies have released PP-based powders (special formulations) for laser sintering processes. The materials offered are mostly offered as a low-cost alternative for PA12 and PA11 [Sculpteo 2021]. 3D-printed PP is a translucent gray-white material with an exceptionally high elongation at break and otherwise has similar properties to injection-molded PP. Its robustness, fatigue resistance, and light weight make PP suitable for form, fit, and function testing. Prototypes and test specimens made of 3D-printed PP have the great advantage of being made of the same material as the final component. Ideal applications include functional prototypes for components with snap-fit hooks or film hinges in automotive parts, as well as food packaging, consumer goods, and medical applications. PP as a powder material enables printing of objects with low bulk density, increased moisture resistance, and good chemical resistance, according to the manufacturers [Anusci 2020]. The 3D printing powder has good recyclability of 70–100 % and good processing stability.

2.6.8.3 Polyvinyl Chloride (PVC)

PVC is a thermoplastic polymer produced by chain polymerization from the monomer vinyl chloride (Figure 2.53). PVC is still the third largest standard polymer in terms of production volume after PE and PP, with approximately 44 Mt in 2018 [Statista 2018, PVC].

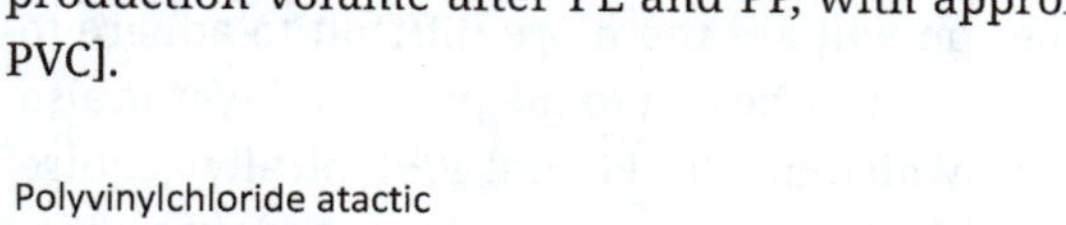

Figure 2.53 Atactic polyvinyl chloride

2.6.8.3.1 Manufacturing process

PVC is produced by radical chain polymerization from the monomer vinyl chloride and is therefore largely atactic and amorphous. Essentially, three different polymerization processes are common:

- **E-PVC:** The oldest process is emulsion polymerization (E-PVC). With the aid of emulsifiers, vinyl chloride in the form of small droplets is stirred into water, where it undergoes radical chain polymerization (first carried out in 1929). Polymer dispersions from this process are used for adhesives or coating, among other things.
- **S-PVC:** In an autoclave, vinyl chloride is liquefied under pressure and water is added. Intensive stirring produces a suspension of very small vinyl chloride droplets in the water, which in turn undergo radical chain polymerization. This is a suspension polymerization (S-PVC). About 90 % of PVC production takes place in this way.
- **M-PVC:** In bulk polymerization (M-PVC), free-radical chain polymerization is carried out directly in liquid vinyl chloride. M-PVC has a very high purity compared to E- and S-PVC. In applications where high transparency is required, such as sterilization films, M-PVC is preferred.

PVC is divided into rigid PVC (PVC-U, unplasticized) and flexible PVC (PVC-P, plasticized). Rigid PVC is used to make pipes and profiles, for example for windows and pharmaceutical films. Soft PVC plays a major role as a cable insulator and is also used in floor coverings, hoses, shoe soles and roof sealants. Plasticized PVC contains up to 40 % plasticizer; rigid PVC basically contains no plasticizer.

The inherently brittle and hard PVC can be adapted to a wide variety of applications by the use of additives, primarily stabilizers to improve the temperature, light, and weather resistance, and impact modifiers to improve the toughness and elasticity. Typical impact modifiers are acrylic polymers and chlorinated PE.

PVC is a thermoplastic polymer with a T_g of 79 °C and is processed in a temperature range of 160–200 °C. At these temperatures, however, a decomposition process already starts with the formation of hydrogen chloride (HCl), which must be counteracted by additives (stabilizers). The addition of suitable stabilizers is a precondition for safe application and processing of PVC. Traditional stabilizers for PVC are based on stearates or carboxylates of heavy metals such as lead, cadmium, tin, barium/zinc, calcium/zinc, and calcium/aluminum/zinc. Nowadays, such heavy-metal-containing thermostabilizers have largely been replaced by hydrotalcite (a magnesium aluminum hydroxycarbonate).

2.6.8.3.2 Features

PVC can be easily dyed and absorbs hardly any water. It is resistant to acids and alkalis and has a limited resistance to ethanol, oil, and gasoline. PVC is attacked by acetone, diethyl ether, tetrahydrofuran (THF), benzene, chloroform, and concentrated hydrochloric acid. Rigid PVC can be machined well, while flexible PVC is difficult to machine. At temperatures of 120–150 °C, PVC can be thermoformed.

PVC burns with a yellow, strongly sooty flame and extinguishes quickly. Due to its high chlorine content, PVC is inherently flame retardant, unlike other engineering plastics such as PE or PP. However, if fires occur in presence of PVC plastics, toxic hydrogen chloride gas is produced, which is highly corrosive.

PVC is a good insulator. The formation of dipoles and their constant realignment in the AC electric field results in high dielectric losses compared to most other insulators. Due to the high strength of the cable sheath and the good insulating properties, low-voltage PVC cables are very suitable for installation under plaster or outdoors.

2.6.8.3.3 Applications

One of the major advantages of PVC is its high weather resistance to sunlight, water, and air. This is why PVC is used mainly for durable products in outdoor weathering applications. PVC can be manufactured in a wide range of colors and decorations, and is very inexpensive. Chlorine, in terms of mass composition, is the main raw material of PVC. Chlorine is produced in excess as a "forced by-product" in the electrolytic production of caustic soda from NaCl (with 60 Mt one of the largest industrial products) and would otherwise have to be disposed of.

Solid PVC is used, for example, in garden and camping furniture and window frames. PVC films are used, for example, as sheathing for water cores in waterbeds, as artificial leather, as pond liners and roofing membranes in the construction sector, and as floor coverings (vinyl). PVC is often used as a flame-retardant cable sheathing (insulation material for electrical cables), as an electrical switch box or as retractable conduit for cables. Foamed PVC in sheet form is used as a carrier material for advertising media, mainly due to its light weight, good color stability, and easy processing.

2.6.8.3.4 Recycling

A take-back system exists for PVC. The main materials collected are flooring, roofing membranes, window profiles, electrical cables, and PVC pipes. The establishment and expansion of recycling structures is based on a voluntary commitment by the PVC industry [VinylPlus 2021]. Technically, PVC can be partially recycled, but as with most polymers, this involves downgrading. Raw material recycling of PVC is currently not carried out for cost reasons.

Besides landfilling, energy recovery of PVC is the most widely used option for its disposal. However, this requires significant measures to treat the waste gases (chlorine, heavy metals). The chlorine content of PVC is completely converted to HCl during incineration and is removed from the flue gas as part of the mandatory flue gas cleaning process to well below the legally permissible emission limit value. The scrubbing liquid is neutralized mainly with burnt lime, and the resulting calcium chloride can be safely disposed of. The hydrogen chloride can also be separated from the flue gas as hydrochloric acid and reused in chemical production. Five waste-to-energy plants operate on this principle in Germany.

The SOLVAir® process (formerly NEUTREC® process) from Solvay offers another option. With the help of sodium bicarbonate, sodium chloride is recovered and purified in the flue gas scrubbing of incinerators. There are plants for the treatment of sodium-containing reaction products in Italy and France [VinylPlus 2021].

2.6.8.3.5 PVC in Additive Manufacturing

In principle, PVC can be processed well in extrusion-based processes. Products with good flame retardancy, high chemical resistance, high impact strength, and overall good mechanical properties are obtained. It is particularly suitable for technical applications and components in the field of electronic equipment due to its low moisture absorption and good insulation properties. Suitably stabilized PVC filaments can be processed at <220 °C and with a build chamber temperature of 60–80 °C.

Despite the favorable raw material properties and the ease of processing, however, the use of PVC in conventional AM extrusion-based processes is still very uncommon, as many users fear the decomposition of PVC with the splitting off of hydrochloric acid during prolonged exposure to high extruder temperatures. More importantly than with most thermoplastics, processing should also take place only in well-ventilated rooms or in printers equipped with air filters. Although PVC is readily available as a powder, there is currently no known application in powder-based manufacturing processes.

2.6.8.4 Polystyrene (PS)

PS is a transparent amorphous or semi-crystalline thermoplastic. Amorphous PS is a widely used, low-cost standard plastic that is used in many areas of everyday life. PS grades occur in a number of modifications: as standard atactic PS, as syndiotactic PS (sPS), as extrudable PS foam (XPS) and as expandable/foamable PS (EPS).

In addition to pure PS, there are a number of important styrene copolymers: rubber-modified polystyrene (high-impact PS, HIPS), styrene-butadiene graft copolymers (SB), styrene block copolymers (SBC), styrene-acrylonitrile copolymers (SAN), acrylonitrile-butadiene-styrene copolymers (ABS), polyacrylonitrile-styrene-acrylate (ASA), crosslinked polystyrene (PS-X) and styrene-butadiene copolymers (SBR rubber).

2.6.8.4.1 Production

PS is mainly obtained by radical chain polymerization of styrene, but can also be produced by other chain polymerization processes (anionic, cationic, and Ziegler–Natta). PS can exist – depending on the production process – in different stereochemical variants (Figure 2.38). The most important is the atactic amorphous PS (Figure 2.54). Atactic amorphous PS is a low-cost plastic with a market volume of 11.2 Mt in 2018 [Ceresana 2019]. Syndiotactic PS has been used only for specialty applications. Isotactic PS can be produced but is of no industrial interest due to its low crystallization rate.

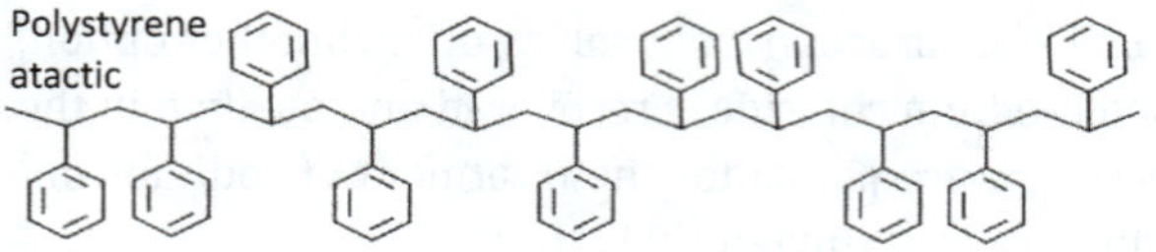

Figure 2.54 Atactic polystyrene

The finished plastic is sold as granules to be processed mainly as extrudable PS (XPS) into plastic parts or containers (e.g. food packaging with aluminum heat-seal lids). Expandable PS (EPS) is produced with gas inclusions during polymerization into solid beads. The beads are transported to the processor. When the beads are steam-heated to slightly above 100 °C, the gas expands and the thermoplastic inflates. The edges of the bubbles fuse, and a molded solid is produced. Depending on the molding template, everything from simple sheets to geometrically complicated molded parts is accessible.

2.6.8.4.2 Features

Unmodified amorphous PS has a T_g of approx. 100 °C. Pure PS is hard and brittle and relatively permeable to oxygen and water vapor. The density of compact PS is between 1040 and 1090 kg/m^3. EPS has a density between 15 kg/m^3 (perimeter insulation) and 90 kg/m^3 (ski helmet), and has a very low thermal conductivity.

PS can be colored in all colors. Solid PS has a tendency to show stress cracking. It is not very resistant to heat; above 55 °C, aging starts to accelerate, which is why PS can only be used in applications up to 70 °C. Foamed PS is white and opaque. Compared to solid PS, it has lower mechanical strength but higher elasticity.

2.6.8.4.3 Chemical Properties

Atactic amorphous PS has good resistance to aqueous alkalis and mineral acids, but is UV-sensitive. PS can be dissolved in common solvents, e.g. with dichloromethane, and welded almost seamlessly.

Syndiotactic PS (sPS) crystallizes sufficiently quickly and is highly suitable for processing by injection molding. Because of its excellent resistance to chemicals, hot water, and coolants, and its good electrical insulation properties, it can be used in the electrical engineering, automotive, and filter technology industries e.g. for application in coolant circuits. When reinforced with 30 % glass fibers, sPS can resist high temperatures and is dimensionally stable up to just beyond 260 °C, making it ideal for applications in electrical engineering in which welding is required.

PS is physiologically harmless and approved without restriction for food packaging.

2.6.8.4.4 Fire Behavior

PS burns with a bright yellow, strongly sooty flame. The fire behavior of EPS is characterized by the fact that it softens at temperatures slightly above 100 °C and then drips off burning. Rigid PS foam that is flame-retardant due to additives is classified for building applications in accordance with EN 13501-1 and classified in building material class E.

2.6.8.4.5 Copolymers

Polystyrene has a good property profile as a material when transparency, surface quality, and rigidity are required. Its range of applications is also considerably extended by copolymers and other modifications (blends, e.g. with PC and syndiotactic PS). The brittleness of ordinary PS is overcome by elastomer-modified styrene-butadiene copolymers. The copolymer of styrene and acrylonitrile (SAN) is more resistant to thermal stress, heat, and chemicals than the homopolymer and is also transparent. ABS has similar properties to SAN, but is even tougher and opaque.

HIPS (high-impact polystyrene) consists of a coherent PS matrix and a rubber phase dispersed in it. It is produced by polymerization of styrene, in the presence of polybutadiene dissolved in the styrene monomer.

Styrenic block copolymers (SBCs) are produced by anionic block copolymerization and are more widely used thermoplastic elastomers (TPEs).

2.6.8.4.6 Applications

The low shrinkage tendency of PS during production enables products with very near to net-shape to be made from PS. Furthermore, very fine contours, edges, and straight surfaces can be produced from PS, enabling the production of precision-fit technical components.

PS is approved in special grades for food packaging, for example as yogurt pots or foam trays. In electrical engineering, PS is used because of its good insulating properties. It is used (as HIPS) for the production of switches, coil bodies and housings of electrical appliances, for mass-produced articles (in the past, for example CD packaging), in precision engineering, and for sight glasses.

EPS is commonly known as a lightweight white packaging and insulation material. Rigid EPS foam sheets can be cut from a block to almost any thickness. PS foamed into shape is widely used as an ultra-lightweight packaging material, as well as for hard hats, solid life jackets, and surfboards.

Loose, free-flowing PS foam beads of about 2–6 mm diameter are used as filling for beanbags, vacuum mattresses in rescue services, road construction, for loosening heavy soils in gardening and landscaping, and sometimes as buoyancy agents in lifting shipwrecks.

Another method of producing PS foam is extrusion (to form XPS). The starting material of PS granules and blowing agent is foamed by heat and at the same time continuously pushed out through a defined opening and cooled. This produces a homogeneous, fine-pored rigid foam that usually has a closed surface and a closed-cell structure. It is classified as impermeable to air, water, and water vapor. Due to its high compressive strength and low water absorption (closed porosity), XPS is used, for example, in the insulation of buildings against the ground (perimeter insulation).

2.6.8.4.7 Recycling and Disposal

Various mechanical recycling processes are currently available. Clean PS waste can theoretically be used to obtain PS re-granulate after shredding and extrusion (mechanical recycling). Single-variety PS recyclates change their properties only slightly, even after repeated processing.

EPS granulate can be used for bonded EPS fill, EPS recycling boards, as lightweight aggregate for concrete or for increasing the porosity of masonry bricks. In reality, however, EPS is hardly ever recycled due to possible contaminations and mixing of grades. Part of the EPS waste is thermally recycled.

2.6.8.4.8 PS in Additive Manufacturing

Due to its brittleness, PS is only used in the form of copolymers in 3D printing by means of extrusion processes. HIPS is a rubber-modified thermoplastic polymer with significantly improved impact strength compared to PS. HIPS filaments are easy to print and can be used either as a direct polymer for the production of 3D-printed objects or as a support material for combination with ABS. The material has a low water absorption capacity and has good strength and heat resistance. Due to its low warpage effect, larger components can be made from it in a heated build chamber at about 80 °C. HIPS is significantly more mechanically capable than PLA. The material has very good insulating properties and high toughness, being roughly comparable to ABS in mechanical performance. The density of HIPS is approx. 1.07 g/cm^3 and it softens at 95 °C.

HIPS can be easily printed at 200–235 °C with a build platform heated to 80–110 °C. Compared to other printing materials, such as ABS, PLA, and PETG, the paintability and post-processability of HIPS are particularly advantageous.

Although HIPS is classified as harmless to health, processing of HIPS should only take place in well-ventilated rooms. Heating HIPS can produce fumes that can irritate the eyes and lungs. HIPS, like all PS-based materials, has only limited resistance to weathering and UV light, and must be additivated accordingly if it is to be used outdoors.

HIPS can be finished mechanically (grinding and polishing) and chemically. Chemical post-processing takes advantage of the good solubility of HIPS to achieve surface post-processing in suitable solvent vapors. HIPS can be painted with acrylic paints and bonded with common PU and PVA adhesives. If the material is used as a support structure in combination with ABS, complete dissolution in a limonene extract (D-limonene from lemon or orange oil) is possible.

HIPS is used in 3D printing in a variety of applications. It can be used in the hygiene and food sector, but also as a support material, housing, safety containers, transport containers, and in all applications where high impact strength is required.

ASA, chemically a copolymer of acrylonitrile, styrene, and acrylates, has similar properties to ABS, but better resistance to UV weathering. Nevertheless, some complications can arise during printing, which is why it is recommended to print at higher build platform and build chamber temperatures. The printing parameters are very similar to those of ABS, with 80–120 °C build platform temperature and 220–260 °C extrusion temperature. Also in the case of ASA, care must be taken to use 3D printers with a closed housing or to print only in well-ventilated rooms due to possible decomposition reactions during extrusion.

PS made from suspension polymerization is a suitable material for use in powder-based AM processes such as SLS due to its low cost and good post-processability. High-resolution parts can often be obtained based on amorphous thermoplastics in powder-based AM processes, but they typically have only low mechanical strength (see also PC and PMMA). The reproduction accuracy can be so high that the printed parts can be used as a lost model for metal casting. Due to the sharp softening point (T_g) of PS and the high melt viscosity, it is not possible to produce dense structures in the SLS process without risking the sintering together of excess powder in the build chamber. The build chamber temperature must be kept within a narrow tolerance just below the melting temperature of PS, i.e. below 100 °C. If sintered PS parts are to be used as components, it is recommended that the resulting highly porous structures are stabilized by resin infiltration, e.g. by means of epoxy resins [EOS 2021].

2.6.8.5 Acrylonitrile-butadiene-styrene (ABS) Copolymers

ABS copolymers are thermoplastic terpolymers in which side chains of styrene-acrylonitrile copolymer (SAN) are covalently attached to a polybutadiene backbone (Figure 2.55). The impact strength of SAN can thus be improved to more than 10 times that of pure PS. The largest share of the ABS production volume of approximately 11 Mt in 2018 is mainly used for automotive interior parts and for applications in electrical engineering and electronics.

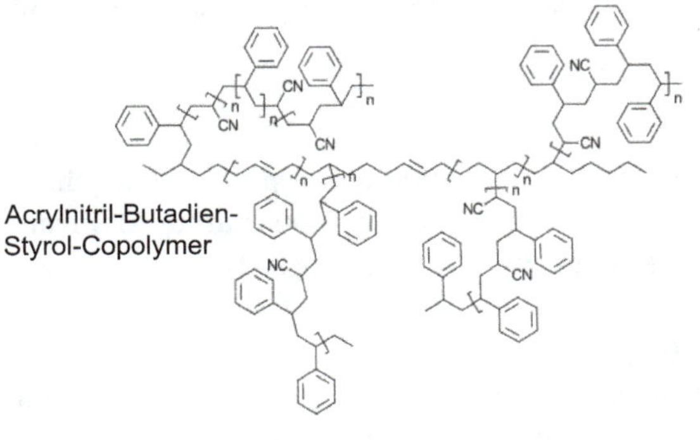

Figure 2.55
Structural formula of ABS copolymer

2.6.8.5.1 Manufacturing Process

To produce ABS, a copolymer of styrene and acrylonitrile (SAN) is attached to a polybutadiene-based main chain by radical chain polymerization grafting. Since the two components are not soluble in each other, microphase separation occurs, which increases the impact strength of the material due to the intercalated soft polybutadiene particles in the continuous, brittle SAN main phase. The proportions of the monomers used vary from 5–30 % butadiene as well as 15–35 % acrylonitrile and 40–60 % styrene.

2.6.8.5.2 Features

The ABS copolymers (ABS) obtained in this way are multiphase products with a density between 1.03 and 1.07 g/cm^3. Although ABS is amorphous, most grades are opaque because some of the individual phases are larger than the wavelength of visible light and therefore scatter. Transparent products are obtained by simultaneously grafting the rubber with PMMA. The operating limits of ABS are approximately between –40 and 85 °C for continuous use, and up to 100 °C for short-term use. Chemical resistance is good and comparable to that of SAN.

ABS is not weather-resistant, and is sensitive to oxidation due to the unsaturated polybutadiene building blocks. For this reason, polybutadiene rubber is replaced by polyethylene-propylene-diene rubber (EPDM) in some grades. If required, finished parts can be coated with polyvinylidene fluoride (PVDF) to achieve higher chemical and outdoor resistance. To produce chrome-plated plastic components, the surface can be galvanized.

ABS is compatible with many other thermoplastics. Technically important are blends with PC, PA, and TPU. In PC-ABS blends, PC increases the toughness and heat resistance compared to ABS. On the other hand, the comparatively low melt viscosity of ABS improves the processability compared to pure PC. ABS-PAblends reduce the high dimensional changes typical of PAwhile increasing the low-temperature impact strength. They have good flow properties and allow the production of finely structured matt surfaces. Examples of applications for these modified variants can be found in mechanical and automotive engineering, but also in sanitary applications, as fittings in galvanized versions and, to a not inconsiderable extent, as impact-resistant general-purpose plastics.

2.6.8.5.3 Recycling

If separated by type, ABS can be remelted and reused without any problems. Mechanical processes are available for sorting and separation that can separate ABS from common waste mixtures to a high degree of purity [EU-Recycling 2021].

2.6.8.5.4 ABS in Additive Manufacturing

ABS copolymer is a thermoplastic polymer widely used in industry, known primarily for its good impact strength at low temperatures and its low weight. The plastic is also used in extrusion-based processes in 3D printing as a very popular and cost-effective classic ("the 'UR' FLM material") and is, next to PLA, one of the most commonly used materials.

Available in the form of 3D printing filaments in a variety of colors, ABS is attractive to many industries looking to 3D-print functional parts or prototypes. Although it is more difficult to print than PLA, ABS remains a very popular material for 3D printing professionals due to its high impact and broad temperature resistance (between –20 and 80 °C). It is opaque, offers smooth and glossy surfaces, and can be easily welded with acetone and smoothed in acetone vapor.

ABS can be processed at an extrusion temperature between 220 and 260 °C. When printing ABS, the use of a heated build plate (between 80 and 130 °C) is mandatory and a heated build chamber is very advantageous. ABS shows significant warping, which is why large or highly filled components easily detach from the build platform. It is therefore advantageous to additionally fix the component to the platform with an adhesive. In the professional sector, ABS plastic is processed in heated, enclosed, and air-filtered build chambers to ensure the correct build chamber temperature and to trap odors and particles.

HIPS is particularly recommended as a support material for ABS, since it can be printed at approximately the same extrusion and build platform temperatures and, unlike ABS, is readily soluble in D-limonene.

With the right printing conditions, printing with ABS succeeds very well, so that only minor post-processing is required. This saves time, especially when manufacturing complex parts. ABS can be mechanically reworked and painted and is very durable. As with all 3D printing processes, for high-precision dimensioned prints, users must plan for shrinkage in the CAD design. For ABS, this means designing a part approximately 1.5–2.5 % larger. The desired density of the printed object correlates directly with the shrinkage here.

ABS is not UV-, sunlight-, or weather-resistant and is easily embrittled if unprotected. Although it only absorbs moisture from the air in small quantities, ABS must be stored dry due to the high printing temperature compared to PLA and the resulting bubble formation. ABS has been an integral part of the available and successfully used materials since the beginning of 3D printing. However, error-free part printing with ABS is often reserved for professional use, since the necessary prerequisites for ABS printing (build temperature, platform bonding, pre-drying of the material, and air filter) are often lacking in the hobby sector.

Some manufacturers have developed special ABS variants that improve one or more disadvantages of the material. For example, there are variants that do not emit any odor during printing or exhibit less shrinkage when cooling. These are marketed as “ABS+” or “ABS Pro” (see for example [3Dprima 2021]).

ABS is also the most widely used substrate material for mixed plastics (both particles and chemical blends), along with PLA and PETG. Since ABS is mainly used for functional printing, these are usually blends that improve certain properties. Since the blends are opaque, they are seldom used for optical purposes, but mainly for the functionality to be achieved. By adding conductive particles, ABS can be made conductive to a limited extent. A blend with PC (also called “PC+”) is also popular This blend reduces the very high processing temperatures and shrinkage of pure PC (approx. 290 °C processing temperature, heated build platform up to approx. 140 °C) to a more moderate 240–270 °C extruder temperature and 110–120 °C heated bed temperature. This means that ABS-modified PC can also be printed on many non-professional FLM printers.

Printing of ABS powders by laser sintering is not known.

2.6.8.6 Polymethyl Methacrylate (PMMA)

PMMA, often called acrylic glass or Plexiglas®, is an amorphous thermoplastic with excellent optical properties. PMMA is characterized by high hardness and excellent weather resistance. With approximately 3.7 Mt produced in 2018, PMMA is a specialty product among standard polymers [IHS Markit 2018].

2.6.8.6.1 Production

PMMA is routinely produced by radical chain polymerization of methyl methacrylate, either pure or as an emulsion or suspension. PMMA produced in this way is atactic and completely amorphous (Figure 2.56). The pressure, temperature profile, and duration of the polymerization process have an influence on the average length and architecture of the resulting polymer chains. This in turn has a direct impact on some physical and chemical properties, which may vary depending on the production process.

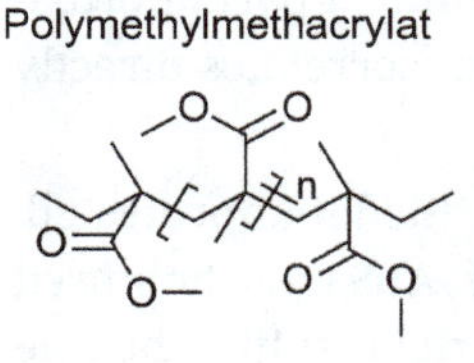

Figure 2.56 Structure of PMMA

2.6.8.6.2 Features

PMMA crackles when it burns, and does so with a yellowish flame, a sweetish odor, a dripping flame, and without residues. PMMA is amorphous with a T_g of approx. 105 °C. PMMA can be polished and machined well due to its brittleness. It can be cut and engraved very well with lasers. Due to its high hardness and a modulus of elasticity of >2700 N/mm², it is more scratch-resistant than most other thermoplastics. PMMA can be bonded well by gluing or welding and transmits light better than mineral glass. Furthermore, it can be easily colored and is largely resistant to acids, alkalis of medium concentration, as well as to gasoline and oil. However, polar solvents such as ethanol, acetone, and benzene attack PMMA. For this reason, acrylic glass surfaces must not be cleaned with alcohol or solvents, as this may result in stress corrosion cracking.

If co-monomers like alkyl or aryl methacrylates are used in the polymerization, the properties of the copolymers obtained, e.g. the heat distortion temperature or the T_g, can be influenced. Via copolymerization with multifunctional acrylates, toughness as well as chemical and temperature resistance can be improved by crosslinking.

With a T_g of approx. 105 °C, the application limits of pure PMMA are –40 to 75 °C continuous temperature and up to 100 °C for short periods. PMMA exhibits good electrical insulating properties, but as a result also has a high electrostatic chargeability and a tendency to attract dust. The medium strength, combined with the high stiffness, scratch resistance, transparency and weathering resistance, as well as an impact strength six times higher than that of silicate glass, clearly indicate the main field of application for PMMA as a glass substitute.

2.6.8.6.3 Processing

A distinction must be made between "cast" high-molecular PMMA, which belongs to the "thermoelastics", and PMMA molding compounds, which behave like typical thermoplastics. Thermoplastic PMMA can be processed by extrusion, injection molding, thermoforming, welding, bonding, and machining. Special processes include mono- and biaxial stretching to improve mechanical properties (impact strength).

2.6.8.6.4 Applications

PMMA is used in the following areas:

- **Optics:** Ophthalmic and watch lenses, magnifiers, lenses, prisms, diffusing screens, light-conducting fibers/threads and elements, plates with LED colorings matched to the individual wavebands, and optical storage devices
- **Electrical engineering/electronics:** Flat screens, diffuser plates, television set screens, plate elements in photovoltaics, switch parts, dials, and luminaire covers
- **Automotive industry:** Bus skylights, caravan windows, reflectors, rear and indicator light covers, and aircraft glazing
- **Construction:** Traffic signs, advertising signs, protective glazing at bank counters, gymnasiums, and laboratories, soundproof walls, lights and luminous ceilings, light domes, washbasins, bathtubs, bathroom units, shower enclosures, and WCs
- **Other:** Models, liquid level indicators, protective glazing of machines, vending machines, measuring and drawing instruments, resins for casting sensitive showpieces (biological objects), in mixtures of dental fillings, and dentures.

2.6.8.6.5 Recycling

Since the market prices for PMMA are high compared to other standard polymers, the recycling of PMMA has been of interest for a long time. However, grinding and remelting of used PMMA parts results in severe quality losses, especially in the optical properties that are often crucial for PMMA applications. In addition to thermal recycling by incineration, PMMA can also be recycled by feedstock recycling in the form of thermal depolymerization. In this process, the polymer molecules can be returned to the monomer, which can be converted back into high-quality PMMA by polymerization [Röhm 2021].

2.6.8.6.6 PMMA in Additive Manufacturing

PMMA can be processed in 3D printing with extrusion-based processes at 200–250 °C with a build platform temperature of approx. 80–100 °C. PMMA shows a rather high tendency to warp on the build platform, but special build platform materials or adhesives can help here. During the printing process, it is important to provide sufficient ventilation to minimize unpleasant odors that occur during melting.

Due to the high brittleness of the material, PMMA components can be machined, ground, and polished to improve their optical properties. When finishing, care should be taken not to drill too close to the edge of the object, as this could cause parts to splinter. Bonding and welding are other possible post-processing methods for components printed from PMMA. Colored highly transparent filaments are available, and coloring the otherwise transparent plastic is also feasible without problems.

PMMA can also be processed in powder-based processes. Depending on the manufacturing process (suspension polymerization), PMMA can be obtained as a very fine, very round, and uniform powder. PMMA is mainly used in binder jetting (BJ) printing by bonding the powder layer by layer with a precisely applied PMMA-solubilizing polymerizing adhesive. The resulting products show a somewhat rough, grainy material surface, which also has a rather low abrasion resistance. The wall thicknesses should not be less than 1.5 mm, as the model still has to be cleaned, mechanically compacted and infiltrated with a binder after printing to ensure final stability.

Despite the initial roughness, quite high imaging accuracy is achieved after post-processing, and fine details can be reproduced as long as the minimum wall thickness is maintained. Other advantages of the process are that the build process is highly efficient and scalable, there is no shrinkage, and unbonded powder can be 100 % reused, as the materials are not subjected to any thermal stress during the build process.

After gluing and infiltrating the powder, the material is hard and brittle and should not be exposed to high mechanical stress. Printed objects made of this material are mainly used for illustration purposes, for artwork, and for creating prototypes with a high level of detail. Examples of applications for PMMA in 3D printing are therefore mainly components that require high weather resistance and good post-processability, such as in the fields of electrical engineering, optics, medical technology, food technology, and art and design objects.

2.7 Processing Methods for Thermoplastics

Thermoplastics were originally processed primarily by injection molding, which is why they are also known as injection molding compounds, in contrast to thermosets, which are also known as compression molding compounds. The second most important processing method for thermoplastics is extrusion. Both processes belong to the so-called primary molding processes. Other common processing methods include blow molding, film blowing, hot caulking, film calendering, free forming, and deep drawing. These methods belong to the forming processes. In addition to primary forming methods, thermoplastics can also be processed and joined mechanically. Mechanical processing includes, for example, sawing, milling, grinding, polishing, turning, planing, as well as the joining methods of bonding and welding.

2.7.1 Injection Molding

Injection molding is a fully automatable process with high reproducibility. It is particularly suitable for the production of complicated molded parts as mass-produced articles [Chemie 2021]. For this purpose, the plastic is melted in a plasticizing and conveying unit, for example an extruder, and injected under pressure into a mold. The injection volumes vary from grams to kilograms. The cycle time of the molding process depends on the cooling or solidification rates of the thermoplastics, and varies from seconds to minutes, depending above all on the wall thickness of the components.

Multi-component injection molding is suitable for the production of multicolored products or for the combination of different plastics (for example hard and soft plastics) in one molded part. It can also be used to produce movable articulated joints in a single step [Maschinenbau-Wissen 2021c].

The principle of **injection compression molding** is that the plastic melt is injected into a larger cavity, which is then mechanically reduced in size. The compression molding process reduces the residual stresses in the molded part that occur during rapid cooling and crystallization [Kunststoffe 2021].

In **foam injection molding**, a foaming agent is added to the plastic melt in order to achieve foaming of the melt in the underfilled mold [Maschinenbau-Wissen 2021d].

2.7.2 Extrusion

In **extrusion**, plastics are continuously melted and discharged through a shaping die. The plastic is then cooled to produce semi-finished products such as profiles, sheets or films. Output rates can exceed 1000 kg/h [Maschinenbau-Wissen 2021b].

Extrusion blow molding is a process for producing hollow bodies from thermoplastics. A special form of blow molding is so-called stretch blow molding [Thielen et al. 2006]. The production of returnable plastic bottles is done by stretch blow molding. In this process, a preheated plastic blank (an elongated, thick-walled, cylindrical hollow body) is controlled and expanded into a desired shape using a so-called blow die and compressed air.

Blown film extrusion, as the name implies, is used for the production of films, which are mainly used in the field of food packaging. The way blown film extrusion works is that during the process, molten plastic is discharged from a ring-shaped die and the resulting tube is inflated directly with the help of air. This produces a melt tube 10–20 m high, the layer thickness of which is controlled by the blowing pressure and the haul-off speed. The already solidified hose is then collapsed by a pair of squeeze rolls. This produces a so-called "layflat", which then runs to a winder and is wound onto roll cores [Maschinenbau-Wissen 2021a]. A distinction is made between single-layer films, known as monoblow films, and multilayer films, known as coextrusion blown films. The latter offers the possibility of combining positive properties of several plastics in one film. This type of film is produced by coextrusion with the aid of several extruders and a specially adapted extrusion head in which the various plastic melts are selectively layered on top of each other during extrusion. The mode of operation of blown film extrusion is otherwise unchanged.

2.7.3 Thermoforming

Thermoforming describes the production of three-dimensional molded parts from sheet-like semi-finished plastic products (films and sheets) under the action of heat and compressed air or vacuum. In principle, this forming process can be carried out using a variety of process technologies. In the forming of thermoplastics, it has become established that the mechanical energy required for forming is introduced by vacuum and/or compressed air as well as by mechanical stretching aids. Forming is preferably carried out in the rubber-elastic state at the transition of the heated polymer to the melt, in order to obtain a stable layer thickness without cracks.
The primary molding processes extrusion and injection molding, on the other hand, are carried out in a temperature range in which the materials are present as low-viscosity melt.

A characteristic feature of forming processes is that the **semi-finished products** must first be brought from room/storage temperature to forming temperature, which is why "thermoforming" has become established in international usage as the generic term for all process variants. The range of parts produced by this process extends from packaging containers (quantities of up to 100,000 pieces/h) to large molded parts, such as swimming pools measuring up to 8 × 4 × 1.50 m (1 piece/h). An important area of application for these processes is the production of large molded parts for the automotive industry (dashboards and interior door panels) [Hopmann & Michaeli 2017]. The thermoforming process is preferably used to process higher-molecular-weight amorphous thermoplastics (PS, ABS, SAN, PMMA, PC, and PVC) and fast-crystallizing semi-crystalline materials such as PP and PE.

3D printing often provides individual templates (negatives) for thermoforming in small batches. A prominent example is the use of 3D-printed negatives as a template for thermoforming a foil for the production of transparent tooth-correction splints, so-called aligners. Aligner therapy is an orthodontic treatment method for the largely invisible treatment of mild to severe tooth misalignments that uses a sequence of individually manufactured, thin and transparent plastic splints.

With the aid of a special simulation process - starting from the actual condition of the scanned dentition - a previously determined treatment objective is represented three-dimensionally and divided into individual treatment phases. Individual splints are then produced for each of these phases. The dental splints are produced in a two-stage process. In the first step, a simulation of the tooth position to be achieved is created in the form of a jaw (upper/lower jaw with teeth) as a thermoforming template. The simulation serves as a base for manufacturing the jaw template using a resin-based AM process (SLA or DLP). A transparent thermoplastic polyester film is then thermoformed over the printed part (the template), cut to size at the edges, and post-processed (for example, cut edges are rounded). This is how the aligner is produced [Tindera 2017].

2.7.4 Extrusion

Extruders are conveying devices that uniformly press solid to viscous masses out of a shaping orifice (extrusion die) at high pressure and temperature according to the basic operating principle of the Archimedean screw. This process is known as **extrusion**. Basically, extruders can be divided into two process principles: processing and compounding extruders. Processing extruders (usually single-screw extruders) are mainly used for heating, plasticizing, and transporting masses. Processing extruders (e.g. co-rotating, close-combing twin-screw extruders, Buss kneaders etc.) are used for chemical and/or physical modification of materials (reacting, mixing, degassing, etc.). There are extruders with one, two, or more screw shafts. Extruders with two screws are divided into co-rotating and counter-rotating twin-screw extruders.

Single-screw extruders are suitable for almost all plastics. The task of the extruder is to convey the material against the resistance of the die (extrusion nozzle) and to melt it in the process. In the co-rotating twin-screw extruder, the screws rotate in the same direction, in the counter-rotating one in the opposite direction. In single-screw and co-rotating twin-screw extruders, conveying and pressure build-up are effected by friction between the rotating mass and the stationary barrel wall (friction conveying). The plastic melt is thereby pushed from the helical screw flights to the outlet nozzle.

In the case of counter-rotating twin-screw extruders, a distinction is made between the parallel and the conical design. In the parallel design, the screw diameter is constant over the entire length of the processing unit. In conical machines, on the other hand, the screw outer diameter and the screw distance from the feed zone to the discharge zone decrease steadily (see Figure 4.4). Twin-screw extruders allow better mixing and degassing of plastic compounds compared to single-screw extruders.

Single-screw extruders are used in 3D printing primarily in the area of extrusion-based additive processes. Whenever large throughput volumes or inexpensive standard plastic pellets (or pellet mixtures) are to be processed, small single-screw extruders are used to enable plasticization and conveying of the plastic pellets (pellet extruder, cf. [Tubbesing 2018]).

The advantage of extruders as conveying units, compared to the usual FLM printers fed with filaments and films, is particularly in the area of conveying very soft materials (TPE, silicones, rubbers, and pastes) or highly filled or fiber-reinforced materials (PP, PA, etc.). Higher throughput rates can be achieved with extruders, which is why they are often used for printing larger components mounted on industrial robots. In the production of large components, the price advantage and better availability of the pellets compared to filaments also speak in favor of their selection as the preferred construction material.

The disadvantage of pellet extruders lies in the higher bearing load of the die head, since an extruder must always be placed in the immediate vicinity of the extrusion opening. Accordingly, these devices must be designed more robustly in order to absorb the acceleration processes during directional changes of the print path [Dyze Design 2021], or alternatively the movement is transferred to the build platform (cf. [ARBURG 2021]).

An important difference of pellet or granulate extruders is the longer residence time of the plastics at the high temperatures in the melt in the extruder barrel compared to filament extrusion and also compared to classical melt processes. Extrusion-based additive processes achieve significantly lower throughputs than are common in injection molding or profile extrusion. As a consequence, the material is subjected to greater thermal stress and must be stabilized accordingly by additives.

2.8 Formulation of Thermoplastics

The formulation or compounding of thermoplastics describes the process of adding additives (fillers, pigments, plasticizers, additives, etc.) to achieve the desired property profiles. The formulation and blending of plastic molding compounds in suitable mixing units allows the design and manufacture of compounds with "tailor-made" properties over a wide range. Blending or compounding is mainly carried out in twin-screw extruders and includes the process operations of conveying, melting, dispersing, mixing, degassing, pressure build-up, and extrusion (ejection).

The objectives of compounding are numerous, and depend on the desired properties of the subsequent component. However, compromises often have to be made in the sum of the requirements, since, as is so often the case when solving technical problems, individual requirements are sometimes diametrically opposed. The most important goals of compounding are the improvement of the base polymer or that of the resulting workpiece in terms of:

- Mechanical resistance (strength, modulus, and/or impact strength)
- Thermal resistance (heat deflection temperature or heat aging)
- Chemical resistance (flame retardancy, hydrolysis resistance, or weather resistance)
- Processing properties (viscosity, adhesion, crystallization rate, or shrinkage)
- Optical properties (color, transparency, or surface quality).

In this process, the addition of reinforcing materials and fillers, plasticizers, polymer blends, and impact modifiers are used to adjust properties such as modulus, tensile strength, elongation at break, impact strength, and heat resistance. By adding pigments and additives, many other property characteristics can be adjusted, such as color, surface finish, crystallization rate, processing properties, and also heat aging, oxidation, discoloration, and weathering resistance.

> The widespread and necessary compounding with fillers, especially for the refinement of standard and engineering plastics, as well as the complex additivation, for example with flame retardants and pigments, make material recycling considerably more difficult when it comes to recovering unmixed and pure raw materials for further processing. As an alternative, thermal recycling by pyrolysis [Scheirs & Kaminsky 2006] and, if necessary, chemical recycling by solvolysis or hydrolysis, can be used to produce recyclable polymers or building blocks and monomers that can be specifically dissolved or distilled off from the contaminated mixtures [Vogel et al. 2020]. In the current discussion on microplastics, fillers and additives play an important role in addition to polymers, as bioaccumulative properties are suspected for many of these substances [Fath 2019].

2.8.1 Color Setting

The color desired by the customer is adjusted via the addition of pigments, which is often done by means of so-called masterbatches. This often results in the first conflicts of objectives, since certain color settings influence not only the mechanical properties, but also the classification, for example as a medical material, the recyclability, the weather resistance, and other properties of the plastics [Bastian & Hochrein 2018].

2.8.2 Flame Protection

Flame retardants can be added to prevent inherently flammable plastics from igniting. Many of the previously widely used highly effective brominated flame retardants (polybrominated biphenyls (PBBs) and polybrominated diphenyl ethers (PBDEs)) are, due to new toxicological findings, now on the banned list, along with polychlorinated biphenyls (PCBs) [Weber 1998]. Instead, they have been largely replaced by mainly inorganic and halogen-free flame retardants, such as aluminum hydroxide (ATH), magnesium hydroxide (MTH), halogen-free phosphorus compounds, and borates. The advantages of the brominated and chlorinated (halogenated) flame retardants, especially in combination with antimony oxides, were their good compatibility with plastics and their high efficiency. The search for effective, compatible, non-toxic, non-bioaccumulative flame retardants is still ongoing today, as there continues to be a growing market and need for these products, especially in construction, electronics, batteries, metal replacement transportation, and mobility [Anon 2017].

2.8.3 Stabilizers

An important additive class besides flame retardants are the stabilizers [Maier & Schiller 2016]. There are essentially the following reasons for the stabilization of polymers:

- Temperature-initiated chain degradation during processing: This can be caused by excessive shearing of the material or by excessively long residence times with simultaneous thermal and mechanical stress in the processing machines. The degradation can be largely prevented by preconditioning (for example drying) and stabilization designed for a short-term load.
- Temperature-initiated chain degradation in the application: Plastic parts that are subjected to high temperatures in the application, e.g. in the engine compartment of a motor vehicle, must be specially stabilized for these conditions.
- Improving weather resistance: Plastic parts in outdoor areas are exposed to severe stress from UV radiation, oxidation, and hydrolysis. These can be compensated to a certain extent by special stabilizers. Depending on the base polymer and stabilization, these effects can be halted for different lengths of time.

> For use in AM by means of FLM processes, oxidation stabilizers are of somewhat lower importance for material stabilization during processing than in classic screw extruders, since the residence times are short. This statement expressly does not apply to the use of pellet/granulate extruders. Oxidation stabilizers are also crucial for use in powder-based processes, such as SLS and HSS, and should be adapted to the average residence time and temperature in the building space. In this case, both yellowing stabilizers as well as oxidation stabilizers are important.

2.8.4 Processing Aids

The role of processing aids is to improve the processing and mixing properties of polymers. Demolding aids, for example, facilitate demolding in the injection molding process, and dispersing aids contribute to the uniform incorporation of fillers [Scheirs & Kaminsky 2006]. The product properties in the application are typically not affected by processing aids.

In AM, processing aids should, for example, reduce the extrusion pressure or increase the extrusion speed, but at the same time not have a negative impact on layer adhesion. The development of processing aids for 3D printing is still in its infancy. In powder-based processes, in addition to flow agents (such as silica), which act on the surface to minimize powder clumping and provide better powder flow or more uniform powder application, fillers (carbon black or titanium dioxide) are used to increase laser absorption or optimize the laser scattering on the particle. In particular, flow aids, which support a more uniform powder application by accumulating on the surface of polymer particles, can, in return, make the sintering properties, such as the merging of polymer powder particles in the melt, more difficult. The polymer must first flow around the flow aids to form a continuous and mechanically stable film.

Nucleating agents (crystallization accelerators)

During cooling, polymers crystallize out of the melt. This crystallization is normally triggered by dust, the vessel wall, or by auxiliary substances such as dyes, etc. If crystallization around the seed occurs slowly and uniformly, spherulites are formed. The larger these spherulites are, the less favorable is the mechanical behavior (toughness) of the plastic. Therefore, an increase in the number of nuclei is generally desirable. Fast and homogeneous crystallization throughout the plastic melt additionally shortens the demolding time in injection molding and improves the dimensional stability of the shaped extrudate in extrusion, and thus overall increases processing efficiency. In order to specifically influence the crystallization rate, so-called nucleating agents are added to the polymer in the melt [Scheirs & Kaminsky 2006].

In 3D printing, these nucleating agents are rather counterproductive for most common polymers, as they have a negative influence on ply adhesion and increase shrinkage in extrusion-based processes. At the same time, they do not act fast enough to positively influence bridging behavior in FLM printing. Nucleating agents are also rather undesirable in powder-based processes because, depending on the build speed, they can lead to inconsistent solidification in the build process and thus to inconsistent shrinkage. In tendency, they reduce the processing window by forcing a higher build temperature to compensate for crystallization that is too fast.

2.8.5 Fillers

The terms “fiber-filled plastics” or “particle plastic composites” are used for an extremely heterogeneous group of materials. The effect of fillers depends not only on the chemical interaction with the plastics, but also on the type of fillers, their geometric dimensions, the particle volume fraction, the distribution quality, and the particle spacing. In particular, the distribution quality of the fillers is determined to a large extent by the mixing process used.

In plastics technology, fillers are used to optimize the property profile of plastics. They are incorporated by compounding, which is why the filled polymers are also referred to as plastic compounds. Optimization often focuses on increasing stiffness, reducing shrinkage, and improving surface appearance. Furthermore, a targeted increase in thermal or electrical conductivity is also possible. If the aggregates only account for a volume share of less than 5 %, as is typically the case with so-called masterbatches, we speak of additives. Important fillers for thermoplastics are carbon blacks, titanium dioxide, talc, chalk, mica, and the class of fibrous fillers.

Since most standard plastics are not suitable for structural applications due to their insufficient intrinsic material strength, reinforcement with fibers or filling with functional particles represents an economical variant for obtaining structurally capable materials with sufficient properties based on inexpensive standard plastics. Due to the variability of the fillers and plastics, the properties of these composites can be specifically adjusted within a wide range (tailor-made materials). In many cases, this has made it possible to replace metallic materials with cheaper plastics for applications in the lightweight construction sector and for the white goods and automotive industries.

2.8.5.1 Particle-filled Plastics

In contrast to fiber-reinforced composites (FRCs), the particles in particle-filled plastics have hardly any reinforcing effect on the strength under tensile, compressive, and flexural stress. Depending on the filler volume fraction, however, an increase in the modulus of elasticity is achieved, with the elongation at break of the filled plastics then often decreasing significantly.

Basically, a particulate plastic composite is a combination of any particles of different geometry and origin, and a plastic matrix that encapsulates the fillers and enables force transmission between the respective composite partners by means of adhesive or cohesive interactions.

The added particles can be inorganic or organic particles that have regular or irregular shapes and generally serve to stretch the polymer (cost containment) or functionalize it (electrical and thermal conductivity, hardness, abrasion, and surface quality).

An additional distinction is made between macro-, micro- and nanoparticles, which are distinguished in particular with regard to filler content, particle size, and spacing in the polymer composite.

Particle-filled plastics typically provide an increased specific modulus (based on density), higher hardness, and improved tribological properties compared to the matrix material. This is associated with a generally reduced elongation at break and, depending on the particle geometry and processing conditions, also with increased anisotropy (directional dependence) of the mechanical properties.

Inorganic fillers are, in analogy to reinforcing fibers, often provided with a so-called sizing and/or an adhesion promoter (see also fiber matrix adhesion), which improve the bonding of the particles to the plastic matrix by forming an interface between matrix and particles. The length-to-thickness ratio for most filler particles is close to 1, so that breakage of the particles is rather rare, and debonding from the matrix is the dominant damage mechanism.

> The basic rule for all particle-filled plastics is that the fillers perform a supporting task (e.g. increase in modulus) or a functional task (e.g. increase in thermal or electrical conductivity) in the plastic–filler composite. Good filler bonding to the respective plastic matrix supports these tasks. In the case of semi-crystalline plastics, fillers also frequently act as crystallization nuclei and influence the crystallization behavior (crystallite size and crystallization rate).

2.8.5.2 Carbon Black

Industrial carbon black is used to over 90 % as a filler in the rubber industry (mainly vehicle tires). Other applications include use as a black pigment for printing inks, inks, paints, and for coloring plastics (especially to improve UV protection), as conductive carbon black in the electrical industry, and as a raw material for engineering ceramics and electrode material.

2.8.5.3 Chalk and Talc

The oldest natural inorganic fillers for plastics are talc, a phyllosilicate with the general formula $Mg_3(OH)_2[Si_4O_{10}]$, and chalk ($CaCO_3$), which are added essentially to dilute costs. Using such naturally occurring fillers with a low carbon footprint can improve the environmental sustainability of compounds. By using talc and chalk the modulus of elasticity, heat deflection temperature and continuous service temperature of a compound can be increased to some extent, while the shrinkage is reduced in proportion to the filler volume content.

Compared to polymeric fillers, the high density of these fillers (>2.5 g/cm^3) produces a weight handicap. Usually, talc (7 to approx. 55 wt%) and chalk (10 to 40 wt%) are added to particle-filled plastics, such as PP. At higher filler contents, depending on the particle size and geometry, a significant deterioration in tensile strength and elongation at break is observed, since the number of potential flaws in the plastic filler composite increases.

In the plastics industry, talc-based reinforcing materials or fillers are used to a greater extent in the reinforcement of polyolefins, such as HDPE or PP for use in the automotive or construction industries. Talc-reinforced PP compounds have been used in a wide range of applications since around 1975.

2.8.5.4 Mica

Another natural mineral filler is the layered silicate mica. Mica exhibits a layered and irregular structure and therefore evokes not only a filler-dependent but also a direction-dependent reinforcing effect (length/thickness ratio up to 50) in the plastic. Due to its low electrical conductivity, mica is of particular importance for technical applications as an insulator material. Mica-filled plastics often also have better temperature resistance than unfilled polymers [Ehrenstein & Pongratz 2007].

2.8.5.5 Particle-reinforced Thermoplastics in Additive Manufacturing

In AM, the use of highly filled thermoplastics is still an exception. Up to now, extrusion dies based on copper and brass have been used in common extrusion-based processes. These are optimized for high heat transfer and are not suitable for handling abrasive fillers. However, alternatives made of hardened and/or coated steel are already available. The viscosity of the melt of highly filled thermoplastics increases considerably, which in turn results in higher extrusion temperatures or a higher pressure build-up at the extrusion die. Most fillers can act as crystallization nuclei, which leads to rapid crystallization. This has the disadvantage that the ply layer adhesion in the component product can be negatively influenced. It is therefore essential to adjust the build and extrusion temperatures when using filled materials. The tendency of particulate fillers to cause higher brittleness of the components product further exacerbates this fundamental problem of additively manufactured components.

On the other hand, filled thermoplastics can have the benefits of lower absolute shrinkage, higher heat resistance, cost savings, and better flame resistance. There are therefore good reasons to look for optimized application conditions, improved extrusion heads, and new conveying technologies, such as the newly emerging pellet/granulate extruders, in order to further expand the use of highly filled thermoplastics in AM using extrusion-based processes.

In powder-based processes, the use of filled thermoplastic powder materials is common in order to optimize the shrinkage of the plastics as well as the advantages mentioned above, and (in particular) the energy absorption efficiency and the scattering of the lasers in the SLS process. Common fillers used here are silica to improve the flow behavior of the powders, carbon black, talc, glass beads, and (more recently) ceramic and carbon short fibers. The use of rheologically active fillers (e.g. nano fillers, fine particles, and anisotropic fillers), which tend to accumulate on the surface of the polymer particles (pickering), is, however, limited to a rather small percentage by weight. Less active fillers, such as glass beads, can be added at higher dosages.

2.8.6 Fibrous Reinforcing Fillers

Short-fiber-reinforced thermoplastics are a group of materials whose properties bridge the gap between unreinforced and long-fiber-reinforced polymer materials. A decisive advantage is the processability of these materials by injection molding.

During the injection molding process, a specific material- and filler-dependent fiber orientation is generated in the part or test specimen. Both the degree of orientation and the orientation distribution of the fibers depend on the flow conditions in the melt, the geometry of the molded part, and on fiber-specific parameters such as fiber length and fiber volume fraction. Common fibers used here are natural fibers, polymer fibers, glass fibers, carbon fibers, and ceramic fibers. In contrast to inorganic fibers, organic natural fibers (cotton fiber, rayon fiber, hemp fiber, etc.) and polymer fibers (polyaramides, polyacrylonitriles, polyesters, polyamides, polyethylene, polypropylene, and polyvinyl alcohol) are limited in processing conditions in that they must not be destroyed or melted during the molding or extrusion process. Accordingly, processing should take place below the softening point of these fibers.

With fibers, the mechanical, thermal, and dynamic properties of thermoplastic polymers can be increased enormously. Manufacturers of technical injection-molded components can choose between compounded short-fiber pellets, pultruded long-fiber pellets, direct fiber injection molding, and a wide variety of combinations of fibers and matrix materials.

Key issues in the production and design of fiber-reinforced products often relate to fiber distribution, fiber length and degree of fiber damage (fiber breakage), fiber orientation and the resulting orientation-dependent part shrinkage, weld line strength during melt confluence in injection molding, and the selection of suitable mixing and processing parameters such as mixing time, screw geometry, shear rate, processing temperature etc.

The use of fibers thus brings significant advantages, but also significantly increases the effort involved in preparing the compounds and designing the parts, taking into account the complex interdependencies between processing and material parameters.

2.8.6.1 Fiber-reinforced Thermoplastics in Additive Manufacturing

In AM with fiber-reinforced thermoplastics, a distinction is made between two fundamentally different processes. One of these is the processing of classic short-fiber-reinforced raw materials in extrusion-based processes. As described above, this results in a significant reinforcement of components in the deposition plane and, due to the orientation of the fiber during the extrusion process, in the extrusion direction. As discussed, when processing short-fiber-filled thermoplastics, care should be taken to ensure that the equipment is designed for the application. This means that the temperature capability and abrasion resistance of the print head must be set up to discharge the higher viscosity melts and abrasive fiber filaments. On the other hand, the processing of long fibers in AM opens up completely new possibilities for the targeted reinforcement of components, preferably in the load direction of the component. The long fibers can be used in the form of long-fiber-filled filaments, or in a parallel plastic and fiber application process, or as a 3D application of long-fiber-reinforced thermoplastic tapes. This makes it possible for the first time to create design- and load-oriented component geometries that were unthinkable before the advent of AM as a product of industrial processes.

2.8.6.2 Glass Fibers

Glass-fiber-reinforced plastic (GFRP) is a fiber composite made of a plastic and glass fibers. Both thermosetting plastics (e.g. polyester resin (UP) or epoxy resin (EP)) and thermoplastics (e.g. polyamides, polyolefins, etc.) can be used as a basis. Compared with long-fiber-reinforced composites made of carbon fibers and thermosetting materials, GFRP has a high elongation at break and a high elastic energy absorption in combination with a suitable plastic matrix, but a relatively low modulus of elasticity. Even in the fiber direction, this is lower than that of aluminum. It is therefore not always suitable for components with high stiffness requirements, comparable with metallic materials.

Short-glass-fiber-reinforced parts are used primarily as cladding or are used in injection molding processes due to their good moldability and great design freedom. The components usually exhibit quasi-isotropic behavior, since the short fibers are randomly distributed. During injection molding of short-fiber-reinforced thermoplastics, a weakly pronounced orthotropy to the injection direction can occur, with the fibers orienting themselves along the flow lines. The addition of short glass fibers to thermoplastics improves their stiffness, strength, and (in particular) their behavior at high temperatures. The creep of short-fiber-reinforced thermoplastics is lower than that of the base material.

2.8.6.3 Carbon Fibers

Carbon-fiber-reinforced plastic (CFRP) is a composite material in which carbon fibers are embedded in a plastic matrix. The matrix serves to connect the fibers as well as to fill the gaps. Epoxy resin is usually chosen as the matrix, but other thermosets and thermoplastics are also possible as matrix materials. CFRPs are used especially where high stiffness is required and the increased cost of carbon fibers is accepted. Examples include bicycle frames, fishing rods, rowing boats, sports accessories, and to an increasing extent compressed gas containers. In 2015, global production of CFRP was 91,000 t, with annual growth of about 12 %. However, a large proportion of this went into thermoset processing [Allied Market Research 2018].

Compared to materials such as steel, carbon fibers have a significantly lower density (by a factor of ~4.3). Their weight-specific stiffness in the fiber direction is, depending on the fiber type, approximately up to twice that of steel. This results in a very stiff material with the appropriate filling grades, which is particularly suitable for applications with a main load direction where a low mass combined with a high stiffness is important, such as in vehicle or aircraft construction.

> **!** As a filament for extrusion-based AM, CFRP is used in particular for the production of strong but lightweight tools and parts due to its high strength [3D-Filament 2021]. Long-carbon-fiber-reinforced parts can be produced using parallel melt and fiber deposition processes (e.g. [Markforged 2021]) as well as processed as tapes [Denkena et al. 2016].

2.8.6.4 Ceramic Fibers

Wollastonite is a naturally occurring calcium silicate. The use of wollastonite in plastics is a borderline case between reinforcement and filling. Due to its rod-shaped crystal structure, a reinforcing effect can be achieved by admixture. Ceramic fibers are mainly used as reinforcing fibers in fiber-reinforced plastic composites or as heat protection material. The physical properties and, consequently, the areas of application are similar to those of glass fibers.

2.8.6.5 Natural Fibers

Natural-fiber-reinforced plastic (NFRP) is a fiber composite made from a plastic (e.g. polyester resin, epoxy resin, or polyamide) and natural fibers. NFRP belongs within the class of biomaterials and biocomposites. It is used both as a long-fiber-reinforced plastic and as a short-fiber-reinforced plastic. NFRPs provide greater stability than pure plastics by virtue of the natural fibers incorporated. The main area of application is in vehicle construction, especially in the design of trim components in the vehicle interior.

A variety of natural fibers can be used for the production of natural fiber composites. The fibers differ in shape and length as well as in their chemical composition and mechanical properties. This, in turn, affects the processing possibilities of the fibers and the properties of the composites obtainable from them. The fibers most commonly used for the production of fiber composites are flax and hemp fibers, as well as subtropical and tropical fibers such as jute, kenaf, ramie, or sisal fibers. Wood fibers are also used as a filler and reinforcing material in fiber composites. Prior to use, these fibers are often specifically modified by ultrasonic pulping to achieve reproducible technical properties through standardizable machining processes. To better distinguish them from other natural fibers and due to their importance in terms of volume, composites with wood are referred to as wood–plastic composites (WPCs).

> A type of WPC also exists in 3D printing in the form of the so-called "wood filament". Usually PLA is taken as the base material and wood fibers are added in varying concentrations [Theias 2021]. With wood filaments, the focus is on the look and feel. The printing temperature usually has to be reduced due to the lower temperature resistance of wood compared to normal PLA.

2.8.6.6 Polymer Fibers

Polymer fibers not only have a high aspect ratio (length/diameter), but also a high orientation dependence in their properties. Their modulus along and across the fiber differs significantly. At high processing or application temperatures, organic fibers melt or decompose. However, this temperature limit for processing and use of polymer fibers can vary greatly depending on the fiber.

The best-known polymer-reinforcing fibers are the polyaramides. Lightweight parts subject to abrasion and impact stress (e.g. the protection of the leading edges of aircraft tailplanes against hailstorms or the walls of kayaks) are preferably made of aramid fiber-reinforced plastics (AFPs). Strength and stiffness are somewhat better than, for example, in E-glass, which is commonly used for composite applications. Other properties include good damping capacity, non-flammability and excellent chemical resistance.

The processing of laminates is often difficult due to the high toughness of the fibers. Special tools (Kevlar shears) are required to cut fabrics. Polyaramide short fibers are also available as fiber fillers for thermoplastic injection molding applications. However, the incorporation and distribution of the polyaramide fibers is also complex here.

> Parts reinforced with polyaramide fibers can be used in AM analogously to materials reinforced with carbon fibers using a parallel melt and fiber deposition process [Mason & Gardiner 2020]. The use of ceramic fibers and natural fibers in 3D printing has not yet been widespread, with the exception of wood fibers as fillers in thermoplastic filaments. However, an increasing use of polymer fibers in AM is to be expected.

2.8.6.7 Thermal Stability and Weathering Resistance of Plastic Fiber Composites

Fibers have different heat resistances. Aramid fibers, for example, already suffer initial strength losses at elevated room temperature. E-glass fibers show strength losses from 200 °C upwards. In contrast, carbon fibers are not thermally affected until 1000 °C.

Glass and carbon fibers do not absorb moisture, and improve the barrier effect against water vapor in laminates. Synthetic polymer fibers, such as aramids, are an exception. Their moisture absorption can be quite significant, depending on the degree of drying and the type of fiber. Polyaramide fibers also exhibit anisotropic swelling behavior with respect to the fiber axis. Measurements show high swelling in the radial direction and contraction in the axial direction. Accordingly, polymer fibers should be well encapsulated by the plastic or duromer matrix, or used only in applications exposed to a small variation in ambient humidity.

The weathering resistance of fiber-reinforced composites (FRCs) is mainly investigated under outdoor weathering conditions. Aging or degradation occurs through molecular chain breaks caused by radiation mainly in the UV-B range (280–315 nm). Synthetic fibers in particular, such as aramid, are subject to rapid photochemical degradation unless protected by colored topcoats/varnishes. The residual strength of a polyaramid part can drop below 25 % within a year, depending on the radiation intensity [Niederstadt 2017].

2.9 Weather-resistant Thermoplastics

Plastics used in outdoor applications can be exposed to massive weathering. Therefore, these applications require weather-resistant plastics that can withstand the negative effects of UV radiation on the material's appearance and mechanical properties. Inherently, weather- and UV-resistant plastics can be recognized by the fact that the following changes in appearance usually do not occur:

- Yellowing of the plastic
- Washing-out of colored plastics
- Fading of the surface of the plastics
- Stress cracking.

Furthermore, none of the following changes in mechanical properties are recorded:

- Brittleness development
- Reduction of strength, elasticity, and hardness.

UV resistance of plastics is usually achieved by adding additives such as UV stabilizers, radical scavengers, black coloring (usually with carbon black), or a protective coating (e.g. with a paint or a metal coating (by e.g. metallization, galvanization, or a metal-vapor deposition processes). The addition of carbon black offers an inexpensive and usually very effective way of producing UV-resistant plastics.

For applications with outdoor exposure, aliphatic polymers that do not contain aromatic building blocks or easily hydrolyzable or oxidizable groups are recommended (PA12, PA6, aliphatic polyurethanes, aliphatic polyesters with a high T_g, or PE), and in particular the material group of chlorinated and fluorinated polymers (e.g. PVC, PTFE, and PVDF), since they exhibit particularly high UV and weathering resistance even in their uncompounded state.

2.10 Electrically Conductive/Insulating Thermoplastics

The conductivity of materials depends to a large extent on the presence and density of mobile charge carriers, such as free (delocalized) electrons or ions, in the conductive medium. If materials with a low electrical resistance are present, they typically have many freely mobile charge carriers. They are then referred to as conductive or conducting. If this is not the case, the material is a non-conductor or insulator. The conductivity of plastics can be altered to a large extent by using fillers. A conductive plastic can be produced from a non-conductive plastic, for example by adding conductive carbon black or metal powders. By contacting the fillers in the compound (a filler-volume-dependent percolation boundary), a conductive path is formed through the plastic. Due to their variable electrical properties (from complete insulation to electrical conductivity), plastics are indispensable as materials in electrical engineering and electronics [Weiß 2005].

2.10.1 Electrically Passive Insulating Plastics

Unmodified plastics often offer inherent electrical insulation properties. These electrically insulating plastics exhibit high or complete resistance to electrical current. They are typically used in a variety of applications where there is a need to inhibit electrical conductivity to prevent injury or property damage. Electrically insulating plastics are commonly used in applications such as switches, lighting fixtures, electrical cables, and control panels. Typical representatives include PE, PP, PVC, PVAC, EVA, PC, PET, ABS, PS, PA11, and PA12. Crosslinked rubber materials such as EPDM, EVA, and CR are used in elastic cables.

The electrical conductivity of a material, when a direct current is applied, can be described by the combination of (specific) volume resistance, the (specific) surface resistance, and the insulation resistance [Schönhals 2015]. Plastics are classified according to their specific surface resistance. The surface resistance describes the resistance of the plastic to the current flowing through the surface. This value is expressed as the quotient of the applied voltage U (in V) and the generated current I (in A) via Ohm's law ($U = R \cdot I$).

2.10.2 Electrically Active Conductive Plastics

Unmodified plastics generally offer electrical insulation properties. However, thermoplastics can be modified by formulating them with fillers in such a way that various electrically conductive, antistatic, or static dissipative properties can be achieved. Antistatic plastics are used to prevent static charging and thus electrostatic discharge (ESD). Static dissipative plastics can be used to ensure the reliable handling of extremely sensitive components, significantly reducing damage to electrical components or their destruction during production. Electrically active materials are also used in mechanical engineering, for example in materials handling or explosion protection. In these applications, the targeted dissipation of electrical charges is an important factor in preventing explosion-triggering discharges. Antistatic plastics offer a number of advantages over other materials, for example in terms of lighter weight, weather resistance, and good resistance to mechanical stresses.

2.10.3 Conductive Plastics in Additive Manufacturing

The targeted design of the conductivity of components in three dimensions, e.g. the printing of three-dimensional conductive tracks and circuits and the targeted shielding of components against electromagnetic radiation, opens up a large number of possible new applications. New components become accessible in this way in the areas of medicine [Reinhardt & Anhalt 2015], wearables, tags and antennas [UMSICHT 2019], load control and failure analysis [Willberg & Rädel 2017], as well as system integration of electronic circuits in mechanically loaded components in vehicle and aircraft construction, in automotive electronics, in the use of LEDs and OLEDs, in photovoltaics, and also in the development of modern battery technologies [Gutberlet 2020].

Conductivity can be achieved by various methods:

- Multimaterial extrusion of highly filled polymers (with metal powder/graphene/carbon nanotubes)
- Precise insertion of metallic wires into components printed by thermoplastic extrusion
- Co-extrusion of conductive fibers/wires in reactive pastes [3Dnatives 2019]
- Application of doped intrinsically conductive polymers (ICP) (thiophene/polypyrrole/polyacetylene/polyaniline) as pastes by paste extrusion (PAM) [3Druck.com 2016]
- Printing of conductive nanoscale metal pigments in dispersion by the inkjet process [Fraunhofer IKTS 2021] or the aerosol jet process [AMFG 2020].

In addition to electrical conductivity, good thermal conductivity combined with simultaneous electrical insulation can be of interest in applications like complex lamp sockets and housings. The principles for setting the thermal conductivity and the application of thermally conductive plastics using 3D printing are analogous to those for electrically conductive plastics. Instead of electrically conductive fillers, only thermally conductive fillers such as aluminum oxide, boron nitride, or aluminum nitride [Fraunhofer INT 2019] are specifically used here. For combined electrical and thermal conductivity, fine metal powders, graphene, and carbon nanotubes can primarily be considered as fillers.

2.11 Fire-retardant Thermoplastics

Plastics consist predominantly of carbon and hydrogen atoms and thus principally possess a high thermal energy content. This benefits the thermal recyclability of used plastics, but turns into a problem in the case of fire. Not all plastics are combustible to the same extent. Plastics such as PE and PP are chemically structured like long-chain mineral oil products (e.g. mineral oil waxes) and are also similarly combustible. Aromatic plastics and plastics with a higher heteroatom content as well as plastics with higher chemical binding forces, on the other hand, are less combustible.

The characteristic value for flammability is the limiting oxygen index (LOI). It indicates the oxygen content of the air in percent that is needed for the plastic to burn. The normal oxygen content of the air is 21 %. Plastics with an LOI lower than 21 % burn very easily, such as polyethylene (LOI = 17 %). At the other end is PTFE (LOI = 95 %), which is thus considered virtually incombustible. Polyamides are in the range 21–25 % and are thus to a certain extent inherently self-extinguishing.

Within a polymer group, the same chemical structure leads to basically similar flammabilities. Differences in fire protection ratings result from the use of flame-retardant additives. There are two different approaches. On the one hand, the flammability of the plastic compound itself can be reduced, e.g. by counteracting the exothermic (heat-generating) combustion process with an endothermic (heat-absorbing) chemical process. Much more effective, however, are additives that secrete gases at high temperatures and displace the oxygen in the air around the product, so that the fire is extinguished at the outset.

Using flame-retardant additives to achieve self-extinguishing materials can also have a negative effect in the event of a fire, for example through the release of toxic gases. Thus, the use of fire-retardant plastics must be well considered. In electrical equipment, the case is clear. Areas where short circuits can occur must be fire-protected. Criteria such as voltage and current, operating time, and whether an operation is supervised or unsupervised are decisive.

In particularly critical application areas such as airplanes or trains, one needs to consider not only how the ignition of a fire can be avoided, but also what happens in the further course after ignition. Specifically, in enclosed spaces smoke development and smoke toxicity are critical. Plastics that are inherently non-combustible due to their chemical structure, such as polyetherimide (PEI), have an advantage here. Unformulated PEI has an oxygen index of 47 % and meets almost all fire protection standards, including those relating to smoke density and smoke toxicity.

2.11.1 Fire-retardant Thermoplastics in Additive Manufacturing

If fire-protected components are to be produced in AM, e.g. for electronics housings or lamp components, then the use of fire-protected thermoplastics is preferable. Formulations that work well in injection molding, however, do not always have to be easily processable in powder-based processes or in extrusion-based processes. Here, it often makes sense to use tested 3D printing materials with inherently good flame retardancy such as polycarbonates or polyimides for smaller series productions by FLM-based processes. If the applications with regard to product sizes are larger, and there needs to be a further cost optimization, then granulate/pellet extruders can be used with a broader range of suitable engineering plastics formulations. In many cases, however, it will be necessary to specifically develop plastic formulations for AM if the demand to fulfill high flame retardancy requirements needs to matched with good mechanical properties and good processability.

Another aspect of flame retardancy concerns the risk of fire in the manufacture of AM parts. In most powder-based processes, the build space is blanketed with nitrogen, argon or CO_2, which significantly reduces the risk of fire, since a critical oxygen concentration (LOI) must be present for fire to occur. In the extrusion-based processes, only very small quantities of a plastic are heated to the point of melting. A fire hazard is more likely to come from technically unsafe electronic components in very low-cost DIY printers, as well as from badly maintained extruder heads/hotends. Here, too, the risk of component failure is more likely to occur in the printer and not in the plastic itself. Especially when using technically unreliable devices, it is therefore advisable to use them in an environment that is poor in other fire-promoting substances (e.g. a metal shelf or a stone slab).

2.12 Foamed Thermoplastics

Foams are artificially produced materials with a cellular structure and low density. Almost all plastics are suitable for foaming. Foams can be compressed, i.e. their volume can be significantly reduced by pressure. Foams are characterized by a very low density (referred to here as volumetric weight) and a low thermal conductivity combined with a low mechanical strength. Due to their internal structure, foam components are almost free of residual stress.

The compressive "strength" of a foam is expressed by the compression hardness (e.g. for mattresses a compression hardness of 50 equates to 5.0 kPAat 40 % compression; this is most common, but inaccurate), or by the indentation hardness (less known but more meaningful). The density is expressed in kg/m^3 or g/L. The density, e.g. RG 35, corresponds to 35 kg/m^3, and the compression and indentation hardnesses are only conditionally dependent on each other. A heavy foam need not be a firm foam. However, if the same material is foamed in different densities, then the density usually correlates well with the compression hardness.

2.12.1 Foaming Process

A distinction is made between physical, chemical, and mechanical foaming of polymers:

- **Physical foaming:** The material is foamed by a physical process, usually involving injecting a gas under pressure into a melt, which then expands at a reduced pressure and solidifies in a foamed state by cooling in the mold (injection molding) or in the air/between boundaries (extrusion).
- **Chemical foaming:** A blowing agent, usually in the form of a so-called masterbatch granulate, is added to the plastic granulate in the extruder. When heat is added, a volatile component of the blowing agent splits off in the melt, causing the melt to expand under reduced pressure and solidify as a foam.
- **Mechanical foaming:** In this process, air is stirred at ambient pressure into the resin or paste to be foamed. The foam solidifies by crosslinking or gelling of the resin/paste (impact foams).

2.12.2 Foam Extrusion

In the production of PP e-foam (this is also possible for polyethylene, polystyrene, PET or some biopolymers), molten polypropylene is mixed with a blowing gas (e.g. pentane, CO_2) in the extruder under high pressure. As it emerges from a perforated nozzle, the plastic expands to 20–50 times its original volume. The resulting foam strands can be shortened to foam particles of approx. 2 to 8 mm diameter by rotating knives in a water ring pelletizer or underwater pelletizer. They are then separated from the water, dried, conditioned in silos and processed into foam moldings in the so-called foam molding process.

The molding process is used to process foam particles (EPS, EPP, EPE) into molded foam parts. Foam beads with diameters of approx. 1 to 8 mm are blown into porous aluminum molds and sintered by means of steam (approx. 1.2 bar for EPS, approx. 3 bar for EPP). After the cooling and stabilization phase, the newly created molded parts can be demolded. This manufacturing process makes it possible to produce thermoplastic foam parts in the density range between approx. 12 kg/m^3 and 300 kg/m^3 with very homogeneous density distribution over the entire foam part. Direct foaming processes using chemical or physical blowing gases in injection molding do not allow this due to the limited cooling rate for larger wall thicknesses. The molding process is very energy-intensive, since the mold (die) and parts of the molding machine (steam chamber) have to be heated up and cooled down again for each cycle (shot).

Applications for EPP molded parts include insulated containers, heating insulation, reusable transport packaging and, increasingly, automotive parts such as bumper cores, sun visors, crash pads, toolboxes, or leisure articles.

The foam extrusion of particle foams is advantageous compared to particle foams produced in an autoclave. They are less expensive to produce, since one process step is eliminated (autoclave treatment), and in addition, a simple and fast color change is possible.

2.12.3 Thermoplastic Foam Casting

Foam structures can be created in injection-molded parts by adding blowing agents to a range of thermoplastics, including those with fillers such as glass or carbon fibers. Depending on the type and consistency, the blowing agents can be added to the granules at loadings of a few percent or pumped into the melt in the barrel in a liquid state under high pressure. The resulting molded parts have a foam structure at the core and a largely compact outer skin toward the surface (integral foam). The process is used, for example, to stabilize thick-walled areas in injection-molded parts with a supporting foam core, or to reduce the weight of a component without having to reduce the wall thickness.

2.12.4 MuCell Process

In the MuCell process, an inert gas (either nitrogen or carbon dioxide) is added to the thermoplastic melt and mixed in the melt cylinder to form a single-phase solution. When this single-phase solution is injected into the cavity of the injection or blow mold, or when it is extruded and exits the nozzle, the mixture experiences a pressure drop. Gas nucleates evenly over the component. A large number of tiny cells are created, which grow and fill a defined volume, resulting in a density reduction in the part.

2.12.5 Foamed Thermoplastics in Additive Manufacturing

Extrusion-based 3D printing processes can be used analogously to foam extrusion for the production of foamed thermoplastics. A chemical blowing agent is added to the thermoplastic filaments, which is first heat-activated in the extruder head (hotend) and then leads to the discharge of a foaming filament strand at the extruder nozzle. The degree of foaming can be adjusted within certain limits by the temperature and dwell time in the extruder head.

The foamed extrudate strands can be arranged on top of each other and in typical structures just as in the standard FLM process. However, it is important that the slicer takes the degree of foaming into account to avoid over-extrusion [Drucktipps3d 2019].

Analogous to the production of foamed profiles based on rubber (solidification of the foam by heating/vulcanization after extrusion), TPE (solidification of the foam by cooling after extrusion), and foamed gaskets (formed in-place foam gasketing, FIPFG) [MarketWatch 2021] using 2K polyurethane reactive compounds, robotic applicators can also be used in 3D printing for reactive extrusion (of pastes) or spray application [Fomicom 2021] to produce 3D foamed components.

2.13 Thermoplastics With Food Contact

In order to ensure the marketability of food contact materials and articles, it must be ensured that they neither release substances of health concern to the food, nor influence its taste or odor in an undesirable way. Thermoplastics used for the manufacture of products that come into contact with food must comply with strict limits with regard to the migration of ingredients. These are laid down for example in Regulation EU No. 10/2011 and in FDA guidelines (Code of Federal Regulations 21 CFR181.32 and Food Notification 190).

Thanks to their light weight and ease of processing, plastics for food contact can replace metals in a variety of applications. They are used, for example, in plants and equipment for meat, fish, and poultry processing, and dairy, bakery, and confectionery production. Typical thermoplastics with food contact approval and food applications include the following plastics:

- PET is resistant to oils, fats, alcohols, and dilute acids, and is widely used for microwave trays, beverage containers, and cups
- PE is the most widely used plastic and is highly resistant to acids, alkalis, oils, and greases. It is well-suited for packaging materials, freezer bags, and containers
- PVC has a high resistance to acids, alkalis, alcohols, and oil, but is preferentially used in the “non-food” sector

- PP has good barrier properties against grease and moisture, and is often used for containers and films
- PS is known as "Styrofoam®" when foamed up, and is often used for bases, trays and containers for fish, meat, fruit, vegetables, or to carry food in restaurants and snack bars
- PC and PAare used in water dispensers, drinking bottles, microwave dishes, and kitchen appliances. Due to the suspected hormone-like effect of monomeric bisphenol A, its use in the food and drinking-water sectors is now declining.

2.13.1 Additive Manufacturing of Food Contact Products Based on Thermoplastics

In principle, all thermoplastics that have food approval for injection molding or extrusion applications are also suitable as raw materials for the AM of components with food approval. In AM, it is important to note that they must be processable in additive processes.

Another problem area for both powder-based and extrusion-based processes is the AM of waterproof and airtight products, such as are important in the packaging sector. Tightness can, however, be achieved by various downstream processes. Of course, only processes that do not bring the materials into contact with problematic substances or change the materials to such an extent that suitability for use with food would have to be reassessed are useful.

Suitable processes include, for example, infiltration methods with approved materials, surface smoothing processes using chemically unproblematic smoothing agents (steam, alcohol, or acetone), as well as post-sintering processes in chemically inert media that are not problematic for the food industry (salt process [YouTube 2022]).

2.14 Bioplastics and Biocompatibility/Degradability

Bioplastic is the term used to describe all plastics produced on the basis of renewable raw materials (such as starch, oils, cellulose, bioethanol, etc.). In principle, almost all raw materials for common petroleum-based plastics can also be produced on the basis of bio-based raw materials through a combination of chemical, pyrolytic and/or fermentative processes, although in some cases at considerably higher costs and with a higher energy input. If the necessary energy can be obtained from regenerative energy sources, then a CO_2-neutral production of plastics is possible. In a narrower definition, bioplastics are referred to as those plastics that can be obtained by simple refining processes of natural raw materials (e.g. viscose from cellulose). Another definition refers to bio-based plastics as those that rot in nature or are compostable and thus do not accumulate in the long term.

> Recently, more and more bioplastics have been used in AM alongside PLA [Wieland 2018]. The common aim here is to improve the impact on the environment with regard to the CO_2 balance of the raw materials and with regard to the possible accumulation of plastics in the environment. Typical examples for bioplastics in AM are PLA as well as PA11 and, depending on the manufacturing process, PETG and PE.

2.15 Thermoplastics in Medical Applications

Plastics are among the most widely used materials in medicine. The advantages of plastic in this field are many. A suitable plastic does not absorb odors, and is resistant to water and other liquids. In addition, plastic can be molded in a variety of ways to adapt to specific conditions. Plastics are therefore used in medicine in almost all areas. The history of plastics in medicine goes back many years, and classic reusable materials were gradually replaced by the lighter plastics decades ago. A major role was played by the fact that plastics are so cheap to produce that they can be thrown away after a single use (significantly improving the hygiene and safety of use). Plastics in medicine can be very well adapted to the individual needs of doctors, nurses, and patients (see also hygiene articles, syringes, etc.). In addition, plastic is very inexpensive for apparatus construction in medical technology and production. It is easy to handle, mostly kind to the skin, and (if needed) “unbreakable”.

The suitability of plastics for medical applications depends to a large extent on the field of application. The following types can be distinguished:

- Medical devices and their housings
- Plastics that have contact with blood or body fluids
- Plastics with direct short- or long-term skin contact (orthotics, prostheses, and aligners)
- Plastics that remain in the body (e.g. dental applications, bone nails, plates, stents, sutures, breast implants, and bone implants)
- Plastics that only come into contact with body fluids for a short time, such as surgical instruments.

For all these applications, there are positive lists of plastics that have already been tested or test criteria that need to be met [Reinhart 2018].

Biocompatibility in medicine is a question of compatibility with the human organism. This must be certified in accordance with ISO 10993 parts 1–20. Certified biocompatibility for greater patient protection is for example attested by certified organizations such as TÜV Rheinland in Germany. By means of coatings, even plastics that are not inherently biocompatible can achieve biocompatibility.

2.15.1 Use of Additive Manufacturing with Thermoplastics in Medical Applications

Especially in medicine, 3D printing has provided enormous progress early on. Many applications of plastics have not only been significantly improved by this innovative technology, but also made much cheaper. However, the decisive factor here is (above all) the benefits for patients resulting from individually adapted medical products, which goes under the heading of "personalized medicine". 3D printing has already proven its practicality here, particularly in the following areas:

- Production of prostheses
- Production of orthoses
- Production of surgical instruments
- Dentures
- Tooth correction
- Hearing aids
- Implants
- Splints and bone nails
- Corrective lenses.

Internal as well as external prostheses, such as artificial hip joints, dentures, hearing aids, and arm and leg prostheses, can be highly customized using 3D printing. This provides patients with near-perfect replacements for damaged or missing body parts. Orthotics, or temporary skeletal supports, can likewise be produced in much better quality. Patients no longer have to settle for standard products, which are always a compromise in the end. Instead, 3D printing provides them with products that are tailored to their needs with millimetre precision.

In surgery, 3D printing has proven its worth not only in the production of precisely fitting implants. The production of precisely adapted surgical tools and aids has proven to be a boon for the attending physicians. In particular, difficult-to-implement surgeries can be performed much more precisely and gently for the patient with individually manufactured cutting, injection, and suturing tools.

Dental implants and hearing aids can be created directly on the basis of the patient's individual anatomical data using AM. This results in customized products that are quickly available and ideally adapted to the patient's needs.

In the field of prostheses and orthoses, plastics based on PA12, PA11, ABS, PETG, and TPU are in use. The polyamides are mainly used in powder-based processes, while ABS, PETG, and PC are mainly used in extrusion-based processes. An exception is TPU, which is used both as a powder and as a filament. Surgical instruments are mostly manufactured on the basis of PA12 and PEEK, and implants are made from PEEK and PLA. UV-curing thermosets are mainly used for dental prostheses and tooth correction.

2.15.2 Organs From the 3D Printer

One high-profile area of 3D printing is the production of true-to-scale, high-precision images of organs. Once the physician has spatial data of an organ using advanced imaging techniques (e.g. tomography), they can have a 3D printer produce a true-mold model of the organ. This gives them a much better impression of the individual structure of the organ. Surgeries can thus be better planned and healing success is more likely. Great efforts are being made to enable the production of implantable organs from the 3D printer. The idea is to use the printer to produce a collagen model of the organ and then "plant" it with stem cells. The goal is cheap, rejection-free, readily available replacement organs. This would suddenly make any form of organ donation superfluous. Even severely damaged hearts, lungs, livers, kidneys, and pancreases could be easily and completely replaced with this procedure. However, it will probably be several years before this technology is available – but many experts already agree that it will one day become the standard.

References

3Druck 2016: Hyrel 3D introduces Syringe Delivery System extruder series. 3Druck.com, 3 February 2016. *https://3druck.com/drucker-und-produkte/hyrel-3d-stellt-syringe-delivery-system-extruder-vor-5441698* (retrieved 4 April 2021)

3D-Filament 2021: Carbon fiber filament comparison. *https://3d-filament.info/testberichte/filament/kohlefaser-vergleich* (retrieved 4 April 2021)

3D HUBS 2021: Post-processing for SLS and MJF 3D printed parts. Knowledge base. 3D HUBS B.V. *https://www.hubs.com/knowledge-base/post-processing-sls-and-mjf-printed-parts* (retrieved 4 April 2021)

3Dnatives 2019: Anisoprint and the 3D printing of composites with continuous filaments. 3Dnatives.com, 3 July 2019. *https://www.3dnatives.com/de/anisoprint-interview-010820191* (retrieved 4 April 2021)

3DPrima 2021: Buy filament for your 3D printer cheap. *https://www.3dprima.com/de/filament-resin/filament/?page=2* (retrieved 4 April 2021)

3M 2021: 3D printing with 3M™ Dyneon™ PTFE. 3M Deutschland GmbH. *https://www.3mdeutschland.de/3M/de_DE/entwicklungs-und-spezialmaterialien/3d-druck* (retrieved 4 April 2021)

Abts 2020: G. Abts, Plastics knowledge for beginners. Fundamentals, properties and recycling of polymer materials. Carl Hanser Verlag, 2020

Acmite 2012: Global Fluoropolymer Market. Acmite Market Intelligence. April 2012. *https://www.acmite.com/market-reports/chemicals/global-fluoropolymer-market.html* (retrieved 4 April 2021)

AGVU 2021: Material fraction polypropylene (PP). AGVU – Arbeitsgemeinschaft Verpackung + Umwelt e. V.*https://www.agvu.de/de/polypropylen-pp-147* (retrieved 4 April 2021)

Alexandre et al. 2020: A. Alexandre, F. A. Cruz Sanchez, H. Boudaoud, M. Camargo & J. M. Pearce, Mechanical Properties of Direct Waste Printing of Polylactic Acid with Universal Pellets Extruder: Comparison to Fused Filament Fabrication on Open-Source Desktop Three-Dimensional Printers. In: 3D Printing and Additive Manufacturing, 7(5), 2020, pp. 237–247. *https://doi.org/10.1089/3dp.2019.0195*

Allied Market Research 2018: Carbon Fiber Market by Raw Material (PAN-based carbon fiber, Pitch-based & Rayon-based carbon fiber), Type (Continuous, Long, and Short), Form (Composite and Non-composite) and End-Use Industry (Aerospace & Defense, Sports/Leisure, Wind Turbines, Molding & Compounds, Automotive, Pressure Vessels, Civil Engineering, Marine, Pultrusion Misc., Misc. Consumer, Sailing/Yacht Building, and Others) – Global Opportunity Analysis and Industry Forecast, 2018–2025. April 2018. *https://www.alliedmarketresearch.com/carbon-fiber-market* (retrieved 4 April 2021)

AMFG 2020: In the spotlight: 5 electronic components that can benefit from 3D printing. AMFG – Autonomous Manufacturing. 20 January 2020. *https://amfg.ai/de/2020/01/26/im-rampenlicht-5-elektronische-komponenten-die-vom-3d-druck-profitieren-koennen* (retrieved 4 April 2021).

Anon 2017: Global flame retardant market projected to reach US$11.96 billion by 2025. In: Additives for Polymers, 2017(1), 10–11. *https://doi.org/10.1016/s0306-3747(17)30014-3* (retrieved 4 April 2021)

Anusci 2020: V. Anusci, DSM introduces Arnilene AM6002 (P) polypropylene powder for SLS. Boasts balance of properties, food contact approval in Europe. 3 September 2020. *https://www.3dprintingmedia.network/dsms-introduces-arnilene-am6002-p-polypropylene-powder-for-sls* (retrieved 4 April 2021)

ARBURG 2021: ARBURG Kunststoff-Freifformen (AKF) and freeformer. ARBURG GmbH + Co KG, *https://www.arburg.com/de/leistungsspektrum/additive-fertigung* (retrieved 4 April 2021)

Bastian & Hochrein 2018: M. Bastian & T. Hochrein, Einfärben von Kunststoffen. Product requirements – Process engineering – Test methodology (2nd edn). Carl Hanser Verlag, 2018

Beaumont 2015: Solid Density vs. Melt Density, Beaumont Technologies, Inc., April 15, 2015. *https://www.beaumontinc.com/solid-density-vs-melt-density* (retrieved 4 April 2021)

Boissonneault 2019: T. Boissonneault, DSM developing first commercial PBT powder for SLS 3D printing. 3D Printing Media Network, 28 March 2019. *https://www.3dprintingmedia.network/dsm-pbt-powder-sls-3d-printing* (retrieved 4 April 2021)

Bonnet 2009: M. Bonnet, Plastics in Engineering Applications. Understanding and reliable selection. Vieweg+Teubner Verlag, 2009, p. 55, *https://doi.org/10.1007/978-3-8348-9303-1*

Braun 2017: D. Braun, On the social history of plastics. In: Mitteilungen Gesellschaft Deutscher Chemiker, 25, 2017, pp. 268–286

Business Wire 2015: Coca-Cola produces world's first PET bottle made entirely from plants: Groundbreaking innovation unveiled at Milan World Expo. 3 June 2015. *https://www.businesswire.com/news/home/20150603006722/de* (retrieved 21 March 2021)

CAMPUS 2021: CAMPUS® – A material information system for the plastics industry. *https://www.campusplastics.com* (retrieved 21 March 2021)

Ceresana 2018: New study on the global market for thermoplastic elastomers. Plasticker, September 20, 2018. *https://plasticker.de/Kunststoff_News_33502_Ceresana_Neue_Studie_zum_Weltmarkt_fr_Thermoplastische_Elastomere* (retrieved 4 April 2021)

Ceresana 2019: Solid and malleable: Further growth expected for polystyrene and expandable polystyrene market. Chemie.de, 24 April 2019. *https://www.chemie.de/news/1160722/fest-und-formbar-weiteres-wachstum-fuer-den-markt-fuer-polystyrol-und-expandierbares-polystyrol-erwartet.html* (retrieved 4 April 2021)

Chemie 2021: Injection molding. *https://www.chemie.de/lexikon/Spritzgießen.html* (retrieved 4 April 2021)

Covestro 2019: Polycarbonate made from recycled plastics. *https://solutions.covestro.com/de/highlights/artikel/stories/2019/polycarbonat-aus-rezyklierten-kunststoffen* (retrieved 4 April 2021)

Covestro n.d.: V-T diagram of Makrolon M2405 measured on a Göttfert Rheograph 20, Covestro Deutschland AG

Denkena et al. 2016: B. Denkena, C. Schmidt & P. Weber, Automated Fiber Placement Head for Manufacturing of Innovative Aerospace Stiffening Structures. In: Procedia Manufacturing, 6, 2016, pp. 96–104. *https://doi.org/10.1016/j.promfg.2016.11.013*

Doi & Edwards 1988: M. Doi & S. F. Edwards, The Theory of Polymer Dynamics. Oxford Science Publications, Clarendon Press 1988

Dragon 2020: T. Dragon, Direct or Bowden Extruder. 8 February 2020. *https://www.makerline.de/direct-oder-bowden-extruder* (retrieved 31 January 2021)

Drucktipps3d 2019: Great foam beater: Colorfabb LW-PLA. 29 June 2019. *https://drucktipps3d.de/toller-schaumschlaeger-colorfabb-lw-pla* (retrieved 4 April 2021)

Dynisco 2019: Understanding the Value of Melt Volume Rate. AZoM.com, 8 May 2019. *https://www.azom.com/article.aspx?ArticleID=18004* (retrieved 4 April 2021)

Dyze Design 2021: Pulsar™ Large-Scale Pellet Extruder. *https://dyzedesign.com/pulsar-pellet-extruder* (retrieved 4 April 2021)

Ehrenstein & Bittmann 1997: G. W. Ehrenstein & E. Bittmann, Duroplastics. Curing – Testing – Properties. Carl Hanser Verlag, 1997

Ehrenstein & Pongratz 2007: G. W. Ehrenstein & S. Pongratz, Beständigkeit von Kunststoffen. Carl Hanser Verlag, 2007

Elias 2002: H.-G. Elias, Macromolecules Set (6th edn). Wiley-VCH Verlag, 2002. *https://doi.org/10.1002/9783527626557*

Elsner et al. 2012: P. Elsner, P. Eyerer & T. Hirth, T. (Eds.), DOMININGHAUS Kunststoffe. Properties and applications. Springer Verlag, 2012

Endres & Siebert-Raths 2009: H.-J. Endres & A. Siebert-Raths, Technical Biopolymers. Carl Hanser Verlag, 2009 (ISBN 978-3-446-41683-3)

EOS 2021: Polystyrene (PS) for Industrial Plastic 3D Printing. EOS GmbH *https://www.eos.info/en/additive-manufacturing/3d-printing-plastic/sls-polymer-materials/ps-polystyrene-primecast* (retrieved 9 August 2021)

EPFL 2021: Molecular Basis of the Structure and Behavior of Polymers. Part II: Chemistry and Structure of Macromolecules – Design of Polymer Chains. Ecole Polytechnique Fédérale de Lausanne, *https://web.archive.org/web/20040612172217/http://scgc.epfl.ch/load/cours_chim/cwandrey_part-2-1.pdf* (retrieved 4 April 2021)

EU-Recycling 2021: The valuable electronic waste plastics. *https://eu-recycling.com/Archive/14043* (retrieved 9 August 2021)

Fath 2019: A. Fath, Microplastics compact. Worth knowing for everyone. Springer Verlag, 2019

Flynt 2020: J. Flynt, Polypropylene Filament: Properties, Applications, and Best Brands. 23 May 2020. *https://3dinsider.com/polypropylene-filament* (retrieved 4 April 2021)

Fomicom 2021: Polyurethane foam for 3D printing. *https://www.fomicom.com/de/anwendungen/3d-drucken* (retrieved 4 April 2021)

Formlabs 2020: SLA vs. DLP: 3D printing technologies for resins in comparison. Formlabs GmbH, 2020. *https://formlabs.com/de/blog/vergleich-3d-druck-sla-dlp* (retrieved 31 January 2021)

Fortune Business Insights 2019: Polyethylene Market Size, Share & COVID-19 Impact Analysis, By Type (HDPE, LDPE, LLDPE), By End User (Packaging, Automotive, Infrastructure & Construction, Consumer Goods/Lifestyle, Healthcare & Pharmaceutical, Electrical & Electronics, Agriculture, Others), and Regional Forecast, 2020-2027. (2019). *https://www.fortunebusinessinsights.com/industry-reports/polyethylene-pe-market-101584* (retrieved 4 April 2021)

Fortune Business Insights 2023: High Performance Plastics Market to Worth $42.08 Billion by 2029. At a CAGR of 9.2 %. GlobeNewswire, Inc., 25 January 2023. *https://www.globenewswire.com/news-release/2023/01/25/2594894/0/en/High-Performance-Plastics-Market-to-Worth-42-08-Billion-by-2029-At-a-CAGR-of-9-2.html*

Frankland 2017: J. Frankland, Extrusion: The Role of Polymer Density in Screw Design. Plastics Technology, Gardner Business Media Inc, 25 September 2017. *https://www.ptonline.com/articles/extrusion-the-role-of-polymer-density-in-screw-design* (retrieved 4 April 2021)

Fraunhofer IFAM 2021: Metallic 3D printing using fused filament fabrication. Fraunhofer Institute for Manufacturing Technology and Advanced Materials, *https://www.ifam.fraunhofer.de/de/Ueberuns/Standorte/Dresden/Zellulare_metallische_Werkstoffe/fused-filament-fabrication.html* (retrieved 4 April 2021)

Fraunhofer IKTS 2021: Functional inks for applications of printed electronics and sensor technology with application of inkjet, aerosol and screen printing. Fraunhofer Institute for Ceramic Technologies and Systems, *https://www.ikts.fraunhofer.de/de/abteilungen/energiesysteme/werkstoffe_komponenten/werkstoffe_gedruckte_systeme/funktionstinten_fuer_inkjet_aersol_siebdruck_anwendungen.html* (retrieved 4 April 2021)

Fraunhofer INT 2019: Thermally conductive plastics. In: European Safety & Technology, December 2019, p. 102, Fraunhofer Institute for Scientific and Technical Trend Analysis

GDCh 2020: Polymers – All-rounders of the Modern Age. Magazine. Spektrum der Wissenschaft Verlagsgesellschaft mbH, Gesellschaft Deutscher Chemiker e.V., 2020. *ttps://www.gdch.de/fileadmin/downloads/Service_und_Informationen/Presse_OEffentlichkeitsarbeit/PDF/2020/2020_GDCh_Polymere_Web.pdf* (retrieved 31 January 2021)

Goodyear 1844: C. Goodyear, Improvement in India-Rubber Fabrics. Patent specification: US3462A (9 March 1844). United States Patent Office

Gordon 1976: M. Gordon, Book Review on: L. R. G. Treloar, The Physics of Rubber Elasticity (3rd edn). Clarendon Press, 1975. In: British Polymer Journal, 8(1), 1976, pp. 39–39. *https://doi.org/10.1002/pi.4980080107*

Goulding 2021: C . Goulding, SABIC And Additive Manufacturing. Fabbaloo, 16 February 2021. *https://www.fabbaloo.com/news/sabic-and-additive-manufacturing* (retrieved 4 April 2021)

Grand View Research 2016: Polycarbonate Market Size To Reach $25.37 Billion By 2024. 2016. Grand View Research, Inc., *https://www.grandviewresearch.com/press-release/global-polycarbonate-market* (retrieved 4 April 2021)

Grellmann & Seidler 2015: W. Grellmann & S. Seidler (Eds.), Plastics Testing (3rd edn). Carl Hanser Verlag, 2015, p. 601/602 (ISBN 978-3-446-44350-1)

Gu et al. 2020: Y. Gu, J. Zhao & J. A. Johnson, Polymer Networks: From Plastics and Gels to Porous Frameworks. In: Applied Chemistry, Volume 59, Issue 13, 23 March 2020, pp. 5022–5049. *https://doi.org/10.1002/anie.201902900*

Gutberlet 2020: A.-L. Gutberlet, Will batteries be 3D printed in the future? Leistungs-elektronik, 6 February 2020. *https://www.leistungselektronik.de/werden-batterien-in-zukunft-3d-gedruckt-a-1043231* (retrieved 4 April 2021)

Hanselmann 2006: R. Hanselmann, Ceiling temperature, RD-03-00810. In: F. Böckler et al., RÖMPP Online-Enzyklopädie. Georg Thieme Verlag. 2006. *https://roempp.thieme.de/lexicon/RD-03-00810*

Hellerich et al. 2004: W. Hellerich, G. Harsch & S. Haenle, Werkstoffführer Kunststoffe. Properties, tests, characteristic values. Carl Hanser Verlag, 2004

Hopmann & Michaeli 2017: C. Hopmann & W. Michaeli, Einführung in die Kunststoffverarbeitung (8th edn). Carl Hanser Verlag, 2017, pp. 243–245

IHS Markit 2018: Challenges in the Global Methyl Methacrylate Market. 6 September 2018. *https://ihsmarkit.com/research-analysis/challenges-in-the-global-methyl-methacrylate-market.html* (retrieved 4 April 2021)

Industrieverband Klebstoffe 2021: 2.3 Plastics. Entry in "Leitfaden Kleben – aber richtig". Industrieverband Klebstoffe e.V., *https://leitfaden.klebstoffe.com/2-3-kunststoff* (retrieved 4 April 2021)

Induux 2021: Plastic welding. Processes, applications, manufacturers. Induux International GmbH wiki. *https://wiki.induux.de/Kunststoffschweißen* (retrieved 4 April 2021).

Jose et al. 2016: R. R. Jose, M. J. Rodriguez, T. A. Dixon, F. Omenetto & D. L. Kaplan, Evolution of Bioinks and Additive Manufacturing Technologies for 3D Bioprinting. In: ACS Biomaterials Science and Engineering, 2(10), 2016, 1662–1678. *https://doi.org/10.1021/acsbiomaterials.6b00088*

Kaiser 2006: W. Kaiser, Plastics Chemistry for Engineers, Carl Hanser Verlag, 2006, pp. 99–100

Kloke 2018: A. Kloke, Droplets every millisecond. In: Plastics, 11, 2018, pp. 33–39

Kohl & Pešík 2016: O. Kohl & L. Pešík, Testing Equipment Location on Springing Base Plate. In: Applied Mechanics and Materials, 827, 2016, pp. 95–98*https://doi.org/10.4028/www.scientific.net/amm.827.95*

Kossmehl 2010: G. Kossmehl, Bakelite conquers the world. In: Kunststoffe 1/2010, pp. 10–13. *https://res.cloudinary.com/sternwald-systems/raw/upload/v1/hugoprd/ARTIKEL_ATTACH/00240F12_4156B148CED5/aae4edde49679b531b42722b6b4afd976942dd88/KU_2010_01_Bakelit-erobert-die-Welt.pdf* (retrieved 31 January 2021)

Kunststoffe 2021: Injection compression molding. *https://www.kunststoffe.de/a/article/article-254107* (retrieved 4 April 2021)

Leapfrog 2021: 3D Printing with Polypropylene. *https://www.lpfrg.com/guides/3d-printing-polypropylene* (retrieved 4 April 2021)

Liang 1995: J. Liang, Effect of the Die Angle on the Extrusion Swell of Rubber Compound. In: Journal of Materials Processing Technology 52 (1995) 2–4, pp. 207–212

Lüpke 2015a: T. Lüpke, Fundamentals of Mechanical Behavior. In: W. Grellmann & S. Seidler (Eds.): Plastics Testing. 3rd ed. Carl Hanser Verlag, 2015. pp. 92–93 (ISBN 978-3-446-44350-1)

Lüpke 2015b: T. Lüpke, Fundamentals of Mechanical Behavior. In: W. Grellmann & S. Seidler (Eds.): Plastics Testing (3rd edn). Carl Hanser Verlag, 2015. pp. 96–110 (ISBN 978-3-446-44350-1)

Maier & Schiller 2016: R.-D. Maier & M. Schiller, Handbuch Kunststoff Additive (4th edn). Carl Hanser Verlag, 2016

MarketWatch 2021: Formed in Place Foam Gaskets (FIPFG). Market 2021 – Share, Size, Global Future Trend, Segmentation, Business Growth, Top Key Players Analysis, Industry, Opportunities and Forecast to 2026. 15 January 2021. *https://www.marketwatch.com/press-release/formed-in-place-foam-gaskets-fipfg-market-2021-share-size-global-future-trend-segmentation-business-growth-top-key-players-analysis-industry-opportunities-and-forecast-to-2026-2021-01-15* (retrieved 4 April 2021)

Markforged 2021: The "Digital Forge" platform. *https://markforged.com/digital-forge* (retrieved 4 April 2021)

Maschinenbau-Wissen 2021a: Blown film extrusion – Mode of operation and line design. *https://www.maschinenbau-wissen.de/skript3/werkstofftechnik/kunststoffe/426-blasfolienextrusion* (retrieved 4 April 2021)

Maschinenbau-Wissen 2021b: Extrusion – Process and mode of operation. *https://www.maschinenbau-wissen.de/skript3/werkstofftechnik/kunststoffe/392-extrusion* (retrieved 4 April 2021)

Maschinenbau-Wissen 2021c: Multicomponent injection molding. *https://www.maschinenbau-wissen.de/skript3/werkstofftechnik/kunststoffe/414-mehrkomponenten-spritzguss* (retrieved 4 April 2021)

Maschinenbau-Wissen 2021d: Thermoplastic foam molding (TSG). *https://www.maschinenbau-wissen.de/skript3/werkstofftechnik/kunststoffe/400-schaumspritzgiessen* (retrieved 4 April 2021)

Mason & Gardiner 2020: H. Mason & G. Gardiner, 3D printing with continuous fiber: A landscape. CompositesWorld, 15 October 2020. *https://www.compositesworld.com/articles/3d-printing-with-continuous-fiber-a-landscape* (retrieved 4 April 2021)

Menges et al. 2002: G. Menges, W. Michaeli, E. Schmachtenberg & E. Haberstroh, Werkstoffkunde Kunststoffe (5th edn). Carl Hanser Verlag, 2002

Meyer 1987: W. H. Meyer, Polymer alloys. In: Chemie in unserer Zeit, 21st volume, 1987, No. 2, pp. 59–69 (ISSN: 0009-2851)

Minch 1999: M. J. Minch, Book Review on: G. A. Jeffrey, An Introduction to Hydrogen Bonding. Oxford University Press, 1997. In: Journal of Chemical Education, 76(6), 1999, p. 759

Montarnal et al. 2011: D. Montarnal, M.Capelot, F.Tournilhac & L.Leibler, Silica-Like Malleable Materials from Permanent Organic Networks. In: Science. 334, November 2011 (6058): 965–968

Mordor 2021: Polyamides Market – Growth, Trends, and Forecasts (2020-2025). Mordor Intelligence, *https://www.mordorintelligence.com/industry-reports/polyamides-market* (retrieved 4 April 2021)

Niederstadt 2017: G. Niederstadt, Economical and ecological lightweight construction with fiber-reinforced polymers. Design, calculation and qualification (2nd edn). Expert Verlag, 2017

Noriega & Rauwendaal 2010: M. Noriega & C. Rauwendaal, Troubleshooting the Extrusion Process. A systematic approach to solving plastic extrusion problems. Carl Hanser Verlag 2010

Plastics Insight 2021: Polyacetals Production, Pricing and Market Demand. *https://www.plasticsinsight.com/resin-intelligence/resin-prices/polyacetals* (retrieved 4 April 2021)

Pockat et al. 2010: G. R. Pockat et al., Myoglobin Blooming Agent Containing Shrink Films, Packages and Methods for Packaging. Patent Specification: US 2010/0266795 A1. Patent owner: Amcor Flexibles North America Inc., 21 October 2010

P&S Intelligence 2021: Polypropylene Market Research Report: By Type (Homopolymer, Copolymer, Random, Block), Application (Injection Molding, Film & Sheet, Raffia, Fiber, Blow Molding), End Use (Packaging, Electrical, Medical, Consumer Goods, Automotive, Construction) – Global Industry Analysis and Growth Forecast to 2030. September 2021. *https://www.psmarketresearch.com/market-analysis/polypropylene-market-report* (retrieved 4 April 2021)

Rajan et al. 2019: A. J. Rajan et al., Influence of Vapor Smoothing Process Parameters on Fused Deposition Modeling Parts Surface Roughness at Different Build Orientation. In: Materials Today: Proceedings, 22, 2019, 2772–2778. *https://doi.org/10.1016/j.matpr.2020.03.408*

Rauwendaal et al. 2014: C. Rauwendaal et al., Polymer Extrusion. Carl Hanser Verlag, 2014

Recycling Portal 2016: Polyamide-6 recyclables. Marine waste by depolymerization to nylon. 2016. *https://recyclingportal.eu/Archive/27888* (retrieved 4 April 2021)

Reinhart 2018: P. Reinhart, Plastics in medicine: materials, application and processing. DeviceMed, 16 June 2018. *https://www.devicemed.de/kunststoffe-in-der-medizin-materialien-anwendung-und-verarbeitung-a-725299* (retrieved 4 April 2021)

Reinhardt & Anhalt 2015: H. Reinhardt & H.-P. Anhalt, Use of a conductive polymer material for medical and orthopedic applications. Patent specification: DE102011101579B4 (3 May 2015). Patent holder: Otto Bock Healthcare GmbH. *https://patents.google.com/patent/DE102011101579B4/de* (retrieved 4 April 2021)

Richards 2007: R. B. Richards, Polyethylene-structure, crystallinity and properties. In: Journal of Applied Chemistry. 1, no. 8, May 4, 2007, pp. 370–376 (DOI: 10.1002/jctb.5010010812)

Ritzkowski 2019: M. Ritzkowski, Bio-Plastics Europe. In: Detritus, Volume 8, December 2019 (DOI: 10.31025/2611-4135/2019.13894)

Röhm 2021: PLEXIGLAS® solid sheets. Environmental product declaration according to ISO 14025 and EN 15804. Röhm GmbH, *https://www.plexiglas.de/files/plexiglas-content/pdf/sonstige-dokumente/Umwelterkl%C3 %A4rung-EPD-PLEXIGLAS-Massivplatten-de.pdf* (retrieved 4 April 2021)

Schäfer & Rehage 1969: E. E. Schäfer & G. Rehage, On the entropy and energy elasticity of cross-linked high polymers. In: Macromolecular Chemistry and Physics, Volume 123, Issue 1, 28 April 1969, pp. 91–118

Scheirs & Kaminsky 2006: J. Scheirs & W. Kaminsky, Feedstock Recycling and Pyrolysis of Waste Plastics. Converting Waste Plastics into Diesel and Other Fuels. John Wiley & Sons, 2006

Schönhals 2015: A. Schönhals, Electrical and Dielectric Properties. In: W. Grellmann & S. Seidler, Plastics Testing. 3rd ed. Carl Hanser Verlag, 2015, pp. 357–398

Sculpteo 2021: Jet Fusion PP 3D Printing Material. *https://www.sculpteo.com/en/materials/jet-fusion-material/jet-fusion-pp* (retrieved 4 April 2021)

Sichina 2007: W. J. Sichina, Characterization of Polymers Using TGA. 2007. *https://www.eng.uc.edu/~beaucag/Classes/Characterization/DSCParts/TGA_Notes.pdf* (retrieved 4 April 2021)

Sole & Solc 1982: K. Sole & K. Solc, Polymer Compatibility and Incompatibility. Principles and Practice. Volume 2. Midland Macromolecular Meeting 1980. Harwood Academic 1982.

Statista 2015: Polyethylene terephthalate (PET) production worldwide in 2014 and 2020. December 2015. *https://www.statista.com/statistics/650191/global-polyethylene-terephthalate-production-outlook* (retrieved 4 April 2021)

Statista 2018a: Polyvinyl chloride (PVC) production volume worldwide in 2018 and 2025. October 2018. *https://www.statista.com/statistics/720296/global-polyvinyl-chloride-market-size-in-tons* (retrieved 30 March 2021)

Statista, 2018b: Most-used 3D printing technologies in 2018. *https://de.statista.com/statistik/daten/studie/760408/umfrage/meistgenutzte-3d-druck-technologie* (retrieved 31 January 2021)

Statista 2021: Annual production of plastics worldwide from 1950 to 2020. December 2021. *https://www.statista.com/statistics/282732/global-production-of-plastics-since-1950* (retrieved 31 January 2021)

Theias 2021: M. Theias, What wood you print? Wood 3D Printer: How to 3D Print Wood. All3DP, 20 July 2021. *https://all3dp.com/2/wood-3d-printer-all-you-need-to-know-about-wood-3d-printing-2* (retrieved 4 April 2021)

Thielen et al. 2006: M. Thielen, K. Hartwig & P. Gust, Blow molding of hollow plastic parts. Processes, machines, tools. Carl Hanser Verlag, 2006

Tieke 2014: B. Tieke, Macromolecular Chemistry. An introduction (3rd edn). Wiley-VCH Publishers, 2014

Tindera 2017: M. Tindera, Out Of Silicon Valley, A Billion-Dollar Orthodontics Business Built With Plastic And Patents. 25 April 2017. *https://www.forbes.com/sites/michelatindera/2017/04/25/out-of-silicon-valley-a-billion-dollar-orthodontics-business-built-with-plastic-and-patents/#363c86f430c2* (retrieved 4 April 2021)

Trinkle & Friedrich 2001: S. Trinkle & C. Friedrich, Van Gurp-Palmen-plot: a way to characterize polydispersity of linear polymers. In: Rheol. Acta 40, 322-328 (2001). *https://doi.org/10.1007/s003970000137*

Tubbesing 2018: K. Tubbesing, Pellet extrusion lowers large-scale 3D printing costs. 23 May 2018. *https://www.hannovermesse.de/de/news/news-fachartikel/pellet-extrusion-senkt-die-3d-grossdruckkosten* (retrieved 4 April 2021)

Ügdüler 2020: S. Ügdüler et al., Towards closed-loop recycling of multilayer and colored PET plastic waste by alkaline hydrolysis. In: Green Chemistry, 22(16), 2020, pp. 5376–5394. *https://doi.org/10.1039/d0gc00894j*

UMSICHT 2019: Plastics for textile applications. Compounds for additive manufacturing, geotextiles and wearables are created at Fraunhofer UMSICHT. *Fraunhofer Institute for Environmental, Safety and Energy Technology.* Press release, 24 September 2019. *https://www.umsicht.fraunhofer.de/de/presse-medien/pressemitteilungen/2019/k2019.html* (retrieved 4 April 2021)

van Gurp & Palmen 1998: M. van Gurp & J. Palmen, Time-temperature Superposition for Polymeric Blends. In: Rheology Bulletin 67(1), 5–8 (1998)

VinylPlus 2021: Recycling. VinylPlus Deutschland e.V., *https://www.vinylplus.de/alles-ueber-pvc/recycling* (retrieved 4 April 2021)

Vogel et al. 2020: J. Vogel, F. Krüger & M. Fabian, Chemical Recycling. Umweltbundesamt, 2020. *https://www.umweltbundesamt.de/publikationen/chemisches-recycling* (retrieved 4 April 2021)

Vrentas et al. 1985: J. S. Vrentas, J. L. Duda & H. C. Ling, Free-Volume Theories for Self-Diffusion in Polymer-Solvent Systems. I. Conceptual Differences in Theories. In: Journal of Polymer Science. Part A-2, Polymer Physics, 23(2), 1985, 275–288. *https://doi.org/10.1002/pol.1985.180230204*

Waßner & Laun 2000: E. Waßner & H. M. Laun, Extrusionsmassen aus rheologischer Sicht. Behavior in the die. In: E. Brunnhofer, E. & R. Sander (Eds.), The Extrusion Die. Guide to design and handling. VDI-Verlag, 2000, pp. 9–30

Weber 1998: B. Weber, Environmental Research in Municipal Policy – A Bridge between Science and Implementation. In: M. Schmidt & U. Höpfner, 20 Jahre ifeu-Institut. Commitment to the Environment between Science and Politics. Vieweg+Teubner Verlag, 1998, pp. 3–7. *https://doi.org/10.1007/978-3-322-83139-2_1* (retrieved 4 April 2021)

Weiß 2005: C. Weiß, Kunststoffe in der Elektronik. A handbook for practical use. E. G. Leuze Verlag 2005. *https://www.leuze-verlag.de/shop/fachbuecher-elektronikfertigung/kunststoffe-in-der-elektronik-detail* (retrieved 4 April 2021)

Wieland 2018: H. Wieland, Additive Manufacturing from Bioplastics. Innovation Series Biobased Plastics, Osnabrück University of Applied Sciences. Workshop: State of the art and market potentials, Lingen, 17 October 2018. *https://www.hs-osnabrueck.de/fileadmin/HSOS/Homepages/Innovationsreihe-Biokunststoffe/Praesentationen/Additive_Fertigung_aus_Biokunst stoffen_-_Hansjoerg_Wieland_3N_Kompetenzze pdf* (retrieved 4 April 2021)

Wikipedia 2021: High performance plastics. *https://de.wikipedia.org/wiki/Hochleistungskunststoffe* (retrieved 4 April 2021)

Willberg & Rädel 2017: C. Willberg & M. Rädel, Understanding damage – peridynamic modeling of cracks. In: Innovation Report 2017. Institute of Composite Structures and Adaptive Systems, German Aerospace Center (DLR)

Williams et al. 1955: M. L. Williams, R. F. Landel & J. D. Ferry, The Temperature Dependence of Relaxation Mechanisms in Amorphous Polymers and Other Glass-forming Liquids. In: Journal of the American Chemical Society, 1955 (DOI: 10.1021/ja01619a008)

Wohlers 2020: 3D Printing and Additive Manufacturing. Global State of the Industry [Wohlers Report]. 2020 (ISBN 978-0-9913332-6-4)

YouTube 2022: Remelting 3D printed thermoplastics (the salt method). *https://www.youtube.com/watch?v=nRLJ4ylGTFc* (retrieved 24 February 2022)

Zertifizierte-altreifenentsorger 2021: Dispose of scrap tires without worrying. *https://zertifizierte-altreifenent sorger.de/wp-content/uploads/zare_imagebrosch%C3 %BCre_210x297_aLo_SCREEN.pdf* (retrieved 31 January 2021)

3 Special Features of Thermoplastic Processing in Additive Manufacturing

Thermoplastics are plastics that soften when heated to temperatures above their melting point and then begin to flow. This behavior is technically exploited in additive manufacturing to process thermoplastics into the desired part using a layer-by-layer melt deposition process such as fused layer modeling (FLM) or selective laser sintering (SLS). By definition, the melting point of thermoplastics is well above their service temperature. After cooling below the melting temperature, the printed layers solidify and the resulting part is ready for use.

The processing conditions of additive manufacturing with thermoplastics differ significantly from those of classic thermoplastic processing methods with regard to processing temperatures, processing times, thermal and mechanical stress, pressure and cooling rate (see Section 3.1.1). Accordingly, the properties of parts of the same design, which have been manufactured using classical or additive manufacturing methods, can differ very significantly from one another.

In order for additive manufacturing to find widespread use as a new manufacturing technology and not remain in the prototype niche, it must achieve a level of process technology maturity that allows it to reliably produce products of at least equal quality. To make additive manufacturing based on thermoplastics economically viable in an industrial environment, a number of prerequisites must be met. This chapter deals with basic additive production processes, materials, process and design requirements, and measures to enhance process reliability and achieve quality assurance. In Section 3.1 the extrusion-based processes are discussed, and in Section 3.2 the powder sintering processes.

3.1 Extrusion-based Processes (Material Extrusion, MEX)

In this section, the special features of extrusion-based processes (Material Extrusion, MEX) are presented. At first glance, the number of thermoplastic materials available for extrusion-based processes appears to be very large. In fact, however, extrudability is only part of the requirements for a material to perform well in a MEX process for manufacturing functional parts. The basic extrudability of thermoplastics is almost always given, with the exception of a few high-performance and/or high-temperature materials such as UHMWPE (see Section 2.6.8.1), PTFE (see Section 2.6.6) and a few polyimides (see Section 2.6.6.2). For these materials, the melt viscosity above the melting point is so high that they can be processed without decomposition only in special powder sintering processes or from solution. Many materials that can be processed well in classic thermoplastic processes such as injection molding (see Section 2.7.1) and extrusion (see Section 2.7.2) show, after processing in additive manufacturing by means of extrusion processes, only a fraction of the expected mechanical properties depending on the build direction (mostly in the z-direction = build-up direction). One of the reasons for this is the basic layer-by-layer build-up process of extrusion-based additive processes.

The MEX process features a rather untypical process profile for thermoplastics by combining a high extrusion temperature and a low part temperature, with only a low contact pressure and short contact time when welding the deposited layers onto each other. The process must be understood and optimized for each material. It is therefore necessary to take the material- and process-dependent product properties into account when designing a part to be produced by a given additive manufacturing method.

The quality of a part from an extrusion-based additive process depends on various conditions:

- Material properties → softening temperature (T_m, T_g), melt viscosity, properties dependent on cooling rate such as crystallization rate, degree of crystallization, crystallite sizes, temperature- and pressure-dependent free volume changes (shrinkage, warping, tension build up) etc.
- Printing conditions → nozzle geometry, nozzle temperature, extrusion speed, extrusion pressure, build chamber temperature, printing- platform temperature, etc.
- Part design → wall thicknesses, part size, geometric complexity, filler structures, overhangs, support structures, platform adhesive support, etc.
- Process settings (G-code) → layer thickness, layer time, travel speed, retraction speed, retraction length, path logic, build time, etc.

All of these parameters must be coordinated with each other, taking advantage of the geometric degrees of freedom of the additive manufacturing process, in order to achieve an optimum result. There are some common rules and typical problems listed and explained below. However, the process of gathering knowledge and understanding the complexities of the MEX process and implementing measures for optimization is far from complete.

The main extrusion-based processes (MEX) differ fundamentally in terms of the feedstock. In the filament layer deposition (FLM) process, a prefabricated semi-finished product such as a filament is used, while in pellet (or granulate) extrusion a low-cost standard thermoplastic pellet can be used as the feedstock.

3.1.1 Filament Extruder

When a filament is used as the feedstock, the throughput capacity of a print head is controlled by its ability to melt the thermoplastic and push it through the extrusion nozzle. This depends largely on the amount of heat/time that can be transferred from the extruder heating zone (hotend) to the thermoplastic filament and the effective feed force of the extruder.

The greater the effective heat input and thus achievable temperature of the melt is, the lower is the viscosity of the molten thermoplastic. The feed force determines in interaction with the nozzle geometry (nozzle angle and nozzle diameter) and the viscosity of the melt at the nozzle opening the maximum throughput rate. The maximum feed force of a filament extruder can only be controlled within narrow limits [Nienhaus 2020]. If the force transmitted to the filament becomes too high, abrasion and slippage effects occur, or the filament buckles (buckling/kinking) and the force is partially transferred into the extruder walls (Figure 3.3). A large heat exchange surface increases the heat transfer capacity. If the heat exchange surface becomes too large, also the pressure drop across the walls increases according to the equation of Hagen Poiseuille. If the residence time of the thermoplastic in the hotend becomes too large, then the thermoplastic can be thermally damaged.

As a compromise from these specifications, filaments with a diameter of 1.7–3 mm have become the preferred feedstocks. These are brought to the target temperature in short heated tube sections with filament diameter and to the desired extrusion diameter through a tapered die. Typical targeted residence times in the hotend are < 30 s. For particularly high-melting and high-viscosity products such as PEI or PEKK, the throughput rate is therefore rather low, while low-melting, low-viscosity products such as PLA and PETG can be extruded at particularly high throughput rates. The products mentioned have in common that they have a high modulus featuring hard filaments that can convey a high feed force. Lower modulus products such as semicrystalline TPU should actually be eminently extrudable at high volume rates due to their pronounced viscosity reduction in the melt. However, particularly soft TPU filaments buckle easily and can therefore transfer significantly less force to the melt, which is why the extrusion rate drops.

The fundamentally limited force transmission in filament extruders is also one of the main reasons why short and long fiber-filled filaments can only be extruded to a limited extent through die diameters in the range greater than or equal to the fiber length. The fiber fill, if not ideally aligned to the direction of extrusion, can cause significant pressure build-up upstream of the nozzle due to “nesting”, thus significantly reducing the potential volume throughput. Filament extruders are typically designed for throughputs < 100 g/h and are mainly used where the printing of high-resolution parts is important. Exceptions are specially designed high-throughput extruders where the more costly material handling of a pellet extruder is to be avoided or part resolution is critical. These are equipped with special filament advancement systems (e.g. "Quad Pinch"), whereby throughputs of up to 1 kg/h of easy-to-extrude materials such as PLA and PETG can be realized [Dyze 2022]. However, here the nozzle diameter is limited to a minimum of 1 mm, as otherwise the pressure build-up would still become too high.

3.1.2 Pellet Extruder

The extrusion of larger material volumes is made possible by the pellet extruder. Single-screw extruders are usually specially designed for use in additive manufacturing. The choice of the right extruder size and extruder design is based on many different factors and has a decisive influence on the achievable product quality and productivity of additive manufacturing (see also Figure 4.4). The size of an extruder is described by three main factors:

- Screw diameter in mm
- Screw length as a multiple of the diameter
- Drive power in kW

A type 16/15 extruder (2 kW) thus has a screw diameter of 16 mm and a screw length of 240 mm (16 mm × 15). The drive power in this case is 2 kW. The screw diameter specified refers to the outer diameter of the screw (flight diameter). Common dimensions for single-screw extruders for additive manufacturing are between 8 mm and 45 mm, but larger systems also exist. The throughput (e.g. in kg/h) that can be achieved by an extruder with a specific screw diameter cannot be answered in general terms, as the achievable throughput always depends on many other parameters, such as the screw design, the material to be processed and the heating power.

On deep-cut screws with a large free volume (so-called throughput-intensive screws), higher throughput rates can be achieved with low pressure build-up than with shallow-cut screws with lower free volume (so-called pressure-intensive screws). With regard to screw length, it is true to say that longer screw systems improve the homogenizing performance and mixing performance of the extruder, but can lead to molecular weight degradation in sensitive materials due to the corresponding longer residence times. As screw volume (length/depth) increases, so does the time and material required to "purge" the system when changing materials. The screw depth controls the shape of the material that can be used as a feedstock and should always be larger than the diameter of the thermoplastic pellets to be processed.

Another important aspect in extruder design is the maximum screw speed. Small extruders can often be run at a higher speed than large extruders. A high speed in combination with a low screw depth not only reduces the residence time in the extruder it also improves the heating capacity of an extruder due to increasing the temperature by friction of the material on the extruder wall while at the same time increasing the shear rate and thus further reducing the viscosity.

Another aspect that must be taken into account when selecting or dimensioning the extruder is the drive power. The amount of energy required to heat up the material to processing temperature can be used as a first indication for a sensible value of the minimum required drive power. For example, if a polycarbonate feedstock is to be heated from ambient temperature to 270 °C, then a desired throughput of 2 kg/h would require approximately 250 W of pure thermal power. Due to the losses that occur in an extruder and the additional energy required to build up pressure, corresponding extruders are equipped with three to five times this power.

The residence time is the time that the thermoplastic pellets remain in the system from their entry into the extruder to their exit as a melt from the extruder. The longer the residence time, the more the material will be damaged by thermal degradation. In principle, it is advisable to extrude with the shortest possible residence times, unless the required homogenizing capacity or heating distance or heat transfer is not sufficient.

The residence time of a thermoplastic pellet in a screw extruder tends to be higher than that of filaments in filament extruders. The average residence time corresponds to the quotient of the extruder volume filled with melt and the volume flow of the melt. Since a certain amount of back-mixing takes place compared to filament extruders, at least twice the theoretical residence time can be assumed to be safe. Dwell times of approx. 60 s are quite typical. The higher the melt temperatures, the shorter the residence times should be. The processing instructions of the material manufacturer are an important guide here.

Compared to filament extruders, pellet extruders have significantly poorer process behavior when the melt flow is interrupted (melt flow stop behavior). Clean retraction of the material is difficult to achieve with low-viscosity melts in pellet extruders that tend to have large nozzle diameters. Many materials tend to drip and pull threads here. Different manufacturers meet these challenges in different ways, for example by using special proprietary drop stop systems.

Another weak point of pellet extruders is the uniformity of the material extrusion. Gaps, air bubbles, etc., cannot always be completely avoided in simple pellet extruders. Extruders with high volume throughput therefore sometimes use melt pumps upstream of the extrusion nozzle/die to improve extrudate uniformity [Shaik & Schuster 2021].

When selecting or designing an extruder, a number of aspects must be taken into account, all of which are interrelated and must always be considered with regard to the input material. A universal extruder that meets all requirements in the same quality for a wide range of materials therefore does not exist for use in additive manufacturing. Every extruder design is a compromise and must therefore be considered specifically for the application.

3.1.3 Thermoplastic Properties

The selection of the right process conditions in extrusion-based processes depends on the properties of the thermoplastic or the compound to be processed. Various variables have to be taken into account which have a direct influence on the choice of process conditions and the resulting product properties:

- Melting temperature
- Heat capacity
- Melt viscosity
- Solidification temperature/solidification kinetics
- Volume shrinkage

3.1.3.1 Melting Temperature

The melting temperature of a thermoplastic is the temperature at which a thermoplastic begins to flow. For amorphous thermoplastics, this is the glass transition temperature; for semi-crystalline thermoplastics, this is the temperature at which the crystalline segments begin to melt. Every polymer has at least one glass transition temperature. Semi-crystalline polymers have at least one glass transition temperature and one melting temperature. The melting temperature of crystalline segments is always higher than the glass transition temperature of the amorphous segments.

The glass transition temperature and the melting temperature can be determined either by calorimetric methods such as differential scanning calorimetry (DSC, see Section 2.5.3) or by thermomechanical methods such as dynamic mechanical analysis (DMA, see Section 2.5.2). From a suitable DSC measurement, in addition to the melting point, the specific heat capacity of a polymer can be read out over the temperature range up to the processing temperature. In addition to glass transition temperature and melting temperature, DMA also provides initial information on the mechanical properties (modulus, damping) of polymers in a temperature range from approx. −100 °C up to the melting point.

The glass transition temperature and melting temperature are available in tabular form for most common and commercial thermoplastics from various reference works [Menges et al. 2011].

3.1.3.2 Heat Capacity

The specific heat capacity of a thermoplastic is important in determining the energy required to heat a thermoplastic in its solid state, to the melt state at a desired processing temperature. This energy must be introduced by the extruder. The heat capacity of a polymer is temperature-dependent and changes depending on the state of the polymer starting in the glassy state up to the melt temperature and then in the melt up to the decomposition temperature. The specific heat capacity in kJ/kgK can be determined via a calibrated DSC measurement can be quantified from −100 °C up to the decomposition temperature [Hobbs 2002]. The values of typical polymers do not differ dramatically. The specific heat capacity of thermoplastics in the solid state is between 1 and 2 kJ/kgK (for amorphous thermoplastics it is nearer 1 kJ/kgK, for highly crystalline thermoplastics it is nearer 2 kJ/kgK) and in the melt between 2 and 3 kJ/kgK. Specific heat capacities can also be looked up in the technical literature or are often included with commercial products as information for processing [Polymer Database 2022].

3.1.3.3 Melt Viscosity/Melt Volume Flow Rate

A widely used and simple method for evaluating the processing behavior of polymers in the melt is the measurement of the melt volume-flow rate (MVR), formerly often colloquially referred to as the melt volume index (MVI). The MVR is used to characterize the flow behavior of a thermoplastic under "constant" pressure and temperature conditions at a given nozzle geometry (see Section 2.5.5). The temperatures, pressures, nozzle geometries, shear and flow rates applied in the MVR method fit well with the typical temperatures, pressures and flow rates in extrusion-based additive manufacturing, at least for filament-based processes, which is why the given values can be a good orientation for finding suitable process parameters. As a first indication for a relevant minimum processing temperature with a sufficient extrusion rate, a flow rate in cm^3/10 min at a force of 10 kg (typical forces as they can also act on the filament in filament 3D printing) according to DIN EN ISO 1133 and ASTM D 1238 should be greater than 5 cm^3/10 min.

3.1.3.4 Solidification Temperature/Solidification Kinetics

For the solidification temperature/solidification kinetics of polymers, for example after melt strand deposition onto the last layer in an (un)heated build chamber, a calorimetric method such as DSC (see Section 2.5.3) is the method of choice to predict the solidification behavior of polymers. DSC allows reproduction of the temperature drop that a melt strand experiences downstream of the nozzle when applied to the previous thermoplastic layer. In an oscillating DSC, each temperature point can be approached sufficiently quickly, for example, to map the solidification or crystallization kinetics of the thermoplastic.

Plastics that exhibit slow solidification after extrusion often provide good layer adhesion. However, if the melt viscosity is low at the same time, then these plastics tend to deform in the part and cannot be used for the construction of overhangs and thin walls because they cannot support themselves sufficiently. To control the cooling behavior, the air flow at the part fan on the print head can be set higher or lower accordingly. In the case of highly crystalline polymers (polymers with a high degree of crystallization), at high temperatures and with very rapidly crystallizing polymers, it is often difficult to melt them again sufficiently after crystallization has taken place in order to realize good layer adhesion.

Polymers with a low melting temperature of less than 200 °C and a glass transition temperature in the range 60–100 °C are particularly easy to process. A prime example of this is PLA. PLA has a glass transition temperature of about 60 °C and a melting temperature of about 160 °C. It crystallizes relatively slowly, but solidifies again quickly in an unheated building environment due to the high glass transition temperature. Typical PLA filaments exhibit good processing viscosity at a printing temperature of about 180–210 °C. This application temperature is sufficient to heat the not yet crystallized receptive plastic layer back above the glass transition temperature and soften it sufficiently to create an ideal adhesion base. At the same time, the glass transition temperature is high enough to allow a rapid solidification of the material, which accounts for PLA's good bridging properties. For thermoplastics with higher melting temperatures or higher glass transition temperatures, on the other hand, it is necessary to create the right process conditions via adjusting the extrusion and the building chamber temperature in combination with the operation of the part cooling fan while optimizing the application speed/layer time.

In order to achieve good adhesion between a given layer and a layer to be applied, both plastic layers (receiving and new layer) must be within a certain temperature and viscosity range at the contact interface so that sufficient cohesion can be achieved by entanglement of the polymer chains [Seppala & Migler 2016]. For this purpose, the elastic modulus G' in the interfacial layer should not exceed a value of 10^6 Pa for a certain period of time so that the chains can flow into each other in a sufficiently short time. Interfacial temperature and corresponding viscosity and modulus can be simulated well if the general conditions (extrusion strand thickness, melt temperature, build chamber temperature and layer temperature, temperature-dependent melt viscosity, specific heat capacity) are known [Fitzharris et al. 2018].

The application temperatures required to sufficiently melt the final plastic layer, at least in the upper few micrometers, are controlled by the heat capacity and the actual temperature of the layer to be applied and can be determined both empirically and by simulation if the data is known.

The difficulty lies in controlling the construction process in such a way that a certain process window is maintained. Both for small layer build times and for large layer build times, the layers should adhere/fuse sufficiently, while the lower build layers must not be heated to such an extent that they deform under the build pressure of the next layer or the dead weight of the printed part. To achieve this reliably, the build speed and/or extrusion temperature must ideally be varied dynamically. An infrared lamp, for example, can be used to control the part temperature or even the surface temperature of the layer to be applied [Nycz et al. 2020].

A further means to improve interlayer adhesion between the build layers is to control the contact pressure of the extrusion nozzle onto the previous build layer. However, this can only be varied to a limited extent, as otherwise the part can easily become deformed. In the case of large pellet printers (Large Format Printing, LFP), a pressure roller is often used which presses on the applied layer, flattens it in a controlled manner and at the same time improves the adhesion/bonding to the penultimate layer.

3.1.3.5 Shrinkage

The volume of thermoplastics is in principle temperature-dependent (see Section 2.5.2.5). The volume of semi-crystalline thermoplastics decreases particularly strongly when solidifying from the melt due to the formation of crystal structures and when the polymer chains mobility freezes at the glass transition point. From a PVT diagram of a given polymer the relationship between pressure, volume and temperature can be read for any ratio. In additive manufacturing with thermoplastics, the temperature dependence of the volume is the main factor, since the processing takes place under constant ambient pressure. In contrast to conventional manufacturing processes shrinkage due to volume reduction cannot be compensated by sufficient high pressure or repressing material into a mold. Therefore, shrinkage causing warpage phenomena regularly occurs under non-optimal process conditions, which in extreme cases can lead to detachment of the parts from the build platform, severe part warpage or material distortions, geometry loss and printing failure [Erhard 2008, Richter & Quitter 2015]. Even if a printing failure does not occur, uncontrolled and unplanned part shrinkage can lead to a critical loss of functions (geometry, mechanics) of the part, which fundamentally precludes its use in more than pure prototype applications.

In the MEX printing processes strong stresses can arise in a part due to time-delayed shrinkage. The sum of tensile forces (stresses) acting in the l part volume due to uneven local shrinkage can cause the part to warp and bend, and this becomes more problematic the larger and more massive the part is.

Warpage is a time-dependent effect caused by the uneven local cooling and crystallization in a printed part. This is specifically critical between print layers due to the different temperatures between the freshly melted filament and the underlying layers. For a good printing result, the temperature differences between the last print layers must be kept as small as possible or the cooling kinetics must be controlled very well during the printing process so that uneven or uncontrolled local shrinkage can be avoided.

In extrusion-based additive manufacturing, the temperature in the build space can be controlled by various methods (see Section 3.1.4.4).

Proper design for extrusion-based additive manufacturing helps to avoid printing failures but is a field that is only slowly developing, as both the process and the material must be considered in the design. For successful design for extrusion-based additive manufacturing methods, the material shrinkage (absolute value and temperature-dependent kinetics) must be known and taken into account both in the target geometry including part surfaces and part filling as well as in the process control (temperatures of extruder and build chamber, local cooling by part fan, application speed, layer time).

3.1.4 Process Conditions

Process parameters that can be adapted with knowledge of the physical properties of the thermoplastics to be processed are:

- Nozzle temperature
- Nozzle geometry
- Volume discharge velocity
- Build chamber heating
- Build platform heating.

3.1.4.1 Nozzle Temperature

Unlike in profile extrusion, where a material passes through different heating zones in an extruder, the nozzle (die) temperature in additive extrusion-based manufacturing largely defines the final temperature of the thermoplastic. In MEX processes, it is often equated with the hot-end temperature. At the nozzle, the thermoplastic experiences the highest temperature (due to accumulated residence time in the hotend), the point of highest pressure build-up and shear and directly after the nozzle the point of contact of the polymer melt with the external environment (including a strong pressure drop, temperature drop and contact to air flow). The nozzle temperature and geometry has a strong influence on the viscosity of the polymer melt and the temperature should be high enough to achieve the desired flow rate of material. A high nozzle temperature also favors a lower pressure build-up at the nozzle, less spray swelling, improved discharge behavior and supports good interlayer cohesion of consecutive layers.

However, too high a nozzle temperature must be avoided. The nozzle temperature, together with the residence time of the thermoplastic in the heated part of the extruder head (hotend), has a direct influence on the degree of decomposition of the polymer. This can manifest itself in bubble formation, color change and loss of mechanical properties. A melt viscosity at the nozzle that is too low can, depending on the thermoplastic material, cause a poorer melt stability and bridging behavior of the deposited melt strand in the part. It also promotes undesirable run-out (dripping), curling and stringing behavior (see Section 3.1.5.3).

‚Specifically high-molecular-weight thermoplastics with a high melt viscosity may require significantly higher nozzle temperatures than are recommended for these materials in standard thermoplastic processing handbooks.

3.1.4.2 Nozzle Geometry

The choice of geometry and dimension of 3D printer nozzles is always a compromise between the desired build speed and the level of detail of the part. The printer nozzle is predominantly attached directly to the heating block and represents the area with the lowest melt flow channel diameter and the highest pressure build-up. A strand of melt is deposited through the nozzle as a layer onto the build platform or the respective uppermost build layer. The layer thickness depends on many parameters. Important parameters are the nozzle distance to the last layer, the flow speed through the nozzle in relation to the traversing speed, the melt viscosity, the die swelling and the tendency of the melt to solidify at the nozzle.

The nozzle diameter limits the possible layer thicknesses. As a rule of thumb, the layer thickness must not exceed approx. ¾ of the nozzle diameter. This is necessary to generate a sufficient contact pressure that enforces complete wetting of the prior layer. The contact pressure is a function of the ratio of nozzle diameter and nozzle distance to the last layer.

Other influencing factors are the ratio of the volume throughput speed to the traverse speed and the melt rheology. The applied melt strand geometry is a function of the traverse speed, it is rarely compressed and broadened, and is more frequently stretched and thus reduced or flattened. The viscosity of the thermoplastic melt as a function of the temperature change and the shear rate change in and behind the nozzle determines the die swelling (see Section 2.5.4). With greater layer thickness (xy-direction), the number of defects due to incompletely filled areas (holes) increases for the same volume discharge. With higher layer spacing (z-direction), the wall surface of the part is increasingly characterized by a alternating concave and convex surfaces in build direction (z). As a compromise between build speed and part resolution, layer thicknesses of ¼ to ½ of the nozzle diameter are usually selected.

The most common nozzle diameter of FLM (FFF/FDM) printers is 0.4 mm. For the majority of technical parts, smaller nozzle diameters are not required. However, bores down to approx. 0.1 mm are available for most printers. Nozzle geometries larger than 0.8 mm are only used when large parts are to be produced, some of which still undergo ablative or smoothing surface finishing. Nozzle geometries larger than 0.8 mm only make sense if the printers allow for corresponding throughput rates. Examples of this can be found particularly in the area of pellet (granulate) extruders.

The nozzle geometry, throughput rate and traverse speed must be ideally matched in a printer to enable a good print result. In addition to generating the ideal traverse path (minimal travel moves), this fine-tuning of the throughput rate with regard to the multiple extruder hotend movements (print lengths and speeds) that constitute a single print layer is one of the main tasks of modern slicers in combination with the printer configuration software/firmware (e.g. open software available for many microcontrollers like Marlin, Repetier, Klipper, RepRap; dedicated firmware for specific microcontrollers like Smoothieware; proprietary firmware for specific printers like Prusa, Sailfisch).

For special thermoplastic compounds, special demands are made on the nozzles. While for unfilled thermoplastics the main concern is a good heat transfer, which is why soft brass nozzles (see Section 2.5.6) are very suitable in comparison, the choice of a suitable nozzle material is of decisive importance, especially for abrasive fillers. The nozzle geometry must not change in the course of a build-job so as not to jeopardize the part quality. A change in the nozzle geometry can have various causes:

- Expansion due to abrasive filler interaction
- Constriction due to deposition of impurities during thermoplastic degradation and coking
- Clogging due to fillers
- Blockage due to fibers

A change in nozzle geometry is always accompanied by a change in pressure, which is why a pressure sensor at the nozzle or in the material conveying system can provide a quick indication of undesirable changes.

Nozzle expansion due to abrasive fillers is countered by selecting specially hardened nozzles (steel, tungsten, diamond-coated). A lower thermal conductivity must usually be accepted in this case (see Section 2.5.6). The narrowing or clogging of nozzles can only be countered by the quality of the feedstock with, for example, a highly homogeneous filler and particle size distribution.

For fibrous fillers, the rule of thumb is that the average fiber length should be smaller than the nozzle diameter. Nozzles with a diameter of 0.4 mm are often already critically small in relation to common short fibers of 0.2–0.4 mm length. Particularly in the case of long-fiber-filled thermoplastic compounds with typical fiber lengths of up to 10 mm, not only the nozzle geometry but also the fiber feed geometry must be optimal. Fibers should ideally be prearranged parallel to the direction of passage in the melt (e.g. by preorientation in the feed filament) to avoid entanglement and displacement off-axis in the extrusion process. Overall, the possible fiber content of thermoplastics in extrusion-based additive processes is usually significantly lower than in comparable injection molding or profile extrusion processes.

Clogging of nozzle channels can be caused by material build-up on the outside and inside. External build-up can be quantitatively removed by a cleaning process between or during a build process (if external build up continues to happen, the processing conditions must be adjusted accordingly!). Internal build-up can be removed by a purging of nozzle at higher temperatures or higher throughputs or mechanically with a steel needle from the outside as well as the inside. However, repeating nozzle clogging from the inside is a bad sign, indicating severe process instabilities such as an incorrect temperature/dwell time ratio or poor quality or homogeneity of the thermoplastic compound. Such process instabilities can lead to continuously varying material discharge and thus a significant reduction in the printed part quality.

In addition to round nozzle heads, there are developments aimed at size- or shape-variable nozzle openings or special nozzle geometries [Koch 2017]. These developments are always useful if the ratio of discharge volume to part resolution can be positively influenced, e.g. for parts that exhibit extreme variations in wall thickness. Here, a variable nozzle opening parallel to the direction of travel can provide a significant speed advantage, as wider melt strands can be deposited in one step, thus saving print and travel distances.

Changing extrusion nozzles or even full hotends during a printing job is becoming increasingly easy with upcoming new FLM printers. Whereas in the past nozzles were mainly attached to the heating block (hotend) by means of a screw thread, there are now different concepts for integrating the nozzles into the hotend. Nozzle replacement can thus be carried out much more reliably, easily and faster. Figure 3.1 schematically shows various concepts for easier nozzle replacement through increasing print head integration.

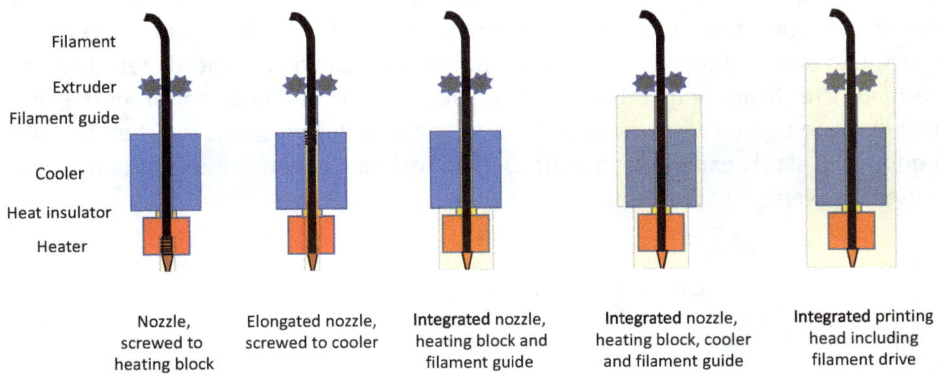

Figure 3.1 Increasing print head integration

3.1.4.3 Volume Discharge Speed

The volume discharge rate in the MEX process depends on four main factors:

- Heat transfer capacity of the print head
- Feed force
- Rheological properties of the thermoplastic
- Nozzle geometry

The heat transfer capacity of the print head is given by the heat transfer surface, the thermal conductivity of the materials used for the hotend, the target melting temperature of the thermoplastic, the maximum heating temperature of the print head as well as its heating power and the dwell time in the heating element and nozzle. These values are typically known or can be easily determined experimentally (see Section 2.5.6).

The feed force describes the ability to transfer force to the thermoplastic melt by means of filament advancement or by means of the extruder screw. If the pressure build-up at the nozzle is too high, the filament can buckle on its way to the extruder. As a result, the conveying resistance increases and slippage occurs at the filament drive. Alternatively, the screw in the pellet extruder stops. If this happens unnoticed even only for a short time, local under-extrusion and quality defects in the parts produced are the consequence (Figure 3.2).

The rheological properties such as melt viscosity versus temperature and shear thinning, etc. (see Section 2.5.4) determine the pressure build-up for a given nozzle geometry. The heat capacity of the thermoplastic over temperature, on the other hand, especially at the transitions from the solid to the liquid state at the glass transition temperature in the case of amorphous thermoplastics and the melting temperature in the case of semi-crystalline thermoplastics (see also Section 2.5), determine the necessary amount of heat that must be introduced into the polymer in order to achieve a suitable processing viscosity.

In a simple experiment, the theoretical volume/mass extrusion speed can be plotted against the experimental volume/mass extrusion speed – in analogy to a mass flow experiment (see Section 2.5.5) – at different specified conveying speeds for different specified print head temperatures and nozzle geometries. The deviation of the real mass extrusion speed from the experimental mass extrusion speed can then be used to infer both stable extrusion conditions and possible compensation factors that can be used for planning a stable process.

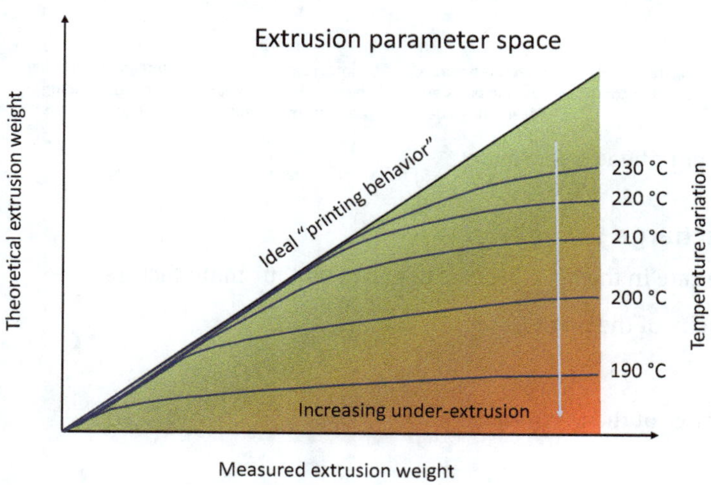

Figure 3.2 Extrusion parameter space

Particularly soft thermoplastics with low Shore D/Shore A values, which melt at low temperatures below 200 °C, are usually limited in their volume discharge speed by the effective force that can be applied to feed the filament into the hotend. Optimization of the filament guidance and the force transmission to the filament are the biggest levers here to increase the process efficiency. When using soft filaments, it is recommended to use so-called direct extruders instead of Bowden extruders (Figure 3.3). In the case of direct extruders, the filament drive is directly coupled with the hotend, and the filament guidance is optimized to prevent kinking or side extrusion.

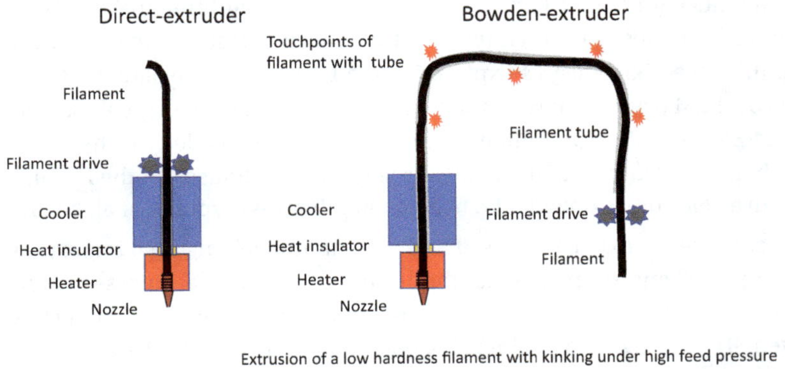

Figure 3.3 Schematic structure of a direct extruder compared to a Bowden extruder

The extrusion of thermoplastics that require particularly high melt temperatures is mainly limited by the heat transfer capacity in the hotend before the nozzle. The reason for this is that at high volume extrusion rates, the residence time is often no longer sufficient to heat the materials to the necessary temperature and bring the melt viscosity into an easily processable range. New hotend designs increase the heat exchange surface specifically of the nozzle.

For the optimization of processes for the production of series parts, where throughput rates and production speed are important, knowledge of the process limits is an absolute necessity. In the prototype case, it is sufficient to identify a stable process range, a so-called process window, in order to obtain high-quality parts for testing.

3.1.4.4 Building Space Heating

As late as 2000 to 2021, the use of build chamber heating for 3D printers, in which the electronic motion control is isolated from the build chamber, was blocked by various patents held by the Stratasys company [Swanson et al. 2004]. This has long hampered the use of high-temperature materials in particular, as well as the use of build chamber heaters in desktop 3D printers. Only a few companies such as AON3D, INTAMSYS, Roboze, and 3DGence were able to develop products for these markets through elaborate workarounds to the problem. As of spring 2021, this technology is now patent-free, resulting in a plethora of new suppliers for extrusion-based additive manufacturing systems featuring a heated build space.

For the consumer market, the existence of US6722872B1 in particular as of 2010 has promoted the use of heated build platforms as an alternative to a heated build space. The heated build platform, now found in most FLM 3D printers, is instrumental in reducing part adhesion problems and alleviating delamination and warpage issues (see Section 3.1.4.4). While the heated build platform is now an integral part of MEX desktop printers, it is not sufficient when it comes to the application of many technically relevant filaments with higher temperature resistance. In addition to ABS, these include polyamides and especially high-temperature thermoplastics such as PEKK and PEI. The significance of Stratasys's expired heated build chamber patents is evident in first low-cost FLM printers now available under open-source design plans that can extrude with PEKK and ULTEM. Dubbed "Cerberus", a three-headed machine can be built for less than $1000 and, similar to the Stratasys design, houses all the low operating temperature electronics outside the heated chamber [Skrzypczak et al. 2020].

A heatable build chamber in extrusion-based additive manufacturing is so important because it allows optimal environmental conditions to be created in each case for the printing of larger and more complex parts with high-temperature-resistant materials in particular. Even 3D printing with engineering polymers such as ASA, PET or PC ideally requires a chamber that can be heated up to 130 °C in order to be able to deliver the best possible quality with regard to the areas of use of printed parts. High-performance polymers such as ULTEM or PEEK also require a build chamber that can be heated up to 250 °C.

Each polymer has a glass transition temperature (T_g). When working with amorphous thermoplastics such as PC, PEI, PEKK or PPSU, it is important to keep the ambient temperature of the material in an optimal range. In many cases, the ideal temperature for the heated chamber is close to the T_g of the polymer to be printed. Around the T_g range, polymers can relax (see Section 2.5.2.3) and relieve internal stresses. Listed below are typical chamber temperatures for some amorphous engineering and high performance polymers:

- ABS: 120 °C
- PC: 130 °C
- PPSU: 225 °C
- ULTEM 1010: 220 °C
- ULTEM 9085: 185 °C
- PEKK-A: 160 °C

If the necessary design temperatures cannot be set, increasingly deformation, warpage, part detachment, layer delamination and reduced mechanical properties are found, depending on the part produced. The cause for this is a non-uniform shrinkage of the materials due to an uneven temperature distribution in the part during the printing process. Shrinkage also affects print adhesion to the build platform and between build layers. If the deposited melt strand cannot transfer enough energy to the last build layer so that it can melt and weld, then the resulting cohesion between the layers is impaired at this point.

The main function of the heated chamber is to keep the printed part at a uniformly stable temperature throughout the printing process. The temperature is selected in such a way that residual stresses caused by shrinkage that is too rapid are prevented and sufficient layer adhesion is ensured. The optimum temperatures for this depend on the material as well as the layer thickness, the build speed and the part geometry. For optimum results, the parameter sets for the slicer must be optimized to some extent. When printing is finished, the temperature in the chamber is slowly and evenly lowered to make shrinkage uniform and to minimize internal stresses to a large extent. This ensures the best possible dimensional stability and preservation of the mechanical properties of the thermoplastic in the product.

In the case of semi-crystalline polymers, the part temperature is maintained at a temperature above the glass transition temperature in a range at which the crystallization rate is so low that the last layer(s) is preferably not yet completely crystallized before the current layer is applied. The optimum temperatures here also depend very much on the compound formulation and the part design to be manufactured and sometimes must be optimized experimentally.

3.1.4.5 Building Platform Heating

The adhesion of the thermoplastic polymer to the build platform has a major influence on the quality of the resulting part. A major influencing factor to improve the adhesion to the build platform is the build platform surface and temperature. Similar rules apply to choosing the ideal temperature of the build platform as to the build chamber. Depending on the thermoplastics to be processed, the temperature should be selected in such a way as to enable a good print part adhesion to the platform that is as reversible as possible.

After cooling the platform to room temperature, the part should be easy to remove due to its thermal expansion coefficient deviating significantly from the platform material. In addition, the temperature of the build platform can also positively influence the cohesion between the first build layers, as the platform temperature transfers to a relevant degree to the next few layers by heat flow through the thermoplastic. Build platform heating was originally designed in combination with closed build chambers as a partial convection-based build chamber heating system to circumvent Stratasys patents. However, this only works well only for flat parts or thermoplastics that have sufficiently low build chamber temperature requirements. Using a build platform for heating a uniformly ventilated and insulated build chamber, build chamber temperatures of up to 80 °C are possible without major equipment effort.

3.1.5 Printing Problems

The following typical problems arise from the combination of thermoplastic properties and process conditions as well as part geometry:

- Lack of adhesion to the build platform
- Lack of interlayer adhesion
- Stringing/oozing
- Delay (warping)
- Under-extrusion
- Over-extrusion

3.1.5.1 Adhesion to the Part Platform

The first applied layer of the printed part is decisive to achieve a good adhesion to the build platform. It must fix the part to the platform against the forces acting on it during the printing process (shear forces, internal stresses, shrinkage). At the same time, the printed part should be as easy as possible to remove from the printing platform after completion without damage. To meet these seemingly contradictory requirements, the printing platform, printing material and process parameters must be well-matched. The combination of printing materials, build platform materials, platform and build chamber temperatures, build platform adhesives and the variation of printing parameters results in an almost unmanageable number of options to influence print platform adhesion. The common measures used to optimize build platform adhesion, most of which are still empirical, are the use of:

- Heated build platforms
- Different build platform materials (tempered glass, PEI, resin plates, powder-coated plates, spring steel plates)
- Removable deformable build platforms (to facilitate easy detachment of the part)
- Adhesive coatings and films (PEI, Kapton, painter's crepe, etc.)
- Pressure-sensitive adhesives (high- and low-temperature, water-soluble, etc.).

Basic recommendations on suitable material and process combinations can often be taken from the information and data sheets supplied with the print material.

In addition to the more "adhesive" issues of build material and build platform compatibility, there are a number of process conditions and slicer settings that support good part adhesion.

Leveling the print bed

Poor calibration of the print bed can lead to locally fluctuating part adhesion due to varying distances between the nozzle and the print bed. The first layer is therefore often run with a smaller layer distance or a higher material throughput to guarantee good contact of the melt strand to the build platform. Visually, this mode of operation can be found as an "elephant foot" on the part and must be removed in critical applications during post-processing.

Cleaning the print bed

A clean and grease-free print bed is the basic prerequisite for controlled adhesion, whereas contaminated print beds can cause adhesion that is either too low or too high (poor release).

Optimizing the print bed temperature

The print bed temperature must be adapted to the material to be processed, either by direct heating or using a build chamber heater.

Building chamber heating

Particularly hard, rapidly crystallizing semi-crystalline thermoplastics (for example polyamide 6/12) with a high melting temperature benefit from a build chamber heater. The particularly strong shrinkage associated with the recrystallization of the thermoplastics can lead to such severe stresses and high warpage in the part that this cannot be absorbed by a good platform adhesion alone. In principle, platform and build chamber heating is recommended for all high-temperature materials (high T_g/high T_m).

Avoiding temperature fluctuations

Large temperature fluctuations on the build platform or in the build space should be avoided in order to minimize local stress differences in the part.

Using special printing plate/flexible printing plate

Various printing plates are optimized for construction materials and enable good adhesion and release in combination with specified build temperatures. When warm, good adhesion is present, and when cooled, the part can be easily detached from the platform. A flexible printing plate further supports easy detachment of the platform from the printed part due to transforming an area adhesion into a line adhesion during flexing the removable printing plate to remove the part.

Using adhesive

Various adhesives support adhesion to the build platform. Particularly suitable are those adhesives which, in combination with an elevated platform temperature, generate adhesion and promote easy detachment of the part from the platform when cooled to ambient temperature. Ideally, residues of the adhesive are water-soluble or can be easily removed from the build platform and/or the part with isopropanol/ethanol.

Activating print bed adhesion

Activating the print bed adhesion functions (skirt, brim) in the slicer increases the adhesion area of the part on the platform. The resulting edges can be easily removed mechanically after the print process.

Optimizing filling level

The greater the material consumption per part/volume (increasing degree of filling), the higher the stresses found inside the printed object. A simulation helps to minimize the part filling by optimizing the part filling level according to the expected part load and with regard to the residual stress. The same applies to particularly large-area parts or parts with a high length in one direction. Here, too, the number of parallel applied lines and lengths of the deposited melt strands should be minimized to reduce part-stress and warping.

Customizing the fill pattern

The filling pattern plays an important role in thermoplastics with high shrinkage. Long build lines in one direction should be avoided, since shrinkage is particularly strong along these lines.

Reducing layer thickness

The layer thickness and build speed together determine the effective shrinkage volume (amount of material that can shrink in a typical cooling time). When the respective layer cools down, significant stresses are generated in the case of temperature gradients in the printed part, which can cause distortion and warpage. In principle, greater warpage is to be expected with higher construction layer thicknesses and to a degree higher construction speed.

3.1.5.2 Interlayer Adhesion

Very good interlayer adhesion (more precise interlayer cohesion) is a fundamental prerequisite for producing functional parts with extrusion-based additive manufacturing processes. Many parameters are relevant for good interlayer cohesion. The rheological state of the receiving layer and the applied layer must match so that molecular "welding" of the layers can take place. They should be liquid at least in the boundary region and have a storage modulus smaller than 10^6 Pa (see Section 2.5.2).

In the case of identical materials, it is sufficient if molecular mixing and entanglement of polymer chains is possible between the previous and the new layer in a depth less than a few micrometers thick. The duration of this condition is dictated by the melting temperature and the temperature-dependent viscosity, as well as by the time required for the thermoplastic to solidify by crystallization or vitrification (see Section 2.5.2.3). The fact that these conditions are not always optimally ensured can be seen in the fact that the part strength in the build-up direction (z) is often significantly lower than the part strength in the layer plane (xy).

The layer-to-layer adhesion of parts made of amorphous thermoplastics is generally more difficult to optimize than the layer adhesion of slowly crystallizing, semi-crystalline thermoplastics. This is due to the fact that the melt viscosity of an amorphous thermoplastic is directly related to the distance from the glass transition temperature and rises exponentially by orders of magnitude when nearing it. The temperature of the interface between last and new layer must thus be higher than the T_g to have any chance to produce a sufficient adhesion between both layers.

The recrystallization of the polymer melt of a semi-crystalline thermoplastic with a T_g typically far below the melting point of the polymer, on the other hand, is a kinetic process which, depending on the actual build chamber- and melt-temperature, may take place with a significant delay in a so-called supercooled melt (see Section 2.5.3), thus allowing sufficient time for molecular mixing and entanglement of polymer chains between layers to take place.

For each polymer there are ideal printing conditions which can be derived from the thermal analysis of the material via DSC/DMA (see Section 2.5.3) and temperature-dependent viscosity/modulus measurements. If the build chamber and platform can be heated, to create a controlled temperature environment, then these conditions can be set optimally. As in the case of platform adhesion, the build speed, the layer thickness and the part geometry also play an important role. However, the conditions for good layer adhesion are partly reciprocal to the conditions for good platform adhesion. Thick build layers and a high build speed – provided that the melt reaches the desired temperature – are definitely positive for good interlayer adhesion. Since the relative surface of the layer decreases with higher layer thickness and the heat transfer to the environment decreases, the heat stored in the application layer can better heat up the previous layer and thus activate it. Especially in large-format granule printing, prints of semi-crystalline and amorphous thermoplastics with high layer thicknesses and with excellent interlayer adhesion can be achieved even in room temperature environments. The corresponding shrinkage and part distortion are controlled here, for example, by using highly filled compounds (with the filler volume contributing very little to shrinkage), which would be difficult to use in filament-based processes.

Good interlayer adhesion is supported by the highest possible contact pressure of the applying layer on the previous layer. In a standard FLM printer, this is achieved by over-extrusion. The melt pressure of the extruded material generates the contact pressure. In larger printers, such as some granule printers, a trailing "spatula" analogous to cement processing can additionally help to compress and solidify the final layers.

It is more difficult to ensure interlayer adhesion between different materials or thermoplastic blends. The same requirements for material compatibility apply here as are already known from two-part injection molding, two-part extrusion and two-part film lamination. In addition to pressure, temperature and melt viscosity, the materials must have a certain degree of compatibility, or an adhesive interlayer must be used.

Materials with exceptionally good compatibility with themselves and many other materials in this regard are thermoplastic polyurethanes. These materials are available in a high variance of hardness ranging from 40 Shore A to 85 Shore D. Due to the often similar polymer building blocks, they exhibit good interlayer adhesion to one another and thus enable parts that can withstand high mechanical loads with high material variability and thus also property variability across multi-material parts.

Sometimes a lack of interlayer adhesion is mistaken for warpage of parts. Although shrinkage forces may well locally exceed interlayer adhesion causing layer separation. Furthermore, part defects and cracks, for example due to omitted layers or layer gaps caused by under-extrusion (see Section 3.1.5.5), can simulate poor layer adhesion.

Every part defect is a potential weak point. Especially in extrusion-based additive manufacturing, the production of a defect-free part is not trivial, especially in the z-direction. Since shrinkage is a fundamental material phenomenon, controlled build chamber temperatures as well as controlled cooling of the part following the build process play an important role to improve layer adhesion and in order to avoid part damage or destruction during the cooling process.

3.1.5.3 Threading (Stringing) and Oozing

Thread formation and leakage are phenomena that have material and process causes. They are often observed when the printer makes an idle run between two build sections.

Stringing

If the viscosity of the molten thermoplastic at the nozzle tip is sufficiently low and the viscosity increase downstream of the nozzle due to cooling is sufficiently high, then it is often observed that a filament is drawn from the nozzle that stretches and solidifies between the last deposition point and the new deposition point (see Figure 3.4).

There are test parts available that help with finding the right temperature, build speeds and retraction conditions to avoid stringing.

Figure 3.4
Test part for optimization of the thread-pulling behavior

Oozing

Leakage is observed when the viscosity of the thermoplastic melt in the nozzle is too low (see Figure 3.5). Gravity causes material to run out of the nozzle when the head slows down or stops and is deposited in an undesirable location. The transition from filament formation to discharge cannot be precisely defined. In both cases, the phenomenon leads to thicker threads where superfluous material is deposited (over-extrusion) and to part gaps (under-extrusion) at the beginning of a new build section, since part of the scheduled material has already flowed out prematurely.

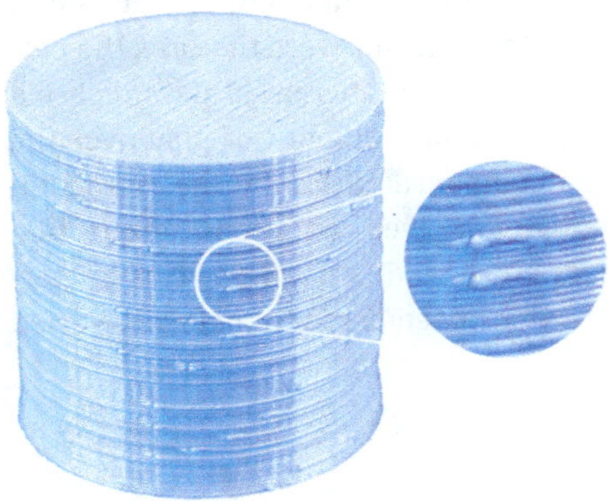

Figure 3.5
Example printout illustrating the effect of oozing

Retraction

Basically, in additive manufacturing, when planning the travel paths through the slicer, it is important to minimize attachment and heel points for the print head. If this is not possible, setting "retraction" is the most common method of minimizing stringing or oozing. When retraction is activated in the slicer, the filament is pulled back a few millimeters each time before or during the extruder travels an empty distance. This pulls the molten plastic up into the nozzle and prevents dripping. As soon as the extruder reaches its next build position, the filament is pushed out again by the retraction length before the extruder resumes printing.

In most slicers, retraction is activated as a function by default. Depending on the material, however, the retraction length and speed must be adapted to the material, the extrusion temperature and also the part geometry. When working with pellet extruders, which often operate with higher throughput rates and nozzle sizes, the retraction must be optimized even more for the material and part, since depending on the screw geometry, an effective retraction is only possible to a limited extent.

Retraction length

The retraction length parameter determines how far the filament is retracted. The retraction length depends on the filament thickness (the thinner, the more retraction), the nozzle geometry (the smaller the nozzle opening diameter, the lower the retraction required) and the melt viscosity of the material in the die channel (the lower the viscosity, the higher the retraction required) and thus the hotend temperature. A larger retraction reduces the probability of material stringing and oozing. During retraction, care must be taken to ensure that no molten material enters the cooling zone, otherwise the conveying of the material can be blocked by solidification of the thermoplastic and adhesion to the wall of the cooling zone.

The retraction length depends on the extruder type. A Bowden extruder generally requires a longer retraction distance, since the distance between the extruder nozzle and the drive wheel for the filament is greater here and the filament thickness and the diameter of the Bowden tube do not match perfectly. If this were the case, the passage resistance of the filament through the tube would be increased too much via static/sliding friction. The filament therefore has a certain amount of play over the "Bowden tube length", which in turn must be compensated for by the retraction length. To find out the optimum retraction length under given conditions (nozzle, filament, material, temperature, extruder type, etc.), pre-made test prints can be performed in which the retraction length and retraction speed are systematically varied at different nozzle temperatures [Johnson 2019].

Retraction speed

The retraction speed determines how fast the filament is retracted. A higher retraction speed accelerates the build process. If the retraction speed is set too high, the filament may break off inside the extruder nozzle and the remainder in the front part of the nozzle will continue to pull threads and run out. Professional slicers often have coast and wipe parameters that can be used to further refine retraction settings.

Coasting (letting the nozzle roll out)

Thermoplastic droplets always occur when the built-up pressure in the extruder nozzle is too high or the viscosity is too low and more plastic than intended emerges from the nozzle.

If the drop occurs at the end of a travel section or at the end of a build position, the pressure build-up at the nozzle is reduced when the coasting setting is activated by switching off the extrusion shortly before the end of a build movement. The coasting value determines the shutdown time before the travel move begins. The value can be varied empirically until no more drops are visible.

Wiping (cleaning nozzle)

The wipe function strips off material residues at the nozzle outlet at the end of a layer before the travel (empty) run begins. Activating the wipe function tends to create better flowing transitions to the previously printed line, especially at the layer jump. Wipe can also help to prevent thread formation because residual material is still wiped off before the empty run.

Temperature

The viscosity of the thermoplastic melt correlates with the heating temperature in the heating head and at the nozzle, which together form the hotend. The lower the melt viscosity, the higher the tendency to ooze out of the nozzle. A reduction in the melt temperature, on the other hand, leads to a higher pressure build-up in the nozzle, lower interlayer adhesion and a reduction in the possible build speed. The optimum material temperature depends on the material and the other pressure settings. However, if thread formation or leakage is observed when retraction is activated, it is generally advisable to lower the print head temperature slightly as a first measure.

Printing speed

The printing speed can also have an influence on thread formation. The time between the heel point and the starting point of an empty run should be kept as short as possible. Modern slicers allow the printing speed to be separated between empty runs and the actual printing (build) process. During the empty run, the movement speed of the head is significantly increased.

Residues on the extruder nozzle

During longer printing processes, residues occasionally form on or next to the extruder nozzle. These can further promote thread formation. It is generally advisable to clean the print nozzle automatically after predefined intervals. Professional printers offer parking positions with nozzle cleaning for this purpose.

Filament conditioning

Damp filaments are one of the biggest problems in extrusion-based additive manufacturing. Care should always be taken to ensure that the filaments used are sufficiently pre-dried. This can be done in specially designed drying ovens or by using moisture absorbers. In addition to bubbling, foaming and clouding, moist filaments often tend to lose molecular weight, which is accompanied by a reduction in viscosity, and increased threading and oozing. Sensitivity to moisture is generally expressed in terms of the moisture absorption or equilibrium moisture content of thermoplastics and is very material-dependent. Moisture absorption is a value that can be taken from the data sheet of a material [CAMPUS 2022].

3.1.5.4 Delay (Warping)

Warpage occurs during printing in the MEX process due to volume shrinkage of the thermoplastic during cooling from the melt to the solid state (see Figure 3.6). Different wall thicknesses, stretched and large parts, high application speeds, high application temperatures and temperature gradients in the part and in the installation space increase the tendency to warp.

Shrinkage

In principle, the shrinkage of a thermoplastic on cooling is a specific and known physical, temperature- and pressure-dependent property which can be looked up in respective material property tables [CAMPUS 2022]. In injection molding and extrusion, shrinkage is already taken into account in the part design, which allows warpage to be minimized. Something similar is also being considered in additive manufacturing, but it is much more difficult due to the poorly controlled environmental conditions and the tendency for additively manufactured parts to be more complex.

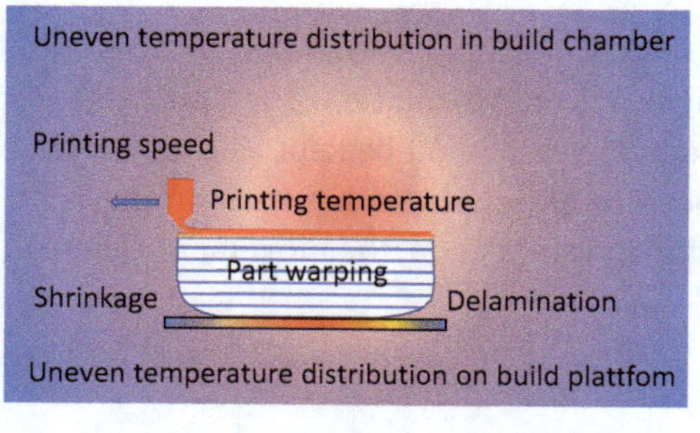

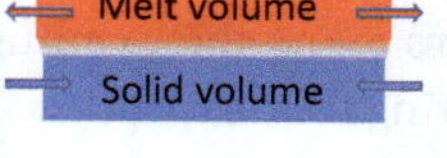

Figure 3.6 Warping of a part due to uneven material shrinkage

The shrinkage behavior is very material-dependent, which is why the problem can vary in severity depending on the material. A PLA part can usually be printed without problems even with larger dimensions using a heated printing plate. In comparison, warpage quickly sets in even with a small print object made of ABS.

Building chamber heating

The greatest lever for avoiding warpage in in parts is a heated build chamber. The thermoplastic can thus cool down in a controlled manner and stresses can relax evenly (see Section 2.5.2.3).

Layer adhesion

Warpage is often noticeable by detachment of layers and especially by detachment from the build platform as the area with the weakest layer adhesion in the build. A heated and well-leveled and calibrated build platform as well as good adhesion (see Section 3.1.5.1) of the part to the build platform can effectively counteract warpage.

Building speed/melting temperature

A slow build process and low melt temperature reduce the tendency to warp, but worsen build efficiency and interlayer adhesion.

Part fan

In order to slow down the cooling process, it helps to operate the part fan at the lowest possible level so that overhangs and bridges due to cooling are still possible, but the thermoplastic has more time to evenly relieve stresses during slow cooling. The settings have to be optimized in a very material-specific way.

Filling density

The tendency to warp increases with the printed volume. Accordingly, a reduction of the filling structure density (infill) can help. However, this measure has a direct influence on the mechanical properties. Furthermore, the type of infill has an influence on warpage. A geometrically complex infill with many changes in the direction of the print head reduces the tendency to shrink, but in turn has a negative effect on the build speed and to some extent also on the part properties.

Part design

In addition to process variations, design changes/adaptations of the part to the manufacturing method can reduce warpage. The rounding of corners or the targeted reinforcement of areas at risk of warpage have a positive effect.

Extrusion parameters

In addition to the infill, parameters such as the extrusion width and height as well as the extrusion multiplier, have an influence on the warpage behavior and may have to be specifically adjusted for a given part design. Low layer heights and a low deposition speed help against warpage.

In summary, warpage in parts produced by the MEX process depends on many material and process parameters. However, with the optimization of the part adhesion to the build platform, the correct selection of build temperature, nozzle temperature and the printing speed, most warpage problems can be solved satisfactorily.

3.1.5.5 Under-extrusion

A common problem in 3D printing is under-extrusion (see Figure 3.7), which occurs when the real extrusion volume is below the planned extrusion volume. Under-extrusion is noticeable by missing or very thin layers as well as depressions and holes in the part. Under-extrusion degrades the appearance of parts and creates voids that can greatly affect part mechanics and lead to early part failure. Under-extrusion occurs when the real extrusion volume is below the planned extrusion volume. Possible causes of under-extrusion are as listed below.

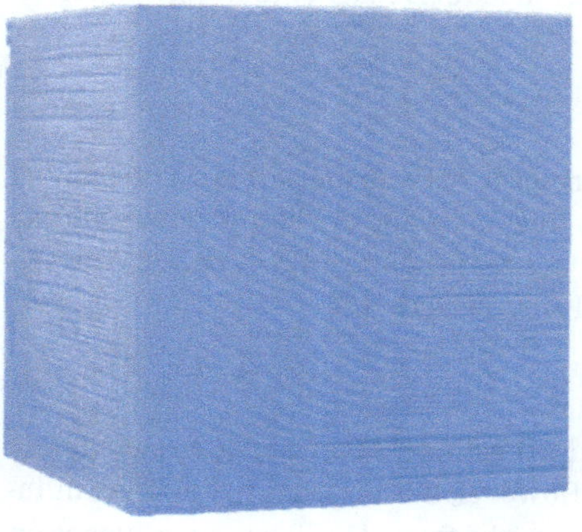

Figure 3.7
Fault pattern of a part exhibiting under-extrusion

Blockage at nozzle or extruder

If there are impurities or residues of filaments in the filament channel or in the nozzle opening, the filament drive must exert more force. If the filament drive can no longer provide the necessary conveying force, under-extrusion occurs until extrusion stops. The transitions from under-extrusion to standstill can be fluid, as the filament drive wheels can slide along the filament or even generate filament abrasion. In either case, the resulting slippage of the drive wheels off the filament strand reduces effective material propulsion. Regularly cleaning the nozzle and nozzle channel typically eliminates the problem.

Pressure build up in the nozzle

The combination of effective propulsive force, nozzle geometry, temperature, and melt viscosity determines the maximum material discharge velocity for each printing process. Above a certain nozzle pressure, the relationship between theoretical and practical filament extrusion velocity deviates from a linear relationship (see Section 3.1.4.3). After that, readjustment can only be made within a narrow range by increasing the feed rate in the extruder: the system is basically limited.

Slip

Continuous slippage of the filament feed unit (slipping of the drive wheels from the filament) in the extruder usually leads to abrasion of the filament and thus to increasing contamination of the feed unit. The power transmission decreases continuously. If the advance unit is not designed to be self-cleaning, under-extrusion increases over time due to increasing slippage.

Jammed filament spools

Filaments tend to jam if they are not wound correctly. If the filament spool jams, additional force must be applied by the filament feed unit, i.e. not only pressure must be built up, but tensile force must now also be applied. This circumstance is often the cause of a briefly effectively reduced material flow, which is expressed in under-extrusion, but can also lead to the build process itself being aborted by e.g. filament rupture (specifically with hard and brittle filaments).

Bowden extruder

FLM printers based on Bowden extruders, whose feeder unit is spatially separated from the heating unit and the nozzle, mostly use low friction tubes for example PTFE or silicone based to guide the filament into the print head. If these tubes are bent or distorted in any way, they cause additional friction with the filament. This additional friction increases the required feed force, which can consequently lead to under-extrusion. In addition to the normal friction on the guide tube, excessive conveying pressure can lead to kinking effects in the filament, which in turn further significantly reduce the effective force for filament conveying. Particularly soft filaments (TPE, TPU) are sensitive to kinking and can only be processed with Bowden extruders to a limited extent. So-called direct extruders are preferred here (see Section 3.1.4.3).

Overheating

If the temperature in the print head is set too high or the retraction is set too high molten material can enter the cooling zone. As the heat increases, the modulus of the filament decreases. It tends to buckle and, in extreme cases, stick to the wall in the cold zone above the extruder. When this happens, the conveying pressure increases abruptly and the extrusion volume decreases or stops.

Feed force

Most FLM printers offer the possibility to adjust the contact pressure (feeder pressure) for the feed unit in the filament advance. This pressure controls the force that the feed unit (drive wheels) can exert on the filament to feed the material. If the pressure is too low or the filament is abraded by the feed unit, the feed unit loses its grip on the filament, which in turn leads to under-extrusion.

Filament diameter

The filament diameter correlates directly with the extrusion volume. Incorrectly set filament diameters in the slicer or fluctuations in the filament diameter have a direct influence on the extrusion volume and can cause both under-extrusion and over-extrusion (see Section 2.5.7).

Print platform calibration

If the print platform is poorly calibrated, the resulting part defects may look like temporary under-extrusion. Detachment from the printing platform is a possible consequence.

Line width

In general, the printing line width should be equal to the diameter of the nozzle. However, a small change can definitely improve the print result. A line width that is too small can look like under-extrusion.

Retraction, retraction distance, retraction speed

If the under-extrusion always occurs at the beginning of a new layer, it may be that the retraction distance of the filament is too large. The retraction operation is associated with less force than the feeding operation. If the retraction speed and the corresponding feed speed are high, it is possible that the feed speed (actual volume feed), due to a higher force required, is not sufficiently matched to the retraction speed (actual volume retraction). As a result, the printer does not manage to extrude enough material in time for the layer at the line start after a previous retraction.

3.1.5.6 Over-extrusion

Over-extrusion is given when the printer extrudes more material than intended (see Figure 3.8). Over extrusion can be recognized by the fact that excess material swells out on the outer sides of the object's construction walls and the surface shows irregularities. The most common reason for over-extrusion is an incorrectly set filament diameter or extrusion multiplier. A printing temperature that is too high can also promote over-extrusion. Leakage and stringing are often perceived as over-extrusion extrusion, but in reality they are always coupled with subsequent under-extrusion.

Figure 3.8
Fault pattern of a part exhibiting over-extrusion

Extrusion multiplier

The extrusion multiplier (usually set to values from 0.9–1.1) regulates the ratio of theoretical and practical extrusion volume in a simple way. The extrusion multiplier is set empirically from the observation of the part quality of test prints.

Z-jump

Material spillage is often combated with the z-jump (hop). A short, rapid lifting of the nozzle from the part reduces the leakage. However, this can also create bumps on the surface of the print layer, which in turn are ideal starting points for subsequent threading.

Heating/melting zone (hotend)

Printers with high throughput rates in particular have a large melt zone in addition to a large nozzle opening. This sometimes accumulates a large supply of molten material, which can easily run out under the effect of gravity through the large nozzle.

Print temperature

If the print temperature is set too high, the material may be too liquid when it exits the nozzle. Too much material may then be discharged locally during printing causing over-extrusion and elsewhere under-extrusion.

Nozzle size

The printer nozzle opening has a major influence on part quality. Generally, smaller nozzles produce higher resolution parts (lower layer thickness). The nozzle diameter also has a major influence on the possible build speed, since the pressure build-up at the nozzle decreases exponentially with the nozzle diameter. Consequently, the volume of melt that can be extruded per time increases for a given conveying force (allowing for builds with higher layer thickness or higher printing speed), assuming that the heat input in the heating zone is sufficient to bring the material to the desired melt temperature.

With small nozzle diameters, the irregularities observable on the part surface due to over-extrusion decrease. However, the permissible overhangs (see Section 3.1.7) and bridging distances also decrease with smaller nozzle diameters, which is why more support structures may have to be planned.

3.1.6 Process Simulation

Process simulation in additive manufacturing must take various parameters into account. Analogous to classic molding processes, the extrusion process can also be simulated in MEX-based manufacturing. This includes – if the properties of the thermoplastic to be processed and the device setup are known – the heat transport capability, the shear- and temperature-dependent melt rheology, the pressure build-up at the nozzle as well as the correspondingly required feed pressure, etc. The results of this simulation process are predictions of maximum material flow rates from the nozzle [Hofstaetter et al. 2015].

A second essential factor for a successful process simulation is the acute knowledge of the properties of the printing device itself. In each case, there is a highly accelerated relative motion between the print head and the part platform. The design of the printer therefore limits the allowed acceleration of the masses involved. The greater the masses or acceleration, the more stable and heavier the printer must be built to avoid natural oscillations occurring and the inevitable resulting construction errors. These can also be simulated and used for optimizing the move/print patterns to make maximum use of the capabilities of a given printer (see input shaping and pressure advance system, Klipper software).

The prediction of part properties from the real production process of MEX based printing is more difficult to access by simulation. This includes taking into account the planned travel paths, the temperature-dependent melt rheology and cooling behavior of the thermoplastic used, the interlayer adhesion and build platform adhesion, the temperature distribution in the build space, and the local shrinkage or resulting warpage. Instead of simulating the product of the real printing process, mostly the part is simulated in its geometrically ideal form on the basis of various additional parameters, such as empirically determined factors of material strength in the xy- and z-directions. Every process detail from the areas of heat transport, material flow and travelpath design has an influence on the validity of these simulation results. As a basic prerequisite for part simulation and a successful printing process, maximum effort should be devoted to creating reproducible process conditions.

Additive manufacturing enables part geometries, filling structures, wall thickness variations, functional groups and surface effects/defects that would be unthinkable in classic parts and for which there is only limited experience as to how and whether these can be simulated with sufficient precision to infer precise part properties. Improving part simulation on the basis of an increasingly precise representation of the real construction process is a major task, but also an opportunity for future developments in additive manufacturing. The goal is to move away from prototypes and towards functional part production.

The simulation of parts is particularly exciting when it is used for automatic geometry or topology optimization. In combination with thermoplastic printing of fiber-reinforced materials or even with processes that allow continuous fiber reinforcement, this enables optimized generative "organic" designs that can now only be envisioned using additive manufacturing processes. Generative design systems are used here that allow designers to describe a design problem simply by goals (functions) and constraints (allowed spatial part volume, weight, etc.) [Glamsch et al. 2019]. This information is then used to simulate in the computer, based on given production process- and material-specific parameter sets, different generative design solutions that satisfy these goals. The designer can then choose between alternative approaches and select design solutions best fitting to a given additive manufacturing process or vice versa.

Some of the problems arising from the deviation of the simulation results and the production reality can also be solved part-empirically by printing critical part zones, testing them mechanically and adapting their design to the necessary properties again iteratively or generatively. This somewhat contradicts the idea of the fully digital process, but it helps in part production for the time being [Covestro 2017].

3.1.6.1 Input Shaping

Input shaping is a technology that effects the travel path of an MEX printing process aiming to cancel out the vibrations due to natural frequency resonances produced during acceleration and deceleration of the print head. Input shaping enables significantly higher printing speeds without compromising on printed part quality.

Typical defects caused by vibrations due to high acceleration and deceleration of the print head are visible on outer walls of printed parts as ringing (also known as echoing, ghosting or rippling) (see Figure 3.9). Besides reducing those effects, input shaping effectively reduces the shaking and noisiness of a printer.

Figure 3.9
Fault pattern "ringing" due to natural vibrations of printer

Input shaping is not a slicing feature but a specific feature of the underlying printer driver software (see Klipper). Input shaping is based on an algorithm. There are several algorithms available that have been successfully implemented in printer driver software with different focuses and strengths (see also *Klipper3d.org*).

As such input shaping requires printer specific measurements before it can be activated. Before being used for MEX based additive manufacturing processes, input shaping was already a well-known technology for optimizing high speed processes involving accelerated masses, e.g. on multiple axis robotics.

3.1.6.2 Pressure Advance System

The pressure advance system accounts for pressure variations by pushing in calculated extra amounts of filament during acceleration and slowing down or even retracting filament during deceleration. With a high acceleration and high pressure advance setting in the slicer software the extruder may not have enough torque to push through the required filament effectively limiting the ability of a printer to move faster. Specially designed high heat transfer/high flow hot ends and nozzles (see also bondtech core heating technology/and e3d revo high flow nozzles) tackle this problem.

The effect of pressure advance on print quality is twofold: it reduces oozing during non-extrude travel moves and blobbing (see Figure 3.10) during cornering/deceleration.

In order to use pressure advance the printer needs to be calibrated and tuned involving the printing and inspection of test objects while iterating the pressure advance setting accordingly. There are several guidelines available on the internet (*https://marlinfw.org/tools/lin_advance/k-factor.html*; *https://www.advanced3dprinting.com/tools/*) to help with this procedure or use one of the available common slicer software that include settings for pressure advance (*GitHub-ollyfg/cura_pressure_advance_setting: A basic plugin for Ultimaker CURA that adds a (Klipper) pressure advance option under materials*).

Figure 3.10
Fault pattern "blobbing" and cornering due to deceleration specifically at directional changes

Be aware the pressure advance system setup is very sensitive to changes in printing speed, materials quality and printer variations. Ideally, the system should be setup and calibrated for individual prints making it most useful when larger print runs or multiple prints of same or similar objects are intended. The pressure advance system does not by itself change the timing or path of the toolhead. A print with pressure advance enabled in the slicer will take the same amount of time as a print without pressure advance. Pressure advance also does not relevantly change the total amount of filament extruded during a print. Pressure advance only affects the extruder output during acceleration and deceleration phases of the print head. The purpose of the pressure advance setting is mainly to provide better print quality (reduce oozing and blobbing) with a given printing setup. However, with the right settings applied, it also helps to improve part quality with faster printing.

3.1.6.3 Klipper

Klipper is a rather new 3D printer firmware. It combines the power of a general-purpose computer with one or more micro-controllers. Klipper is Free Software available e.g. on github (*https://github.com/Klipper3d/klipper*).

- Klipper allows for high precision stepper movements utilizing a simple application processor (such as a low-cost Raspberry Pi) when calculating printer movements. The software calculates precise step times based on the physics of acceleration and the physics of the machine kinematics allowing for more precise stepper movement thereby in tendency providing a quieter and more stable printer operation.
- Klipper is hardware agnostic and works on ARM, AVR, and PRU based micro-controllers. Existing "reprap" style printers can run Klipper without hardware modification by simply adding a Raspberry Pi or similarly powered computing unit.

- Klipper supports printers with multiple micro-controllers. For example, one microcontroller could be used to control an extruder, while another controls the printer's heaters, while a third controls the rest of the printer.
- Klipper supports smooth "Pressure Advance" settings – a mechanism to account for the effects of pressure variations during acceleration and deceleration of printing moves within an extruder. This reduces extruder "oozing/blobbing" and specifically improves the quality of print corners.
- Klipper supports "Input Shaping" to reduce the impact of natural vibrations on print quality allowing for faster printing speeds while still maintaining a high print quality.
- Klipper uses an "iterative solver" to calculate precise motor step times from simple kinematic equations. This makes porting Klipper to new types of robots comparably easier keeping timing of steps of various motors precise even with complex kinematics involved.

3.1.7 Support Materials

Also, additive manufacturing processes have geometric design limitations. In MEX processes based on thermoplastics, overhangs can only be printed to a limited extent – based on only one construction material (see Figure 3.11). The angle of the overhang and the allowed ratio of printing area on the build platform (as adhesion and stabilization area) to the area in a given layer plane are material specific. The faster a material solidifies after extrusion, with good build platform adhesion and good interlayer adhesion at the same time, the larger the overhang may be and the longer a bridge may be (see Section 3.1.8.5.1). The interlayer adhesion and solidification rate can be influenced within limits by the extrusion temperature, extrusion speed and fan settings. Basically, overhangs larger than 45° and bridges larger than 5 mm are typically problematic.

Here, too, PLA stands out as a building material with a balanced ratio of good build platform adhesion, interlayer adhesion as well as rapid solidification after deposition of the melt strand. It is still considered by far the most popular printing material in the private and semi-professional sector for mechanically undemanding functional parts and especially prototyping applications.

Through clever planning and positioning of a design, many parts can be constructed in such a way that they do not require support structures. In some cases, it is necessary to print a model in several parts and then assemble or glue them together in a second step. The advantage is that printers with only a single print head are generally sufficient for this. Furthermore, printing time and material are saved because printing without supports takes less time. This technique works well for some rather simple objects [Phillips 2022]. However, many relevant functional parts cannot be built without support structures. In simple FLM printers with only one print head, the support material is, in principle, the same as the build material.

Figure 3.11
Cube built on corner with support material ("building material")

Advantages:

- Low-cost printer
- No print head calibration necessary
- No compatible second material necessary
- Good material adhesion
- No risk of material degradation in the hot print head between build processes (support-structure/part)

Disadvantages:

- High costs for high-performance materials, as support material equals waste
- Surface/part defects possible especially when detaching tough/elastic materials
- Building material partly not suitable for printing filigree support structures

To some extent, the disadvantages can be reduced by fine-tuning the construction process of the support structures. For example, the printing temperature of the support structures can be lower, the line width narrower and the vertical distance between the support layers of the model larger. As a result, the adhesion strength of the support/framework structures to the part decreases, and the supports can be removed more easily. Increasing the distance between the support structure and the part, which is advantageous for better subsequent removal, tends to impair the surface quality of the part. Necessary support structures in areas that are difficult to access (cavities, etc.) are generally difficult to remove and should be avoided.

Alternative methods for producing parts with large overhangs are always based on the use of printers that can process at least two different materials in parallel or successively. There are various design options here:

- Two materials are fed alternately through the same extrusion nozzle. For this to work, precautions must be taken in the part design. The transitions between the materials must be cleanly separated. This is preferably done in a purge tower, which ensures that only uniform and uncontaminated material is processed as support or construction material with transition areas being disposed of in the waste tower. To ensure that the materials can be fed well through the same nozzle, they should have a similar melting behavior so that the pressure settings do not have to vary too much between the construction and support material. The filament guide required for multi-material printing can be fitted either in front of the print head [Kelly 2022] or in the print head [Samhaber 2021], and can be retrofitted if necessary.
- A printer has two or more independent nozzle heads and filament guides mounted on one print head. This design allows the printer to select independently from the two nozzle heads between the job. The disadvantage here is that the non-active print nozzle head must be continuously tempered to be ready for use. This can cause the material to leak or be thermally damaged. As a consequence, unattractive under-extrusion material type marks can appear on the part. This can also be countered by the slicer settings including by using a smaller waste tower. This ensures that only fresh material from a clean nozzle is used for the part.
- IDEX (independent dual extruder) printers are printers that have two independent extrusion heads that move side by side on one axis and work together on alternating printing of one part or allow simultaneous parallel/mirrored printing of the same. The two print heads work with one build material and an independent support material or with two independent build materials. Disadvantages of IDEX printers, similar to those described previously, are the need for continuous tempering of the permanently operational second print head and the reduction of the effective build space, since the print heads must not interfere with each other during printing.

Common support materials fall into four categories:

- Brittle support material (break-away) provides sufficient structural stability for the build process, but can be easily and quantitatively broken away from the produced part. This leaves the problem of inaccessible voids from which the material is difficult or impossible to remove. A common example of brittle support material is HIPS, which is often used in combination with ABS.
- Tough-elastic support material that builds up just sufficient adhesion to the part material can then be removed well and quantitatively. Here, too, there are sometimes problems with complete removal in areas that are difficult to access.

- Low-melting support material has a melting point and viscosity well below that of the construction material. After the construction process, the part including the support material is heated in an inert liquid or directly in a hot-air oven to a temperature at which the support material liquefies and can flow off. The interaction of a suitable support material and build material is not trivial here since the higher melting temperature of the build material can be problematic when layers are deposited on the support material causing deformations due to excessive liquidification of the previous layer(s). The same applies to materials that require very high build chamber temperatures, which may well go in the direction of the melting point of the support material.
- Soluble back-up material can be removed by a washing or dissolving process following the build process. The soluble support material should be compatible with the typical process conditions of the part material, be sufficiently stable, and have adequate adhesion to the build material. Furthermore, it is necessary that the support material dissolves in a liquid to which the build material is inert. If sufficient openings are provided in the part to ensure unobstructed inflow and outflow of the wash/solution fluid, then soluble back-up materials are the preferred solution for complex part geometries with overhangs.

There are five main types of soluble support materials for extrusion-based additive manufacturing:

- The most widely used water-soluble support materials are polyvinyl acetate (PVA), butenediol–vinyl alcohol copolymer) BVOH and hydroxypropylcellulose (HPMC). These thermoplastics dissolve easily in warm water. However, the water solubility implies the need to process the materials dry, as they are typically highly hygroscopic (water-attracting). Due to their comparatively low melting point and high polarity, the common water-soluble materials are only suitable to a limited extent for support pressure in combination with high-temperature thermoplastics (premature melting) and non-polar thermoplastics (adhesion problems).
- Oil-soluble support materials (e.g. HIPS) dissolve in non-polar oils. Some of these oils may come from biological precursors, such as limonene, which is suitable for dissolving HIPS as a support material and is obtained from orange peel. Be aware, however, that products from biological precursors can also be of environmental concern and require protective measures when working. Limonene, for example, is classified as flammable, hazardous to aquatic organisms and irritant.
- Solvent-sensitive support materials: In principle, material combinations with different stability behavior towards solvents can be combined with each other for construction and support materials. In addition to a direct solution, as in the case of HIPS as a support material and limonene in combination with ABS as a construction material, solvents can also selectively change the properties such as brittleness and consistency of support materials to such an extent that they are me-

chanically easier to separate from the part or fall off the part. Polar thermoplastics (polyester, polyurethanes, polycarbonate), for example, are sensitive to polar solvents such as acetone, ethanol, ethyl acetate, etc. and can be selectively weakened or removed by them.

- Alkali-soluble backings (e.g. various acrylate copolymers) are soluble in warm, mild alkali solutions. The melting point and adhesion properties can be adjusted within a certain range by changing the composition of the acrylate copolymers, which is why they are also suitable to a limited extent as back-up materials for high-temperature thermoplastics. When working with mild alkali solutions, it is also essential to wear suitable protective clothing (gloves, protective goggles, skin protection).
- Acid-soluble support materials: Most polyesters are degraded in warm, slightly acidic aqueous solutions. The polymers do not always dissolve. In some cases, the polymers become brittle or disintegrate into smaller pieces due to stress cracking, for example. Since this process tends to be more time-consuming, it is rarely used for thermoplastic support materials. The advantage, however, is that the polymer fragments are usually not water-soluble and can therefore be separated more cleanly from the water phase, which is why the water phase can be disposed of more easily.

3.1.8 Post-processing Methods

Parts produced by MEX processes often need to be reworked before they are ready for use. Post-processing can address the following problems, which are explained in more detail below:

- Removal of supporting material (mechanical/chemical)
- Surface treatment (mechanical/chemical)
- Functionalization (chemical)
- Consolidation (thermal/chemical)

3.1.8.1 Removal of Support Material

Depending on the material and part, support material can be removed mechanically and/or chemically.

3.1.8.1.1 Mechanical Removal

Support structures can best be removed with fine side cutters, scalpels and tweezers. The attachment point of the supports can be smoothed with sandpaper. Protective equipment such as safety goggles and respirator mask protect against flying parts and dust.

Gloves, goggles, respirator

Gloves should always be worn during post-processing to prevent fine plastic splinters from getting caught in the skin or sharp edges of the plastic or cutting tool from causing injury. Commercially available safety goggles are helpful against splintering and breaking off flying support parts. Machining with sandpaper produces fine plastic powder and sandpaper abrasion. Simple FFP2 or FFP3 masks protect against involuntary inhalation.

Tweezers, side cutter, filament cutter

Support structures from FLM prints cannot usually be removed automatically. They are removed as far as possible by hand at planned predetermined breaking points. In addition, various special tools help to remove difficult-to-reach or stubborn residues. Tweezers can be used to remove very fine structures very close to the predetermined breaking point without damaging the part. Side cutters and small cutting pliers, such as those used as filament cutters, are suitable for coarser structures.

Scalpels

Some support structures must be cut off. Special scalpels should preferably be used for this purpose. The use of "carpet cutters" is generally not recommended!

Sandpaper and polishing felt

Support structures can rarely be completely removed mechanically. Residues often remain on the part surface. These are best removed – if accessible – by processing with sandpaper and/or polishing felt. When using sandpaper, rather fine grit sizes larger than 400 should be used, since the abrasion resistance of plastics is typically rather low and a coarse grit size can easily lead to deep scores in the part surface. As in woodwork and paint finishing, work down from the coarser to the finer grit. Sanding of larger surfaces can also be done advantageously with small hand-held sanders, while sanding of hard-to-reach areas can be done with small multifunctional tools that have abrasives mounted on their rotary heads.

3.1.8.1.2 Chemical Removal

Before employing any chemical post-processing to remove support material, easily accessible and easily removable support material should be removed mechanically. This reduces the necessary amounts of solvent and detergent and the resulting liquid waste. Chemical removal of support material is based on the different chemical resistance of the construction material and the support material in a selected "solvent". Thus, at least two materials and appropriate printers capable of handling multiple materials are used for this purpose. During chemical removal, the support material must dissolve in the solvent or be destabilized by the solvent to such an extent that it detaches itself from the part or can be removed more easily (see Section 3.1.7).

3.1.8.2 Surface Treatment

Surface finishing of an additively manufactured part can be carried out in various ways. Classically, imperfect surfaces can be improved by polishing and painting. Similar polishing equipment, fillers and paints are used as for parts of the same material class (thermoplastic type, etc.) that have not been additively manufactured. If the parts are less complex and the surfaces are rather contaminated by small residues, then purely mechanical and automatable deburring methods such as sandblasting (ceramic or glass bead blasting), vibratory grinding and cryo-deburring can also be used.

A special method of improving the surface quality of additive manufactured plastic parts is the treatment with solvents and solvent vapors (highly volatile solvents such as acetone, methyl ethyl ketone, ethyl acetate, isopropanol, etc.). These smoothen the surface by (repeated) selective dissolving and drying. However, due to the nature of the process, some surface details may be lost. In the case of structural ribs and smaller holes and elevations, this is desirable, but in the case of particularly filigree structures it can also lead to their destabilization, dissolution and destruction. Furthermore, this process must be carried out with volatile solvents under conditions that absolutely avoid "solvent vapor poisoning" as well as reaching solvent vapor concentrations between the critical explosion limits (lower explosion limit (LEL) and the upper explosion limit (UEL) (see also the safety data sheets of the corresponding solvents).

3.1.8.3 Chemical Functionalization

Even the painting of a surface can be understood as a functionalization. In addition to painting/coating, there are also flame treatment, hydrophobizing, hydrophilizing, metallizing and other methods by which an additively manufactured plastic part can be surface-modified in the same way as conventional parts. Functionalization always serves the purpose of specifically changing a surface property and thus introducing relevant properties that are not provided by the building material itself (e.g. conductivity, weather resistance, bondability). Particularly in the context of enormously complex and large surfaces and internal structures, such as those accessible from additive manufacturing, surface functionalization is becoming increasingly important as a value-enhancing method. Examples include the additive manufacturing of medical parts [Jiang et al. 2020] and of parts and reaction vessels for chemical engineering [Parra-Cabrera et al. 2018].

3.1.8.4 Consolidation (Thermal/Chemical)

Consolidation (strengthening, stabilization) of parts can be carried out thermally or chemically.

3.1.8.4.1 Thermal Consolidation

A particularly important method of post-processing thermoplastic parts deals with their structural consolidation through thermal post-treatment. Thermal post-treatment is standard in many additive metal printing and liquid printing processes. In metal printing processes, a binder is often removed by burning (debinding), and the metal powders pre-sintered into an mechanically unstable green body are not combined into a persistent part until the post-sintering process. In the case of liquid resin systems, thermal post-treatment is used to remove previously unconverted reaction groups and to selectively form new, mechanically more stable compounds. Prints made from thermoplastic materials also benefit in principle from controlled thermal post-treatment.

Defects, porosity and insufficient layer adhesion represent possible failure points in a printed part and can in principle be improved by targeted heat treatment (post-sintering). Furthermore, the material can be relieved of stresses and cooled in a controlled manner to produce desired crystallization structures. When a thermoplastic is heated to a point near or above the glass transition temperature, the polymer chains become mobile and begin to relax. This allows stresses to be relieved. If a thermoplastic is heated to a point near or above the softening or melting point (glass transition point for amorphous thermoplastics, melting point for semi-crystalline thermoplastics), then the chains not only become mobile, but begin to flow. In this way, defects can be filled, surfaces smoothed, and transparency and layer adhesion increased (see [YouTube 2022]).

To prevent this softening by gravity from deforming the part, the part should be in a temperature and reaction inert matrix that can flow around and into the part and support it against gravity during thermal post-treatment. This matrix can be liquid (white oils, silicone oils, etc.) or solid (fine sand, table salt, metal powder, etc.), must not react with the part, and should be easily removable/washable from the part after heat treatment [Achten et al. 2021]. This method is particularly advantageous if no functionally stable parts are obtained via the building process itself because the build chamber conditions could not be perfectly adapted to the requirements of the material. A prerequisite for such a treatment is that all cavities of the part are reversibly accessible to the matrix substance. This is basically the case with powder-based parts, since here unreacted powder must be removed from cavities. In the case of extrusion-based parts, the use of this post-treatment must be taken into account in part planning, particularly with reference to cavities, as otherwise the surfaces may collapse. However, this restriction already largely applies to the use of soluble support materials (see Section 3.1.8.1.2).

Typical temperatures for heat treatment start at about the beginning of the glass transition temperature (T_g) and can go up to a few degrees above the melting point (T_m). Temperature and dwell time or cooling rate must be matched to the part and the specific thermal and rheological properties of a thermoplastic.

3.1.8.4.2 Chemical Consolidation

Chemical consolidation of a part can be achieved by various measures. The most common method is infusion with reactive resins, which can smooth surfaces, fill voids, and modify the chemical and mechanical resistance of the part depending on the choice of reactive resin. Typical reactive resins are based on reactive 2K epoxy or polyurethane resin formulations. Often, the surface of a part is also coated, which helps with amending critical surface defects and, depending on the coating, can improve the mechanical and chemical properties of the part. The selection of the right coating depends on the part to be processed.

3.1.8.5 Part Properties

The properties of an additively manufactured part depend on three key factors:

- Building material
- Building procedure
- Part design

The material and processing properties of suitable thermoplastics for extrusion-based additive manufacturing as well as the basic construction methods have already been discussed in detail in Chapter 2. In the following chapter, the effects of part design on part properties will be examined in a little more depth. In part design, there are various aspects that must be taken into account:

- General design recommendations for extrusion-based processes
- The influence of infill structures on part properties
- Load-oriented design

3.1.8.5.1 General Design Recommendations for FLM Parts

General design recommendations for extrusion-based processes address the aspects listed below.

Wall thickness

The minimum wall thickness of a part manufactured by the MEX process depends on the diameter of the filament nozzle. The wall construction should always be a multiple of the nozzle diameter, and in general, however, wall thicknesses below 1.5 mm should be avoided.

Part resolution

The CAD export of a part to the slicer should always be done with high resolution, so that part gaps and movement path errors in the slicer translation of the CAD file are minimized.

Tolerances

In the FLM process, process-related tolerances of approx. ±0.5 % must be taken into account, depending on the part size. However, the minimum tolerances are at least

±0.2 mm in the xy-direction for a standard 0.4 nozzle. With larger parts, the tolerances tend to increase. A major reason for this is the more pronounced shrinkage in FLM printing compared to thermoplastic injection processes, which can operate with high holding pressures in the mold. The tolerances depend largely on the part design and slicer ratio. The print files may have to be modified by design adjustments to account for the real printing results.

Moving parts

If moving parts are created in a printing process, then a distance of at least 0.5 mm should be provided between the individual parts and surfaces.

Minimization of support structures

Removal of support structures can result in damage to the part. Clever part positioning on the build platform helps minimize support structures. Support material is always needed where overhangs become too high an angle or distances under bridge structures become too large. To minimize support material, either the part can be placed on the largest support surface in the slicer or the alignment on the build plate can be iteratively optimized with a view to achieving the least amount of support material required. In general, support material should be positioned so that as little of it as possible is used within the part. Support material that extends from the build platform to the part is significantly easier to remove than support material that is built up between part structures.

The wall thicknesses of the part stabilized by support material should always be thicker than the support material to allow safe mechanical removal. If possible, predetermined breaking points from the support material to the part should be provided in the slicer.

The lower the filling thickness of the support structures, the easier they can be removed. However, care should be taken to ensure that they are sufficiently close to the points where overhangs and bridges are to be stabilized against sagging. The distance between the lines of support material can be increased to make it easier to remove the support. If the distance becomes too great, again the bridges will begin to sag.

The overhang threshold represents the slope in the direction of the horizontal build platform (measured from the vertical) above which the slicer places support (90° = horizontal). Supports can be minimized by changing the value: The higher the value, the less support is required. A common maximum value is between approx. 45° and 65°. The value depends strongly on the material and the process conditions as well as the part geometry and wall thickness. Fast-setting, amorphous thermoplastics allow higher threshold values than semi-crystalline thermoplastics.

Straight-line and non-crossing support structures are easier to remove than supportive contiguous patterns.

Part distortion

Uneven consolidation of part layers in an unheated build chamber leads to stresses and warpage. Accordingly, elongated and flat structures should rather be printed upright, since warpage tends to be lower in the z-direction.

Bridges/overhangs

Bridging refers to the printing distance between two points that have no connection to each other. In principle, an FLM printer can only build up layers on top of a base layer, although a certain continuous overhang is allowed. If a gap between two base layers cannot be closed continuously with an overhang, the printer must print a certain distance in the air. If this distance is not sufficiently underbuilt by support structures, then the filament either sags from a typical overbuilt free distance, or a gap is created. The permitted bridging distance is material- and process-specific and can be optimized accordingly by various measures.

The layer thickness and the nozzle size have a major influence on the bridging distance. Broad and flattened layers in relation to the nozzle diameter offer more contact area in comparison and can therefore stabilize overhangs better. The overhangs also become visually better the flatter the individual material layers are. The layer thickness is determined by the nozzle distance to the previous part layer. By changing the flow rate, the printer can, if it detects a bridge or a critical overhang, simulate a larger nozzle for a short time and increase the part adhesion to the last layer locally in a targeted manner (bridge flow ratio/bridging extrusion multiplier). This setting is often found in slicers in the "advanced" area for advanced users.

In general, bridging is a problem upwards a length of approximately 5 mm, since the deposited thermoplastic strand tends to sag in the melt.

Another decisive factor is the printing temperature. The melt viscosity of the thermoplastic decreases with the temperature at the nozzle. A low printing temperature increases the melt viscosity, thus the stability of the melt strand, which accordingly helps with bridging. However, if the printing temperature is too low, then the printing speed must be reduced because otherwise the thermoplastic will be printed uncleanly or incompletely, which can result in gaps. This can mean that the last layer may no longer be able to bond with the previously printed layer. A slower printing speed while bridging again increases the likelihood of sagging. Generally, a fast printing speed during bridging is preferred.

Likewise, a warm build chamber delays the solidification of the melt strand, while a cold build chamber accelerates the solidification. However, cooling the build chamber is not recommended due to part distortion caused by poorly controlled shrinkage. Instead of adjusting the build chamber temperature, the part fan setting on the print head is used specifically to promote faster solidification of the thermoplastic from the melt for a limited time and locally during the build process. Rapid cooling of the deposited material by the part fan on the extruder is advantageous when printing bridges and overhangs, because the melt strand solidifies faster and sags less. To ensure that the fan is particularly effective, the direction of travel of the print head should ideally be aligned in the direction of the air flow for bridges.

Bridging behavior varies from thermoplastic to thermoplastic and sometimes from filament to filament. Filaments based on the same plastic can differ significantly in properties (see Section 2.6.6.2). Plastics are often blended (compounded) with additives and fillers prior to filament production, some of which improve bridging behavior.

For the empirical setting of suitable printing parameters, so-called temperature towers are useful, which test the printing behavior of materials during FLM printing of bridges and overhangs at different temperatures (see Figure 3.12).

Also the printing speed must be adapted to the part. This is particularly important if the layer time in the part varies greatly. The last layer deposited should always have sufficient time to solidify before the next layer is deposited. If the last layer is still too soft, even perfectly vertical structures and structures with only a small overhang angle tend to sag.

To ensure that filigree structures can also be produced in parts with short layer build times, either the build speed should be reduced or a pause should be inserted between layers. This pause is usually bridged by a waste tower to avoid caking in the extruder nozzle, for example. Layer times should not be less than approx. 20 s. For the bridging itself, the build speed should be set comparatively high in order to quickly achieve anchoring of the extruded strand on the other side of the bridging.

Figure 3.12
Temperature tower to optimize settings for bridging behavior [Steinhausler 2017]

Extreme overhangs are best printed with large nozzle diameters and low layer thicknesses because, as previously described, such a large bearing surface of the deposited melt strands ensures stabilization of the overhang. Some particularly fast-solidifying materials tend to warp upward when overhangs are printed, since they can cool more quickly in this direction and shrink accordingly. To prevent these overhangs from being torn off in the next travel move of the printer, the latter can be raised during empty runs by changing the z-hop setting in the slicer. However, the z-hop setting must be combined very well with the retraction setting to avoid unsightly elevations and threads.

Another helpful setting in combination with sensitive part geometries with overhangs and bridges is "avoid crossing perimeters". If this function is activated, the print head tries to avoid crossing existing part layers as far as possible in mere travel movements.

Overhangs of up to 45° can usually be printed cleanly. With optimized settings and part geometries that are not too filigree, overhangs of up to 70° are often possible. However, the shape of the overhang is decisive here. Linear and concave overhangs are printed more cleanly than convex ones. Printing individual struts is generally difficult because stabilization by other part structures is lacking. There are also ready-made test prints for overhangs that check various parameters and facilitate the empirical selection of suitable slicer settings (see Figure 3.13).

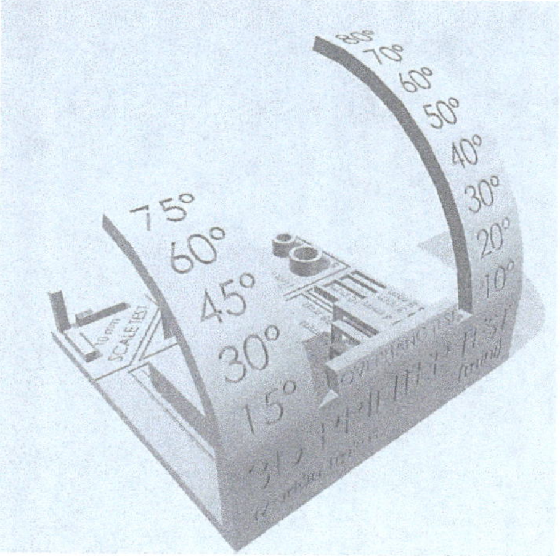

Figure 3.13
Test part for analyzing the printing properties of a material or a slicer setting [Trpkoš 2018]

To avoid supporting critical overhangs, FLM printers are being developed based on multi-axis systems. For example, the extruder nozzle can be equipped with an additional axis so that the application can also be performed at 90° to a given layer by deflecting the nozzle itself by 90° [3DDruck 2017]. Another method adds two axes to the build platform, allowing the part to be aligned with the nozzle, or a six-axis robot is used directly with the extrusion nozzle mounted. In either case, not only can overhangs and bridges be built better, but support structures can be largely eliminated. Six-axis FLM printers can also be used to create vertical reinforcement structures in the strand laydown direction, enabling true load-bearing part design. Printing with multi-axis systems significantly complicates the design of the print path. Existing standard slicer software must be considerably enhanced for this purpose, for example, to enable automated load-path-dependent printer control [Wang et al. 2019].

Raised and recessed surfaces

Raised lettering and details should have a minimum wall thickness of 0.5 mm and a height of 0.5 mm. For recessed details, the line width should not be less than 1 mm and the depth at least 1 mm. Fonts without serifs are preferred for lettering.

Corners

Since the nozzles are circular, the edges and corners are printed at best with a radius the size of the diameter of the nozzle. Thus, objects produced with FLM printers cannot have perfectly sharp 90° edges. If parts are to be printed with corners or edges that are as sharp as possible, then the first part layers are particularly critical. If the nozzle diameter is larger than the layer thickness, then the extruded filament of each layer will be pressed down and deformed. This is good for layer adhesion, but causes the first layers in particular to visibly overhang wider on the unyielding build platform. The overhang of the material of the first layers is known as elephant foot.

At sharp edges, the speed of the print head changes dramatically due to changes in direction. It first drops to zero and then needs time to reach the target speed again. If the amount of extrusion is not precisely adjusted in time to this, then over-extrusion usually occurs at edges and corners (link to pressure advance settings). Ideally, parts manufactured using FLM printing should always be designed with slightly rounded edges.

Drilling

Drill holes should have a minimum diameter of 2 mm. Smaller holes should only be indicated by means of centering or countersinking and drilled out later. Drilled holes should be designed approx. 0.1 mm to approx. 0.75 mm larger than actually necessary (depending on the nozzle size used), since the slicing process and over-extrusion tend to reduce the size of the holes towards the inside. Holes are calculated by the slicer in the center of the printed web curve. Ideally, a 0.4 mm nozzle also deposits a 0.4 mm wide melt strand on the substrate. This then protrudes by half, i.e. 0.2 mm, beyond the hole circle on both sides.

The quality of the STL file also influences the holes. Bore edges are divided into triangles. Viewed from above, the tips of the triangles are then on the diameter, the lines go through the inside of the hole. At low STL resolution, the borehole becomes even smaller. Therefore, inner contours should always be resolved as high as possible.

Part connections

Smaller vertical pins and corresponding holes help the part to be aligned or assembled later if it has been printed in several parts. Larger pins have a diameter of 5 mm or more, and are used to increase the strength of the connection between fillings or outer layers. For particularly small vertical pins, a small fillet can be added at the base. Thanks to the addition of a fillet at the base of the pin, the stress concentration at this point is reduced, and at the same time the strength of the part increases. Regular printer calibration significantly improves the success rate for filigree part sections.

Hole alignment: Hard-to-remove support material in holes complementary to the pins can be avoided by changing the orientation of the print. If designs with multiple holes in different directions are required, then blind holes should be printed first, prioritizing first by smallest and largest diameter and finally by hole size [GSR 2022].

3.1.8.5.2 Resolution

The resolution of FLM printers depends on several factors, which are listed and explained below.

Nozzle size

The diameter and shape of the extrusion nozzle limit the resolution of melt strands in the xy-direction. Stable line resolution is typically equal to and preferably greater than the nozzle diameter. Nozzle diameter for common FLM printers is typically equal to or greater than 0.2 mm. For nozzle diameters smaller than 0.2 mm, the pressure at the nozzle increases sharply and the build speed is thus significantly reduced while the residence time in the hotend is equally increased, both unwanted conditions for stable and efficient printing.

Resolution of the stepper motors of the printer in xy- and z-direction

The resolution of the stepper motors and the stability of the x and y axes further limit the resolution of the print in the xy-direction. The stepper motors in the z-axis independently control the resolution in the z-direction. Motion in the xy-direction is controlled simultaneously by two separate stepper motors, further reducing accuracy, especially for highly accelerated motion and complex motion patterns. A resolution in the range of less than 0.1 mm is possible here only at low travel speeds. FLM printers can mechanically realize a significantly higher resolution in the z-direction than in the xy-direction. Resolutions greater than 0.05 mm are technically feasible with precisely calibrated build platforms while the first layers are particularly critical.

Build speed

The build speed has a major influence on part resolution. The stability of the build frame of the printer against vibration plays a decisive role here, as high forces are generated by the accelerated extruder head, especially during fast build processes. Here, lightweight hotends attached via a filament tube to Bowden extruders have an advantage over the heavier direct extruder and pellet extruder heads. A robust, heavy metal frame that anchors the printer and the print bed to the floor further protects against vibrations and natural frequency resonance. In advanced printers stepper motor control via input shaping can help to increase the printing speed further (see Section 3.6.1).

Part geometry

The part geometry and part positioning on the building platform determines how smoothly an extrusion process can run. Parts that can be built with a continuous and uniform print head movement and have wall thicknesses in integer multiples of the nozzle diameter, as well as allowing sufficient time between the individual layers, are particularly easy to manufacture with high resolution and high printing speeds.

Viscosity and solidification behavior of the thermoplastic material

To enable a high layer resolution in the z-direction multiples smaller than the nozzle diameter of the printer, the thermoplastic to be processed must "have special properties". The discharge viscosity must allow the printer to deposit or press the extruded melt polymer mass, in a thickness smaller than the nozzle diameter, evenly onto the previous layer. If the viscosity of the melt is too high, the discharge mass tends to pulse in the z, but especially in the xy-direction, because the discharge pressure becomes too high. This is particularly the case with very small nozzle diameters. If the viscosity is too low, then unintentional leakage from the nozzle can occur with the consequence of over- and under-extrusion defects. Once again, the property behavior of PLA is exemplary here, which is why filigree prototypes are often constructed from PLA.

Slicer and printer settings

Slicer settings make all the difference in print quality. Even high-resolution FLM printers, print models in poor quality if their settings are not optimized for the print job. The necessary optimizations are many (as discussed above) and very part design dependent. Although if the slicer settings are optimized, then a clear distinction can be made between the production results of high-quality precise FLM printers and simple, inexpensive do-it-yourself (DIY) printers. In addition to the part surface quality, the difference becomes particularly visible with more complex parts that have diagonal or arc-shaped lines or embossing or engraving designs. If the "adaptive layers" function is activated, a compromise between accuracy and printing time in a slicer is achieved. The layer thicknesses are dynamically adapted to the part geometry. Depending on the part design, the layer thickness is reduced at critical points of the object for higher accuracy and increased at non-critical points for faster print time.

3.1.8.5.3 Density

In MEX processes, each part is built up layer by layer from a large number of melt strands. The arrangement of the melt strands in relation to each other as extrusion lines has a decisive effect on the material density and thus on the part strength and the tightness of the part walls. Essential parameters are

- Nozzle diameter
- Layer height
- Line width
- Setting of extrusion factor
- Overlap/air gap.

The nozzle diameter determines the minimum width of the extruded melt strand. The height and width of the melt strand are further specified by the combination of set layer height and extrusion multiplier. The width of the deposited melt strand is usually larger than the nozzle diameter. The layer height is usually selected to be significantly smaller than the nozzle diameter in order to achieve a certain contact pressure on the previous material layer and thus improve adhesion and fill gaps.

If the ratio of line width to nozzle diameter is too high/if the layer thickness is too small, or if the materials are too highly viscous, a too high pressure can build up at the nozzle. If less mass is a conveyed than required, then imperfections in the part result from under-extrusion. The consequence are often air gaps between parallel extrusion lines, which have a negative influence on part strength and density. An air gap is often referred to as a negative overlap.

When adjusting the slicer, such air gaps can be avoided by deliberate over-extrusion, either by adjusting the extrusion multiplier or by placing the build lines closer to each other than the nozzle diameter of the print head. This results in a positive overlap of extrusion lines. The extrusion multiplier is a correction factor which adjusts the amount of material to be extruded with reference to the given parameters of nozzle diameter, layer thickness and travel speed respectively the advance speed of the filament in the filament extruder calculated by the slicer. In the layer direction and in the direction of the extrusion line, an increase in strength can be expected with increasing line width and line overlap, as well as an increased tightness of the part. In general, as the extrusion multiplier and thus the amount of material is increased, the strength of the part increases. However, excessive over-extrusion should be avoided. This usually manifests itself in material accumulation in the uppermost print layer or in extrusion lines swelling out laterally (over-extrusion) and can have a negative effect on surface quality and, under certain circumstances, also on part strength, depending on the direction of loading.

In order to achieve the highest possible part strength, the orientation of the extrusion lines should be selected so that they are deposited as far as possible in the direction of the expected load. In order to achieve the highest possible part surface density when used as a fluid vessel (under pressure), care must be taken to ensure that the extrusion lines are deposited perpendicular to the acting internal pressure of a gas or liquid. Tight parts are preferably produced with a high degree of wall filling so that as few defects as possible occur perpendicular to the pressure effect. Analogous to the general load strength, the z-part strength is also the most critical factor for tightness in pressure applications.

In principle, under ideal construction conditions, a gas/liquid-tight part should be possible with only one extrusion line width. In reality, however, at least two or three layers are necessary, in order to achieve a tightness in the "normal" pressure range of up to one bar.

Dense, FLM based pressure capable parts are increasingly used in research as inexpensive and fast to manufacture reaction vessels [Heikkinen et al. 2018]. The density of additively manufactured products – in terms of part weight – is typically lower than the density of classically manufactured parts while at least in theory also being able to provide better mechanical properties in load direction.

In reality only if the possibility of designing parts with geometrically and load optimized infill structures and fiber reinforcement with non-statistical fiber alignment is consistently realized, can additively manufactured parts – for now – compete mechanically with classically produced parts, especially if not only small series or highly complex parts are considered.

3.1.8.5.4 Dimensional Accuracy

Dimensional accuracy is an essential property when it comes to the additive manufacturing of functional technical parts. For this purpose, the planned part geometry should match the printed part geometry as exactly as possible. Dimensional accuracy is influenced by many factors, the most important of which are listed and discussed below.

Under- or over-extrusion

Both under- and over-extrusion not only cause surface defects (elevations, protrusions and holes), but also directly affect part dimensional stability. Particularly at risk are sharp edges (rounding), filigree structures (stability and dissolution), build-ups and holes (dimensions), and overhangs and bridges (surface steps and holes).

Shrinkage

Shrinkage is a fundamental physical property of thermoplastics. Ideally, it acts uniformly in all build directions. Shrinkage is particularly problematic for large surfaces and structures arranged parallel to the printing platform. Shrinkage should be taken into account already in part design. For critical parts and materials, a printer with build chamber heating should be used to minimize non-uniform shrinking.

Filament quality (ovality, diameter variation)

Irregularities in filament diameter or filament roundness that are not taken into account can lead to transport problems (clogging/slipping) and usually have a negative effect on the effective discharge volume of the melt (typically causing under-extrusion).

Printer resolution

The resolution of the produced part is limited by the nozzle diameter (extrusion line thickness) and the physical printer resolution (xy/z). Adding to this are influences such as build speed (stability of the printer against vibrations) and shrinkage (material/process-dependent).

Part design

When designing parts, the basic capabilities of a printing technology must be considered. For each material-extrusion-based printing technology (MEX) and setup like xyz, delta, number of axes robot-assisted, Bowden extruder/direct extruder, granulate extruder, acceleration speed, different specifications must be taken into account. Basic information on this can be obtained from the printer manufacturer on request and is partly already programmed into the selections and routines oft CAD programs and slicers.

Print bed calibration

Good print bed calibration is not only critical for printing the first layer. If the nozzle is too high above the build platform or too close to it, then this can have a major impact on the next 10–20 layers of a part and lead to complete print failure. Modern printers calibrate the z-axis before each print. Alternatively, support structures can be used to start a part well above the printing platform, so that the printer has the opportunity to compensate for the z-axis itself in the printing process during the first few layers.

3.1.8.5.5 Fill Structures (infill)

Fill structures in additive manufacturing are permanent support structures inside a part. The fill structure is specified by the user in the slicer. A choice can be made between different regular structures and a filling level (spatial density) of the fill structure can be set. The ability to use partially complex, undercutting and particularly load-bearing infill structures is one of the key advantages of additive manufacturing. The infill significantly determines the mechanical properties, the part density and the manufacturing time of a part to be printed.

Fill structures can generally be divided into 2.5D and 3D structures. A 2.5D structure is created by projecting a 2.5D structure into the part space to be filled. Typical examples of 2.5D structures are: lines, zigzag, grid, triangles, tri-hexagonal, honeycomb, cubic, octet, gyroid, etc. (see Figure 3.14). In contrast, 3D structures are arbitrary structures with arbitrary position in space (arbitrary is to be taken literally here; the fillers can be both functional and "decorative"). What all these filler structures have in common is that they have clear units of repetition.

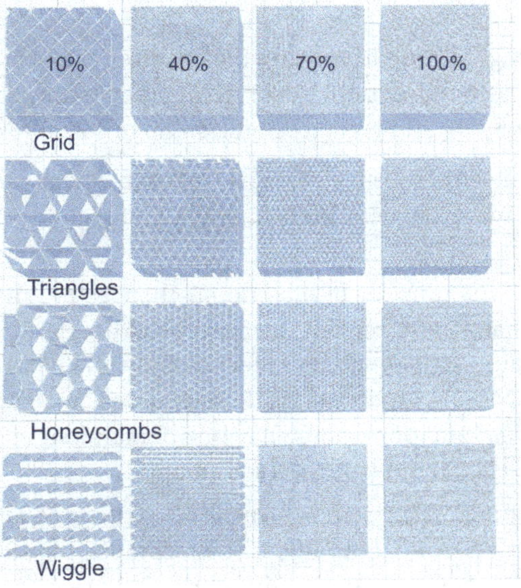

Figure 3.14 Example fill structures and fill structure densities produced with a 0.4 mm nozzle diameter as 40 × 40 × 40 mm cube

Irregular fill structures are another category of fill structures that are becoming increasingly important in modern part design. These originate from optimization calculations, often based on the expected load case of a part. Such fill structures are often characterized by an optimized property-weight ratio and are the object of intensive research.

Like part geometry, fill structures are also dependent on the capabilities of a printing technology. Classic xyz and delta printers can effectively print 2.5D structures, which is why they are also part of all common slicers. The printing of 3D structures and irregular fill structures is best achieved with five- or six-axis printing systems, and using self-supporting print processes such as powder sintering technologies (see Section 3.2).

Load-oriented filler structures become particularly interesting when different materials are used with the aid of multi-head or multi-material printers, thus making local property optimization possible, or when reinforcing fillers (ideally continuous fibers) can be processed in a load-oriented manner. In this way, parts with previously unattainable functional property combinations (strength, elasticity, weight) can be produced.

The common 2.5D fill structures can be roughly sorted into different categories:

- Fill structures with low strength but high building speed: lines, zigzags or grids (50–100 mm/s depending on the material, with input shaping and high acceleration up to 500 mm/s)
- Fill structures with medium strength and medium build speed: grids, triangles or tri-hexagon (up to 50 mm/s depending on material, with input shaping and high acceleration up to 250 mm/s).
- Fill structures with comparatively high strength and significantly reduced construction speed: cube, cube sub-area, octet, quarter cube or gyroid (depending on material up to 30 mm/s, with input shaping and high acceleration up to 150 mm/s)
- Filling structures for elastic parts: concentric, transverse, 3D transverse, wave or wiggle (up to 30 mm/s depending on material, with input shaping and high acceleration up to 150 mm/s).

Basically, the build speed decreases with the number of directional changes per distance unit of the filling structure. At a change of direction the print head decelerates to 0 and then has to be accelerated again. Accordingly, the demands on the design stability of the printer increase in order to avoid vibrations. Rounded design structures such as gyroid/wiggle improve this situation somewhat but demand significantly higher computing speed and data transfer to be processed smoothly.

In the direction of load- and weight-optimized filler structures, 2.5D structures are developing that, depending on the part environment, adjust their complexity and filler density. For example, terminating elements (walls, ceilings/bridging areas) are readily reinforced analogously to support structures directly below an overhang by continuously increasing the fill structure density (number of lines and/or line thickness) or by changing the fill structure geometry itself as required. These functions are often found in the “Advanced areas” of slicers in the “Fill structure settings” section. Some slicers also allow the user to selectively set additional fill structures or to thin out/remove automatically generated fill structures. This enables the designer to exert considerable influence on the optimization of the part design manually and empirically through their experience.

The higher the expected part load, the higher the infill should be. Simple prototypes and figures often get by with 5 to 15 % infill. Parts with higher loads should have 20–30 % infill. In areas that are secured by bolting, for example, infill levels of up to approx. 60 % should be used. If the filler content is selected too high, the risk of warpage in the reinforced part area due to shrinkage increases significantly. Part strength does not increase linearly with the degree of filling. Depending on the load, fully filled areas can be less stable and tough than part areas that are 50–80 % filled, while exhibiting only an insignificantly higher flexural modulus.

In addition to the choice of fill structure, its alignment is also important. Fill structures in xyz-printers like to be aligned with a 45° angle on the build platform, as this allows complex areas to be filled more uniformly and the printer to perform uniform/fast xy travel moves. In this way, build speed can be optimized. However, it may be advantageous to set a different angle of fill structures when aiming to obtain maximum part strength or flexibility.

To achieve particularly high anchorage and force transmission of the filling structure to the part envelope, the internal structure can overlap with the outer walls at points and edges. Under no circumstances, however, should the overlap be set too large, so that the filling does not penetrate so far into the outer walls of the part that it visibly pushes through to the outside. An overlap of 10 % typically provides both good and uniform adhesion between the walls and the infill without unduly risking the infill to push through the outer layer.

Another way to stabilize parts is to increase the part shell thickness. The outer shell of the part is usually printed first and represents the spatial boundary of the internal filler structures. An effective filling structure is only possible if the outer shell of the part can transmit sufficient force, which is why it should have a sufficient thickness. Sufficient shell thickness is also necessary for possible mechanical post-processing/tempering. For more information on infill, see the literature [Gebhardt et al. 2016, Fernandez-Vicente et al. 2016, Grund 2015 and Wipperfürth 2018].

3.1.8.5.6 Mechanical Properties (Depending on Direction)

In principle, MEX-based additive processes generate anisotropic properties. The printing of a mechanically functional part therefore depends essentially on the understanding of the necessary part properties in combination with the capabilities of the printer and suitable printing materials. In accordance with the desired properties, the part must not only be designed to withstand the load, but must also be manufactured in sufficient quality by the given construction process. High tensile and bending forces during part use should be ideally always aligned in the xy construction direction.

Since the material is not applied continuously and stress concentrations form due to the many individual layer connections, the layer strength is typically weaker in the z-direction. If not designed properly small valleys and mountains form between the individual layers during the building process. In the valleys, the stress is concentrated and cracks can form more easily.

The directional dependence of the mechanics of parts is strongly influenced by the "welding conditions" of the thermoplastic melt (temperatures, viscosities, cooling rates, crystallization times, etc.) between the build layers and can be optimized to some extent by targeted thermal post-curing (under normal pressure or elevated pressure). Preferably, post-curing is performed in a supporting, temperature-inert and non-adherent material, to ensure that gravity does not cause further deformation of the softened or melted part during post-curing. The design of a part must take the annealing step into account, as the supporting material must be able to fill all cavities and also drain out of them. Post-treatment temperatures should be slightly above the softening temperature to allow the layers to flow together and sinter. Suitable inert materials are, for example, white oils, waxes, silicone oils and finely ground alkali metal salts, e.g. NaCl, (known as the "salt method" [YouTube 2022]).

3.1.9 Software

In the production of 3D parts using additive processes, different types of software are used, in some cases consecutively, depending on the requirements of the part.

Design Software

If the product shape does not come from a scan file or is predetermined, then CAD (computer-aided design) programs are usually used to create the 3D shape of a product. From the CAD program, file formats such as STL, OBJ, DAE, WRL, X3D, PLY, and 3MF can be exported, which are subsequently converted into control commands for 3D printers in so-called slicers (for a list of common free and paid CAD programs, see [Wikipedia 2022a]).

Topology optimization

In addition to pure design software, certain parts of CAD program families often also allow product optimization processes, topology optimization and product simulations to be carried out. The result can be, for example, load-optimized and space- or weight-optimized part designs. Additive manufacturing, unlike traditional machining techniques, enables the production of parts with more complex geometries whose overall weight can be optimized using a numerical method called topology optimization. Topology optimization is a part of a numerical design that allows, thanks to mathematical formulas, to find the optimal material distribution in an allowed volume to be found, subject to given mechanical property constraints. Thus, topology optimization allows saving material where no stresses/loads are applied to a part. This is done with the aid of special software.

Conventional numerical design methods work by applying loads to an already designed part and evaluating where it will weaken. Engineers then further optimize the design until the part meets the specified mechanical properties. In topology optimization, the mechanical loads, along with the allowed part dimensions, are already part of the input data that allows the software to propose an optimized geometry through continuous iteration and simulation starting from a crude initial part geometry.

Topology optimization and thus effective weight and installation space savings are highly relevant in a number of application areas where masses have to be accelerated. Examples include the automotive and aerospace industries. In the medical sector, it is primarily a matter of optimizing function. Basically, every gram that does not have to be printed saves time. In additive manufacturing, the specifications for the optimization software must be chosen to reflect the capabilities of the desired manufacturing processes. This results in significantly different specifications for the individual additive manufacturing processes and a high level of complexity in setting the correct optimization criteria and limits in the program itself.

Slicer

A slicer program divides 3D objects into many thin horizontal layers, which are then printed layer by layer. A so-called G-code or control code for the printer is created by the slicer for each layer. The G-code is a standardized movement instruction that was originally developed for CNC milling but is now also used for 3D printers. The G-code contains precise movement instructions for the print head, including movements and movement speeds in the xyz-direction. In addition, machine parameters such as print temperatures, extrusion speeds and build temperatures are mapped in G-code (for a good overview of common G-code commands, see [Wikipedia 2022b]). Slicers are sometimes proprietary programs sold with the 3D printers. This comes into play especially when printers have special capabilities such as printing with more than three axes or continuous filament processing. Alternatively, slicers are designed to work with a wide variety of printers. Some of these slicer programs are offered for a fee or as public domain freeware programs (for a good overview of well-known slicers, see [3Druck 2022]).

G-Code Viewer

G-Code Viewers help visualize the printer's travel moves in advance to detect any errors and gaps in execution early on. The viewers are often available as part of slicers or as stand-alone solutions. Some G-Code viewers are also capable of directly controlling CNC mills and 3D printers. Most G-Code viewers are freeware [O'Connell 2022].

3.1.10 3D Printing Materials Recycling

Various recycling options are being discussed in connection with the additive manufacturing of thermoplastics.

Recycling of classically manufactured products from thermoplastics for additive manufacturing

The large quantities of uncolored and comparatively unmixed packaging plastics such as PET, PBT, PLA, COPE and, to a lesser extent, thermoplastics from technical applications such as PA6 from fishing nets can in principle be used as raw materials in MEX processes. For reuse in additive manufacturing to work, the plastics must be largely unfilled, unmixed, clean, dry and either regranulated or in filament form. Reuse as a feedstock in additive manufacturing is therefore preceded by the familiar process steps in recycling:

- Collection
- Sorting
- Crushing
- Cleaning
- Drying
- Regranulation
- Filament extrusion

The quality requirements for material uniformity and processability are high in additive manufacturing with thermoplastics, even with the more robust extrusion-based processes (MEX). Filaments and granules made from recycled materials that achieve this quality are offered at about the same price as freshly produced (virgin) materials. To guarantee the necessary quality, the material is tested for its rheological and mechanical properties before use. Frequently, recycled material must be blended with a relevant proportion of fresh material before it can be used again. The share of recycled thermoplastics in additive manufacturing is still small, but there is a growing demand.

Recycling of waste products/misprints from additive manufacturing

Even with optimized design, extrusion-based additive manufacturing regularly produces waste from support materials or failed prints. These can be easily collected by type, shredded and regranulated or re-extruded as filaments. The mechanical stress on the materials especially in filament-based additive manufacturing is rather low, with the thermal load being determined by the process conditions (nozzle temperature, build chamber temperature, dwell time in the nozzle/heated build chamber).

Materials that can be printed without a build chamber heater only experience a short thermal load of a few seconds in the print head. If the materials are sufficiently dried before printing and the print head operates trouble-free at a "normal" extrusion speed, then the thermal load in the print head is rather low and most materials do not suffer any relevant damage.

More problematic in this context is the process of shredding, regranulation or filament production from the waste. "Home" equipment for shredding and filament extrusion available in the non-professional sector causes a significantly higher mechanical and thermal load than additive manufacturing itself. It is not possible to predict the residual properties, and a property assessment here can only be made empirically. If, on the other hand, the waste is collected and made available to professional recycling companies, then a mechanically good recycled product can be created again by controlled admixing of fresh material. However, the color of the recycled materials is often irreversibly affected by the usual mixing of qualities. The same applies, of course, to the recycling of additively manufactured parts after their service life, in which case the classic recycling processes and requalification measures as described above need to take effect.

Availability of regranulate and refilament

The availability of recycled material for use in extrusion-based additive manufacturing is highly dependent on the scale of consumption of the materials. The most readily available are recycled PLA, PET, and to a lesser extent ABS in filament form. When it comes to recycling plastic, it is also worth looking at independent open-source initiatives, such as the Dutch project Precious Plastic [Precious Plastic 2022], which encourages in various ways reduction the consumption of thermoplastics through targeted recycling. A growing number of companies, which can easily be found on the Internet under the search term “recycled filament”, already offer recycled thermoplastics in the form of filaments. At the Kunststoff-Zentrum in Leipzig (KUZ), the “ReUp-3D-Printwaste” research project funded by the BMWi started in 2021. The KUZ project focuses on the one hand on the recycling and upcycling of printing waste from the field of selective laser sintering (SLS), specifically PA12 waste powder, and on the other hand on support structures and misprints from FLM printing. Here, the main focus is on PLA and its reuse in the form of filaments in extrusion-based printing processes [KUZ 2022].

3.1.11 3D Printing Materials Ageing

In any processing method, the processed material can be changed. The changes can be of a chemical nature, for example induced by thermal stress during processing, or they can be mechanical stresses caused, for example, by high shear of the materials during processing.

3.1.11.1 Thermal Stress

The thermal stress of materials in extrusion-based additive manufacturing does not differ significantly from the stress in classic molding technologies. Factors that influence the thermal damage of the material during the production process are

- Extrusion temperature
- Dwell time in the nozzle
- Formulation parts of the thermoplastic
- Degree of drying of the thermoplastic.

For each thermoplastic or thermoplastic formulation, there are recommended safe processing temperature ranges, which in the professional environment can be taken from the technical data sheets and safety data sheets supplied with the materials. In extrusion-based additive manufacturing, these temperature ranges are often exploited to the maximum in order to achieve a lower melt viscosity, a higher discharge rate and thus a higher assembly speed. At the same time, the residence time in the melt in most filament-based processes is significantly lower than in classic molding processes such as injection molding, extrusion or deep drawing.

The thermal stress in granule-based additive MEX processes is more comparable to classical extrusion processes, since similar extruder temperatures and in some cases even higher residence times are common. The temperature resistance of a material or formulation can be analytically pretested by a thermogravimetric analysis (TGA) () (see Section 2.5.8). Here, information on the thermal decomposition stability is obtained. If a gas chromatography–mass spectrometry (GC-MS) instrument is coupled to the waste gas stream, it is possible to obtain additional qualitative information about the possible decomposition products during thermal processing.

Problems occur when materials are processed above their decomposition temperature [DGUV 2022] or if the formulations contain increased amounts of volatile organic compounds (VOCs) such as plasticizers or monomers (for example, styrene is the largest emission from ABS and HIP [Azimi et al. 2016]). It is imperative to avoid the accumulation of material residues at the nozzle or heating block, where material can remain hot and with air access for uncontrolled periods of time. Such materials can coke up, clog the nozzles and possibly lead to an increased concentration of toxic degradation products. Even in the home, it is therefore essential to ensure that the extruder head is clean and that it is operated in a well-ventilated area or in a closed pressurized enclosure equipped with an air filter.

The normal process temperatures are basically selected so that spontaneous combustion of the material cannot take place. The fact that filament printers in the home nevertheless regularly burn down is often due to the inadequate electrical equipment of the printers, short circuits or unchecked heating of the heating cartridge due to faulty temperature measurement. This results in sparks and temperatures that can cause the printer to ignite, although the extruded thermoplastic often plays a rather minor role here.

3.1.11.2 Mechanical Stress

The mechanical stress on thermoplastics in filament-based additive manufacturing can be considered low compared to classical extrusion and injection molding applications. The shear rates and elongation flows occurring in a common print head can be easily calculated and simulated (see Section 2.5.7). Only low shear rates are achieved in the print head, since the melt strand discharge velocity of less than 100 mm/s is rather low compared to the diameter reduction in the nozzle channel for common filament diameters of 1.7–3 mm and nozzle diameters of 0.4–1 mm.

In pellet-based extrusion processes, the highest shear and stress often occur in the plasticizing unit of the extruder. However, the dependence of the shear on the extrusion speed and the nozzle diameter is comparable to a classic profile extrusion. As a tendency, the residence time is somewhat longer, but the discharge speed at the nozzle and thus the shear at the nozzle opening tend to be lower.

3.1.12 Requirements for the Production Environment

While additive manufacturing processes were primarily used in prototype construction in the past, they have recently also been increasingly used in series production. This industrialization step brings with it new challenges in production planning and control (PPC).

PPC is also referred to as work control and comprises the organizational planning, control and monitoring of production processes under quantity, deadline and capacity constraints. It is an elementary part of order processing and the core of every manufacturing company.

The trend towards customized products in additive manufacturing and shorter product life cycles poses new challenges for production systems and the associated work planning and control.

Additive manufacturing processes represent the "fourth industrial revolution" like hardly any other manufacturing technology [Aiden 2022]. The almost direct production of parts from their geometric data is one of the main reasons for the great potential of the technology, especially for the production of customized products and small quantities. Production programs in additive manufacturing are characterized by high geometric product complexity, variant diversity and short lead times. This in turn creates a need for new solutions for PPC that meet the requirements of Industry 4.0 and cover the possibilities of additive manufacturing.

In extrusion-based processes, the processes, design, material selection, device selection, job placement, job processing time, object removal, object post-processing and object quality control must be mapped. Ideally, in industrial processes, all these process steps are linked together and executed in a spatially as well as temporally optimized manner in a generation system. A build-job can also be a collection of independent parts that are produced in one run through the generation plant. Ideally, there is an integrated software solution for displaying and controlling the entire production process from order entry to order dispatch [Ostermann & Ostermann 2006].

3.1.12.1 Machine Selection

The requirements of the product to be manufactured play the most important role in the selection of the additive process. The process consists of the selection of the specific production machine and post-processing technologies, the materials to be processed, the desired part size and the part resolution. Health and safety should be equally considered prior to purchase of an additive production system to avoid accidents, health hazards, retrofits or even an officially ordered shutdown of the production. Companies must therefore take an early and intensive look at which processes and machines are suitable for them from the point of view of occupational health and safety.

3D printers used in Europe must meet the requirements of the respective applicable EU directive. 3D printers are basically machines in the sense of the Machinery Directive 2006/42/EC (MRL) and must comply with this directive [Ostermann & Ostermann 2006]. The CE mark (Conformité Européenne, European conformity) on the type plate is not a test mark. It is only a declaration by the manufacturer that the applicable EU directive has been applied. Only if the machine also has a GS test mark (Tested Safety), basic safety requirements have been checked by an approved, independent test center. The GS mark is the only legally regulated test mark in Europe for product safety [BAUA 2022a].

In principle, printers that can be operated and controlled cloud-based are preferable for modern production processes. Various manufacturers of professional filament- and granule-based printers offer such capabilities in different variations (see Section 4.1).

The integration of material loading and part removal is becoming increasingly automated. More and more printer farms are emerging that are at least partially automated, demonstrating the applicability of additive manufacturing in volume part production [Gregurić 2019].

3.1.12.2 Workspace

3D printers that operate commercially using the MEX process should generally only be operated in suitable commercial premises by trained personnel. The minimum requirements are a dry, well-ventilated room, sufficient electrical safety and compliance with legal fire protection measures.

3.1.12.3 Occupational Safety

Like any technology or new process, additive manufacturing needs guidance and guidelines for safe and effective use. Against this backdrop, the German Federal Institute for Occupational Safety and Health (BAuA) discusses workplace hazards associated with working with 3D printers [BAUA 2022b].

Technical documentation

3D printers must further in Europe meet the requirements of the applicable EU directive. 3D printers are machines in the sense of the Machinery Directive 2006/42/EC (MRL) and must comply with this directive [Ostermann & Ostermann 2006]. Manufacturers or suppliers must provide operating instructions in the language of the country where the additive production system will be employed. These instructions must contain the necessary information on the topics of safety, installation, operation, loading, maintenance, cleaning and troubleshooting and form an important basis for the instruction of employees.

Risk assessment

If there are residual hazards that cannot be avoided by technical measures, then appropriate warning notices must be attached to the machine and information on this must be described in detail in the operating instructions. There must be a type plate on the machine. Even before purchasing a machine, companies should clarify with the manufacturer which steps are required for successful use (pre- to post-process) and which conditions enable safe working:

- The installation conditions (i.e. defined requirements for temperature, humidity and air exchange rate) must be known and checked for feasibility before purchase
- Maintenance and cleaning work should be user-friendly and safe to perform
- If hazardous substances are required for the work, suppliers must provide safety data sheets in the language of the country the printer is employed.

Once companies have identified and assessed all hazards and defined protective measures, they must instruct employees in the safe use of the 3D printer on a regular basis, at least once a year. Both the risk assessment and the instruction must be documented. This documentation serves to communicate the company's occupational health and safety in a comprehensible and transparent manner and to be able to prove it.

Despite the steadily growing use of additive processes, there are currently hardly any studies available on possible health hazards for employees. The specific hazards associated with working with 3D printers depend primarily on the process used. From the point of view of occupational health and safety, all three production steps, i.e. pre-, in- and post-process, must be considered. Particular attention must be paid for MEX based processes to work in the post-process, which is usually carried out manually and in connection with sharp or pointed objects (deburring, removing support material). During the printing process, there are hazards from moving machine parts, volatile parts and degradation products, and depending on the printing process, the machine or object may be very hot, posing a risk of burns.

When selecting the necessary protective measures, an order of priority, the so-called STOP principle [BG BAU 2019], should be applied:

- Substitution
- Technical measures
- Organizational measures
- Personal safety measures

The best measure is always to avoid the hazard or eliminate it completely (substitution). If this is not possible, the hazard must be kept as low as possible. As a rule, technical solutions (for example, a closed system) are the most effective for occupational safety. They take precedence over organizational regulations and personal and behavior-related protective measures.

The machine must be operated in accordance with its intended use. This includes the following rules:

- Do not intervene in the running printing process without stopping the machine
- Do not remove any covers or protective devices
- Allow objects to cool for as long as possible with the cover closed
- Exercise caution when finishing objects, and when maintaining and cleaning machines

3.1.12.4 Particulate Matter/Volatile Organic Compounds (VOCs)

There are not yet many studies that have looked at ambient exposure to particulate matter or volatile organic compounds from extrusion-based additive manufacturing processes. In the MEX process, a thermoplastic filament or pellet is melted by strong and rapid heating and forced through a die to be successively deposited in thin layers on a build platform and cooled.

The large number of different commercially available 3D printers, the variety of filaments and granulates used, and the variability of process conditions make a general assessment of the hazard potential of this technology for the operator difficult. In principle, the use of devices with a closed installation space and controlled filtered venting is recommended. However, in the DIY sector, many devices with open installation spaces and without air filter devices are also in use. At the same time, a wide variety of materials are used, e.g. acrylonitrile-butadiene-styrene copolymer (ABS), polylactic acid (PLA), polyvinyl alcohol (PVA), polycarbonate (PC), high-density polyethylene (HDPE), high-impact polystyrene (HIPS), polyethylene glycol-coterephthalate (PETG), thermoplastic polyurethanes (TPU), polyamides (PA) and many other polymers.

Manufacturers usually recommend processing temperatures for each filament material depending on the thickness of the filament and intended melt flow. ABS and PLA are currently still the most commonly used filaments in home 3D printing, although other materials such as PETG and TPU are becoming more popular.

It is known that both gases and particles are emitted from hot thermoplastic melts. The emissions depend on the thermoplastics and thermoplastic formulations used. Emissions are particularly evident when processing styrenic thermoplastics such as ABS, PS, and HIPS, with styrene as the residual monomer being most prominent here. Furthermore, volatile ring esters were found during the processing of PA 6, formaldehyde to a small extent during the processing of POM. PLA and other polyesters show comparatively low emissions of lactides and glycols in this context.

In principle, the volatile components found are consistent with the expected emissions of the thermoplastics mentioned and are also found in normal VOC analyses when processing molten thermoplastics. The volume of thermoplastic melted simultaneously at any point in the printing process is comparatively low in extrusion-based additive manufacturing. Accordingly, the measured emissions also reach only comparatively low values, which, cumulatively, can nevertheless be problematic in poorly ventilated rooms [Azimi et al. 2016]. Strongly formulated and plasticized thermoplastics should be regarded with caution, especially if they are offered by less professional/non-certified suppliers as filament or granules.

3.1.13 Quality Assurance

3D printing enables the production of parts in small batch sizes and with high complexities. In extreme cases, no two products are alike. Especially in medical technology and orthopedics, parts are individually designed for the customer. One challenge for companies is quality assurance for additively manufactured products. Personalized medical products in particular are subject to strict safety requirements. Manufacturers must be able to guarantee that there are no defects in the part and that the geometric requirements and the desired material properties are achieved.

The more individualized the production, the more difficult quality assurance becomes. In the case of goods manufactured in large series, it is often sufficient to carry out quality tests on a few randomly selected parts in order to be able to draw conclusions about the entire batch. This is not possible with single-item production. Here, each part must be examined in detail to ensure quality. A solution to this verification problem can be general construction inspection in addition to individual part inspection. To this end, more and more industrially used printers are being equipped with sensor technology to seamlessly monitor the printing process. The sensor data collected can then be analyzed and evaluated using a quality model. In this way, the general hope is that production defects that are not visually detectable on the part can be identified early and reliably [IPH 2023].

3.1.14 Standardization

In additive manufacturing parts are produced layer by layer on the basis of three-dimensional models. With the technological development, rule setting is also gaining in importance. Standards and guidelines for additive manufacturing support the competitiveness of companies and protect users and consumers at the same time. This is done locally on country and European level but also by international standardization at ISO (International Organization for Standardization).

A standard is a document that specifies requirements for products, services or processes. It thus creates clarity about their properties, facilitates the free movement of goods and promotes exports. It supports rationalization and quality assurance in business, technology, science and administration. Standards are developed by those who later apply them. To ensure that the market accepts the standards, broad participation, transparency and consensus are the basic principles. E.g. at the German Institute for standardization DIN. Anyone can submit a request for standardization. All parties interested in a topic are given the opportunity to participate and contribute their expertise.

The application of standards is basically voluntary. Only when standards become the content of contracts or when the legislator makes compliance with them mandatory do standards become binding.

Comprehensive norms and standards for additive methods and processes are in the making. For example, the first standard for quality assurance was recently published. DIN SPEC 17071 defines uniform requirements for equipment, materials, processes and employees [DIN 2019].

The quality concept is based on ISO 9001, while standards from the automotive industry, such as IATF 16949, and from the aerospace industry extend the quality management system.

Relevant standards for extrusion-based processes are:

- DIN EN ISO/ASTM 52903-1:2021-06: Additive manufacturing – Material extrusion-based additive manufacturing of plastics – Part 1: Starting materials (ISO/ASTM 52903-1:2020); German version EN ISO/ASTM 52903-1:2021
- DINEN ISO/ASTM 52903-2:2021-02: Additive manufacturing – Material extrusion-based additive manufacturing processes for plastics – Part 2: Process accessories (ISO/ASTM 52903-2:2020); German version EN ISO/ASTM 52903-2:2020

3.2 Selective Laser Sintering (SLS) of Plastics

In this section we will explore the unique aspects of Selective Laser Sintering (SLS) – also referred to as Laser Sintering (LS) or powder-based fusion (PBF) – which is an additive manufacturing process that uses thermoplastic polymer powders. We will cover the SLS process, part properties, and considerations related to periphery and post-processing. These insights are particularly relevant for mass production using the SLS technique.

SLS is an important additive manufacturing process in the industrial environment, which is based on the selective of thermoplastic powders. Compared to established plastic processing methods such as injection molding, extrusion, and thermoforming, the SLS process offers the possibility of manufacturing polymer parts with a high degree of geometric freedom directly from CAD data and without the need for costly mold or complex support structures. In addition, it offers the possibility of decentralized and economical small-series production down to single individualized parts [Schmid 2015].

Terminology

In the 1980s, researchers at the University of Austin, Texas, first explored SLS, with Carl Deckard playing a significant role in this field, filing the first patent application as early as 1986 [Schmid 2015, Gebhardt 2016, Deckard et al. 1992, Deckard 1997]. Initially, it was believed that the thermoplastic powders used in SLS would only sinter at the surface, resulting in brittle yet highly defined products. This assumption holds true when applying SLS to amorphous polymers with a high glass transition temperature (T_g), including applications like lost cores made from PMMA as used (for example) for the metal casting industry.

Over time, the SLS process has gained prominence, especially when using the material polyamide 12 (PA12). In PA12, the polymer powder particles undergo nearly complete melting. This thorough melting results in a strong interlaminar bond and imparts high mechanical properties to the printed parts. Consequently, the term "laser melting of polymers" has also been coined. However, it is important to note that the term "laser-melting" (selective laser melting, SLM) is typically reserved for metal-based processes. According to guideline VDI 3405, SLS is officially defined as laser sintering (LS), as the term "Selective Laser Sintering" (SLS) is protected by copyright [VDI 2015].

To avoid confusion, from this point we will use the abbreviation SLS, since it has become the predominantly used synonym for the process in public and scientific communications.

Notice:

- In SLS, thermoplastic polymer powders are melted (not sintered)
- The SLS process is also referred to as "laser melting of polymers". The term laser melting (selective laser melting, SLM) is usually dedicated to its metallic counterpart

According to the guideline VDI 3405, proprietary Selective Laser Sintering (SLS©) is also referred to as laser sintering (LS).

What distinguishes SLS?

SLS is a powder-based, layer-by-layer additive manufacturing process that uses laser-induced melting (as shown in Figure 3.15). Here are the key points which characterize the SLS process:

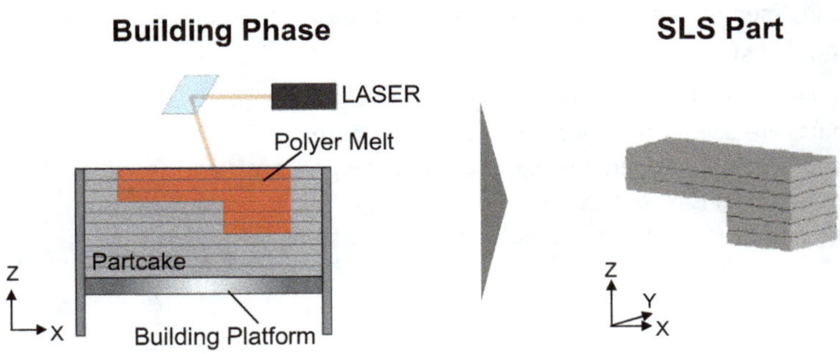

Figure 3.15 Basic manufacturing principle in SLS (© Martin Launhardt)

- **Laser-induced melting**: In SLS, a high-power laser selectively fully melts thermoplastic powder particles. This results in strong bonds both within each layer (planar) and between layers (interlaminar).
- **High mechanical properties**: SLS manufactured parts provide exceptionally high mechanical properties compared to other additive manufacturing methods.
- **Parallel manufacturing**: Unlike many other processes, SLS allows simultaneous manufacturing of multiple parts within a single build space. This feature enables cost-effective production of many small individual parts but can be cost-intensive for large geometries.
- **Built-in support structure**: The part geometry is directly generated in the powder bed. The powder itself serves as the support structure, eliminating the need for additional supports. This reduces post-processing labor and avoids aesthetically unattractive connection points on parts.
- **Surface roughness and powder structure**: During SLS, the entire part geometry is molten and adheres to the surrounding powder. The "rough" powder structure on the part's surface contributes to the characteristic surface roughness of SLS parts.
- **Freedom of design and functionality**: SLS allows increased freedom of design by using the powder bed as a support structure. Parts can integrate functional features such as film hinges, spring elements, snap hooks, movable hinges, or chain elements directly, reducing post processing assembly steps.
- **Process chain considerations**: For economical series production, additional steps beyond SLS are essential. These include a well-designed powder cycle, efficient unpacking of build-jobs, de-powdering of SLS parts, and adapted post-processing with safety precautions.

Advantages of SLS:

- High strength, comparable with injection-molded polymer parts
- Economical production of many small parts in batches
- No support structures
- Savings in assembly steps with high utilization of design freedom.

Disadvantages of SLS:

- Parts have a characteristic surface roughness
- Large parts are usually uneconomical to manufacture
- Hollow bodies require openings to remove the trapped powder
- The process chain is extensive
- Handling fine powder requires adapted safety measures.

3.2.1 Available Materials

The widespread adoption of powder-based additive manufacturing processes, particularly for small-batch production, has led to a significant surge in global sales of polymer powders for the 3D printing industry. According to the Wohlers Report, polymer powder consumption related to 3D printing has grown nearly ten-fold since 2008, reaching \$540 million in 2019. This remarkable growth rate underscores the substantial economic potential of industrial 3D printing, particularly when using powder-based techniques (Figure 3.16).

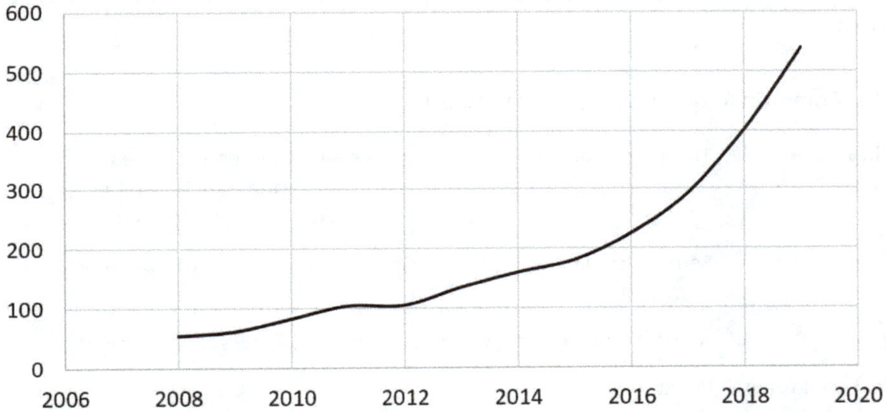

Figure 3.16 Worldwide consumption of polymer powders for powder-based additive manufacturing in millions of US dollars [Wohlers 2020]

The thermoplastic polyamide 12 (PA12) played an especially important role in the recent advancement of SLS. The production process of PA12 already yields a suitable powder for SLS printing. In addition, parts produced using SLS from PA12 exhibit excellent mechanical properties, since the quasi-isothermal conditions provided by the SLS process, perfectly fit the specific needs of PA12 (Section 3.2.2). For these reasons PA12 has long been the primary powder material for SLS, helping the technology to grow and spread to many applications. As SLS gained more acceptance and the technology was further refined and improved, demand for a broader material range has grown. Nowadays, a wide variety of SLS-compatible materials are available, and ongoing research continues to expand the available options.

The basic requirements for polymers suitable for the SLS process are *meltability*, as given by thermoplastics, and the *availability in powder form* (Table 3.1). The maximum process chamber temperature limits the processing of high-melting thermoplastics. The recrystallization temperature of a thermoplastic must be within the possible process chamber temperature. Due to the quasi-isothermal process control it is possible, especially for semi-crystalline thermoplastics (e.g. PA12), to achieve complete melting of the powder particles and thus to create a compact and stable structure (see thermal process control, Section 3.2.2). Amorphous thermoplastic polymer powders are also occasionally used for the process, such as polycarbonate (PC) or polymethyl methacrylate (PMMA), which do not fully melt in SLS due to their thermal properties, but rather sinter together at the powder surface. The parts made from PC or PMMA usually exhibit high porosity and are used, for example, as a lost core in casting molds for metals. If these basic requirements are met, the parameters in Table 3.1 can be used to co-determine the processability and the resulting part properties.

Table 3.1 Common Requirements for Materials in SLS

Fusibility	Thermoplastics	Recrystallization temperature below 200 °C (maximum build chamber temperature for standard machines)
	Semi-crystalline thermoplastics	Good process control comparable to PA12
	Amorphous thermoplastics	Aggravated process management
Raw material form	Powder	Features: ▪ Particle size 50–100 µm ▪ Particles as spherical as possible ▪ Good flowability
Energy input	Absorption coefficient greater than 90 %	

For the processing of a thermoplastic powder in SLS, the powder particles should be smaller than the layer height to be applied. This is usually between 60 µm and 150 µm. If individual powder particles, even if only a few, are larger than the powder layer height to be applied, scoring can occur in the powder bed. A desired average particle diameter is about half as large as the targeted powder layer thickness. In addition, good flowability should be ensured, so that the polymer powder can flow into the process chamber without bridging, and a smooth powder surface is made possible.

A CO_2 laser with a wavelength of 10.6 µm is usually used to induce the energy required for melting the polymer powder. This wavelength is well absorbed by most polymers. If the absorption is insufficient, appropriate absorption aids must be used.

3.2.1.1 Powder Manufacturing Process

There are a wide variety of process engineering methods for the manufacturing of polymer powders. These can be chemical or mechanical processes, and the three most important processes are briefly presented below.

Precipitation polymerization

PA12 is obtained chemically by *precipitation polymerization*, in which individual, approximately spherical particles precipitate from a solution in a reactor and are subsequently collected (Figure 3.17). These can be further fractionated for improved particle size distribution, for example by means of sieves. The resulting powders are usually further formulated with, for example, silica powders to improve the flow behavior (reduce friction between individual particles), pigments (e.g. TiO_2) to achieve a certain color, and fillers such as aluminium, carbon, carbon fibers, glass beads, glass fibers to influence the mechanical properties of a part.

Figure 3.17 Scanning electron micrographs of the powder particles of a PA12-SLS powder of type PA2221 from EOS GmbH [Launhardt 2021]

In Table 3.2 some suppliers of common polyamide polymer powders and PA12 variants for SLS, including the most important material specifications, are presented. Here, typical variations represent a natural-, white-, and dark-colored PA12 powder material. The material color determines the type of laser that can be used for the processing. For example, a white PA12 powder material can only be processed with SLS machines using a CO_2-laser due to the inherent absorption capabilities of the material. Diode- and fiber-lasers, as used in desktop SLS machines or in specially designed industrial SLS machines, need either dark powder material or specially designed absorbing materials in order to provide sufficient absorption and thus processing capabilities. Furthermore, many powder material suppliers provide variations including PA12 variants reinforced with glass beads, carbon fiber, or mineral fiber. The application of flame retardant and halogen-free polyamide materials is attracting increasing interest; however, current materials tend to have very high refreshing rates, which lead to very environmentally unfriendly and expensive manufacturing protocols.

Table 3.2 Commercially Available PA12 Variants for SLS (Based on Manufacturer Specifications)

Provider	Material designation	Most important feature
EOS	PA2200	White, tensile modulus 1,650 MPa, tensile strength 48 MPa, T_m 176 °C
	PA2201	Natural color, slightly translucent, tensile modulus 1,700 MPa, tensile strength 48 MPa, T_m 176 °C
	PrimePart FR (PA2241 FR)	White, tensile modulus 1,900 MPa, tensile strength 49 MPa, T_m 184 °C, flame retardant (halogenated)
	PA2210 FR	White, tensile modulus 2,500 MPa, tensile strength 46 MPa, T_m 185 °C, flame retardant (halogen free), 100 % refresh rate
	PA3200 GF	White, tensile modulus 3,200 MPa, tensile strength 51 MPa, T_m 185 °C, glass bead filled
	Alumide	Metallic gray, tensile modulus 3,800 MPa, tensile strength 48 MPa, T_m 176 °C, aluminium-filled
	PA640-GSL	Dark gray, tensile modulus 3,816 MPa (max. 6,100 MPa in fiber direction), tensile strength 49 MPa, very low elongation at break, T_m 185 °C, hollow glass spheres and carbon fiber filled, very light and heat resistant
	PA2221	Natural color, tensile modulus 1,650 MPa, tensile strength 48 MPa, T_m 176 °C, refreshed optimized (30 wt%)
3D-Systems	DuraForm PA	White, tensile modulus 1,586 MPa, tensile strength 43 MPa
	DuraForm HST	White, glass fiber reinforced PA12, tensile modulus 5,500 MPa, tensile strength 51 MPa, resistant to high temperature
ALM	PA250	White, tensile modulus 1,740 MPa, tensile strength 46 MPa
	PA650	White, tensile modulus 1,700 MPa, tensile strength 48 MPa
Prodways	PA12-L 1600	White, modulus of elasticity 1,600 MPa, tensile strength 48 MPa
Evonik	INFINAM® PA6002 P	White, tensile modulus 1,700 MPa, tensile strength 50 MPa, T_m 187 °C
	INFINAM® PA6003 P	Black, T_m 188 °C

Provider	Material designation	Most important feature
Launhardt	PowderMonkeys™ RePA12 45-20	Anthracite, tensile modulus 1,600 MPa, tensile strength 45 MPa, T_m 180 °C, suitable for desktop SLS systems and fiber lasers
	PowderMonkeys™ SiPA12	Black, tensile modulus 1,750 MPa, tensile strength 48 MPa, T_m 176 °C, suitable for desktop SLS systems and fiber lasers
	PowderMonkeys™ RePA12 50-8 GB	Black, tensile modulus 3,200 MPa, tensile strength 50 MPa, T_m 184 °C, glass bead reinforced PA12, high stiffness, suitable for desktop SLS systems and fiber lasers
	PowderMonkeys™ PA12 40-10	White, tensile modulus 1,500 MPa, tensile strength 40 MPa, T_m 179 °C, economic PA12 for any SLS machine
SinterIt	PA12 Smooth	Gray, tensile modulus 1,470 MPa, tensile strength 32 MPa, T_m 185 °C, suitable for desktop SLS systems
	PA12 Industrial	Gray, tensile modulus 1,840 MPa, tensile strength 52 MPa, T_m 184 °C, suitable for desktop SLS systems
Farsoon	FS3201PA-F	Black, tensile modulus 1,500 MPa, tensile strength 44 MPa, T_m 185 °C, suitable for fiber lasers (FLIGHT®)
	FS3300PA	White, tensile modulus 1,602 MPa, tensile strength 46 MPa, T_m 183 °C, suitable for SLS machines
	FS3300PA-F	Black, tensile modulus 1,820 MPa, tensile strength 48 MPa, T_m 184 °C, suitable for fiber lasers (FLIGHT®)
	FS3401GB	White, tensile modulus 2,644MPa, tensile strength 44 MPa, T_m 184 °C, suitable for SLS machines, glass bead reinforced PA12
	FS3401GB-F	Black, tensile modulus 2,810 MPa, tensile strength 47 MPa, T_m 186 °C, suitable for fiber lasers (FLIGHT®), glass bead reinforced PA12
	FS3150CF	Black, tensile modulus 3,350 MPa, tensile strength 56 MPa, T_m 186 °C, carbon fiber reinforced PA12, suitable for SLS machines
	FS3250MF	Light gray, tensile modulus 6,130 MPa, tensile strength 51 MPa, T_m 183 °C, mineral fiber reinforced PA12, suitable for SLS machines

Cryogenic grinding

Not all polymers are readily available as powder from their manufacturing process like PA12. Consequently, other polymers often need to be converted into powder at a later stage. Various grinding processes facilitate this transformation. For instance, impact mills collide two accelerated streams of polymer granules, mechanically breaking them down due to their high kinetic energy. However, the heat generated during grinding can cause polymers with low melting points to melt and smear. To address this, *cryogenic grinding* is often employed using liquid nitrogen to cool down the material streams before impacting. While effective, this cooling process reduces the economic efficiency of the overall procedure, and can result in powder particles with sharp-edged fractures and reduced powder flowability (Figure 3.18).

Figure 3.18
Scanning electron micrograph of an SLS powder with sharp-edged powder particles using the example of a PA11, Rilsan [Schmid 2015]

Melt atomization

Melt atomization is a process that creates controlled fractures in an accelerated melt flow making use of the material's high surface tension and Raleigh disturbances during the process. This technique is commonly used for metal melts and enables the production of nearly spherical powder particles – for instance, in laser melting of metals.

However, there is a significant difference between metals and polymers. Metals have a high surface tension and low melt viscosity, while polymers exhibit comparatively low surface tension and high melt viscosity. As a result, applying melt atomization to polymers presents challenges related to process settings and stability. Consequently, this powder manufacturing method is limited to only a few specific polymers [Launhardt et al. 2011, Schäfer et al. 2018].

Table 3.3 lists some important commercially available SLS polymer powders from the fields of polyolefins, thermoplastic elastomers, and polyamides (except PA12). A more detailed overview of commercially available polymer powders and their suppliers is summarized in Figure 4.7.

Table 3.3 Commercially Available SLS Polymer Powders Besides PA12 (Based on Manufacturers' Data)

Polymer	Provider	Material designation	Most important properties
TPU	EOS	TPU 1301	White, T_m 138 °C, hydrolysis resistant
	Launhardt	SpheroFlex™ TPU 50	Elastomer variant, natural color, particularly soft, hydrolysis-resistant, UV-resistant, Shore A hardness 55–75, elongation at break 450 %, chemical smoothing up to 700 %
		SpheroFlex™ TPU 80	Elastomer variant, natural color, abrasion-resistant, hydrolysis-resistant, UV-resistant, Shore A hardness 85
		PowderMonkeys™ TPU 50	Black, Shore A hardness 55, 65 and 75, for SLS desktop and fiber-laser systems
		PowderMonkeys™ TPU 80	Black, Shore A hardness 80, for SLS desktop and fiber-laser systems
	Layer by layer	TPU 90A	Black, Shore A hardness 90, T_m 154 °C, elongation at break 250 %
	Prodways	TPU-70A	White, Shore A hardness 86, T_m 105–125 °C, elongation at break 300 %
	Lehmann & Voss	Luvosint X92A-1	Natural color, Shore A hardness 88, elongation at break 500 %
PP	Diamond Plastics	DiaPow PP	Natural white, chemical resistant, low density
	Launhardt	PowderMonkeys™ PP 20-20	Black, chemical-resistant, low density, for SLS desktop systems
	Prodways	PP 1200	Translucent, T_m 140 °C, chemical-resistant
	Layer by layer	PP Flex	T_m 135 °C
HDPE	Diamond Plastics	DiaPow HDPE HX	Black, resistant to oil and gasoline

Table 3.3 Commercially Available SLS Polymer Powders Besides PA12 (Based on Manufacturers' Data) (*continued*)

Polymer	Provider	Material designation	Most important properties
PA11	Layer by layer	PA11	White, T_m 201 °C
	Arkema	Rilsan Polyamide 11	Black, tensile modulus 1,500 MPa, tensile strength 45 MPa, T_m 201 °C
	EOS	PA1101	White, tensile strength 48 MPa, T_m 201 °C
	ALM	PA850	Black, tensile modulus 1,475 MPa, tensile strength 48 MPa
	ALM	PA860	White, tensile modulus 1,475 MPa, tensile strength 48 MPa
	ALM	PA D80-ST	Black, tensile modulus 1,392 MPa
	ALM	HP 11-30	Black, tensile strength 46 MPa
	Fabulous	PA11 Bluecare	Blue, suitable for applications in the food industry
		PA11 Blue Metal Detectable	Dark blue, suitable for applications in the pharma industry
PA613	Evonik	INFINAM PA Polyamide 613	High mechanical properties, HDT 195 °C
PS	EOS	PrimeCast 101	Gray, tensile modulus 1,600 MPa, tensile strength 5 MPa, very low elongation at break, suitable for casting and lost molds
PA6	BASF	Ultrasint PA6	Natural white, tensile modulus 3,700 MPa, tensile strength 66 MPa, T_m 220 °C
PEKK	EOS	PEKK HAT-23	Black, tensile modulus 6,500 MPa, tensile strength 71 MPa, T_m 302 °C, carbon fiber reinforced

The availability of diverse polymer powder materials for SLS has significantly expanded recently. These comprise elastomers, flame-retardant materials, sphere- or fiber-reinforced polymers, and food- and pharma-certified materials that can typically be processed using standard SLS systems. Certain high-temperature polymers require specialized SLS systems for processing – notably, PA6 and PEKK fall into this category.

A class of materials specifically designed for SLS, including some new polypropylene (PP) grades and selected polyurethane-based thermoplastic elastomers (TPU), are driving the growth of additive manufacturing in large-volume applications. Thermoplastic polyurethanes, in particular, offer excellent elastic properties, making them suitable for seals, bellows, dampers, orthotics, pipes, wearables, and even complex structures such as car seats (Figure 3.19).

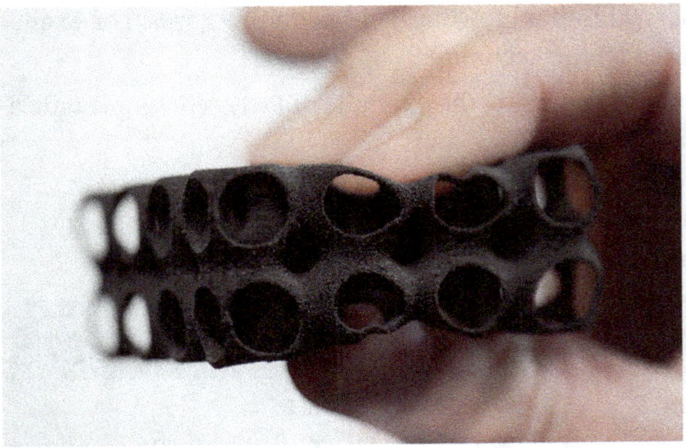

Figure 3.19
Example of a soft SLS-TPU part with a Shore A hardness of 65 (© Launhardt GmbH)

3.2.1.2 Powder Testing Methods

When considering the suitability of a thermoplastic powder for SLS and the general incoming quality assurance, several factors come into play:

- **Flowability:** The flowability of the powder is crucial for good processing properties. Ideally, the flowability should be comparable to that of proven materials like PA12.
- **Bulk density:** Achieving detailed and dense SLS parts requires powders with excellent flowability and high bulk density.
- **Particle size and distribution:** Particle size and distribution impact parameters such as minimum layer thickness, surface roughness of the SLS part, and density.
- **Hausner factor:** Ensuring consistent flowability is essential for a successful SLS application. The ratio of tap density and bulk density, known as the Hausner factor, serves as an indicator of powder flowability [Schmid 2015].

Especially, the Hausner factor provides for structural powder materials such as PA12 a good indication of the powder quality. The Hausner factor represents the ratio of tap density and bulk density, and thus a good indication of the powder flowability, which is essential for the processing of polymer powders in powder bed fusion processes.

In a nutshell: The Hausner factor – an indicator of flowability

Hausner factor = Tap density/Bulk density

- HF ≤ 1.25: High flowability
- 1.25 ≤ HF ≤ 1.4: Reduced flowability
- 1.4 ≤ HF: Cohesive

It is important to be aware that this is only an indication, and cannot fully provide information about the actual processability in powder bed fusion processes. For example elastomers cannot be evaluated by this technique.

Table 3.4 shows the most common values using the example of PA12, which provide a good basis for processing in SLS.

Table 3.4 Common Test Methods for Evaluating Thermoplastic Powders for the SLS process (Adapted From [Launhardt 2021])

Powder property	Test method	Standard	Values using the example of PA12, type PA2221
Flowability	Flowtest with flowmeter	DIN EN ISO 6186	6.4 s (Ø 25 mm) 21.0 s (Ø 15 mm) 61.2 s (Ø 10 mm)
Powder bulk density ρ_{bulk}	Bulk density tester	DIN EN ISO 60	0.45 g/cm^3
Tap density ρ_{tap}	Measuring cup and scale	DIN EN ISO 787-11 and ASTM D7481	0.51 g/cm^3
Hausner factor (HF)	HF = ρ_{bulk}/ρ_{tap}	According to Schmid	1.13
Particle size	Laser diffraction/ optical measurement	DIN EN ISO 9276	$D_{v,90}$ =88 µm
Particle shape, e.g. circularity	Laser diffraction/ optical measurement	DIN EN ISO 9276	Bean-shaped
Absorbance α_{powder} in the wavelength range of the laser	Integrating sphere measurement	DIN 5036	98.3 %

Particle shape and size are typically described statistically, because powders rarely consist of uniformly shaped and sized particles. Determining these characteristics often involves techniques such as automated microscopy or laser diffraction. Optically, the projection area of a powder particle is assessed, and the area-equivalent diameter (CE diameter) is calculated and evaluated (Figure 3.20).

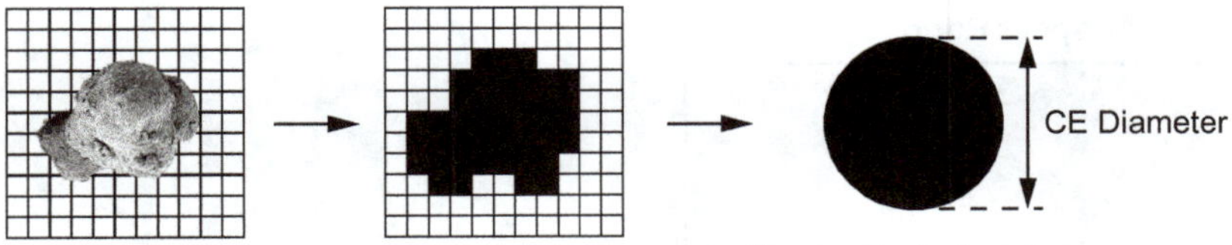

Figure 3.20 Optical evaluation method of the circle-equivalent particle diameter (CE diameter, cf. [Launhardt 2021])

Based on this evaluation, the particle size distribution can be can be represented logarithmically (Figure 3.21). Here, in particular, the significant difference between volumetric and numerical representation can be seen. Numerically, a considerable proportion of small powder particles is detected, which is not of significant importance volumetrically.

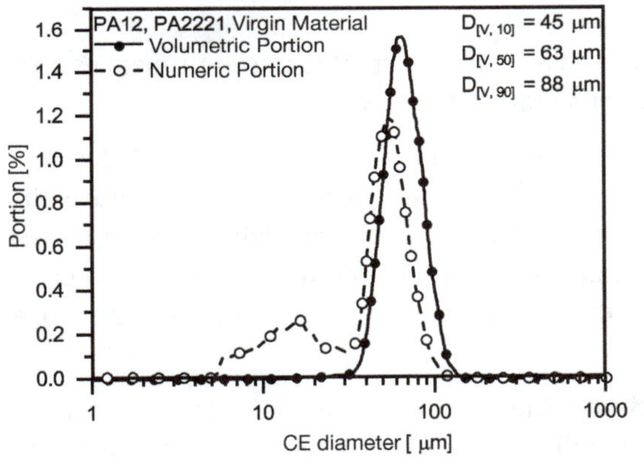

Figure 3.21 Volumetric and numerical powder particle size distribution using the example of PA12, type PA2221 from EOS GmbH [Launhardt 2021]

For the characterization of SLS powders, the volume-based mean particle size specification (denoted $D_{v,90}$ or D[v,90]) has become established.

> $D_{v,90}$ = 88 µm means 90 vol% of the powder particles have a mean diameter smaller or equal to 88 µm.

It is also common practice to define particle shape in terms of shape factors, such as aspect ratio or the sphericity. These statistical values provide insights into particle shape and basic processability. For instance, in Figure 3.22, the aspect ratio of a TPU powder is shown with a value of 0.99. This suggests that the particle shape in the projection plane is nearly circular, allowing us to assume spherical particles.

These fundamental investigations provide guidance, but alone, they cannot definitively determine successful processability or unsuitability for the SLS process. Interestingly, some commercial powders have particle shapes that appear unfavorable based on analysis, yet they can still be effectively processed.

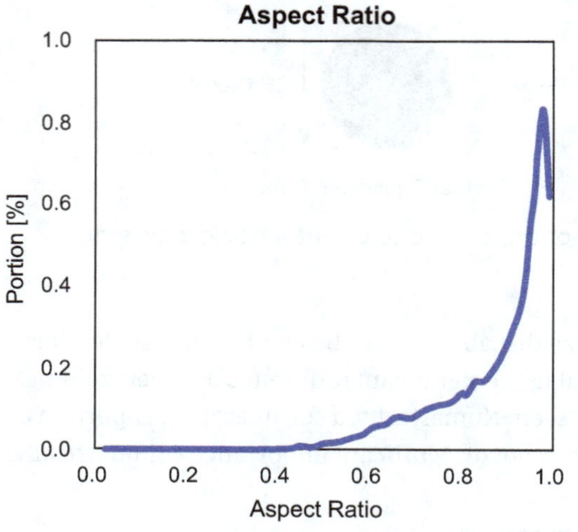

Figure 3.22
Particle size distribution using the example of SpheroFlex™ TPU 50 (© Launhardt GmbH)

In addition to the discussed powder characteristics, thermal material properties also significantly impact the processability of a material in SLS.

- **Crystallization temperature:** The crystallization temperature must be below the maximum build space temperature of the SLS system. A decisive temperature difference between melting and re-crystallization temperature further ensures quasi-isothermal process control. The greater the difference between melting and crystallization temperature, the larger the theoretical process window and the more stable the process control.
- **Heat capacity and heat conductivity:** The addition of mineral fillers can lead to deviations in specific heat capacity and heat conduction within the powder bed. These deviations influence melting behavior and affect achievable detail accuracy for intricate part geometries.

For more details on thermal process control, refer to Section 3.2.2.

3.2.1.3 Thermal Aging

Polyamide 12 (PA12), in particular, exhibits increased thermally induced aging due to its chemical properties. When subjected to prolonged thermal stress, PA12 undergoes molecular post-crosslinking, significantly impacting its melt viscosity. This viscosity, in turn, affects achievable accuracy during the process.

The consequence of this phenomenon is the "orange peel effect", characterized by irregular sink marks and bulges on the part surface (as shown in Figure 3.23). The severity of these defects depends on part geometry and positioning. Parts with high wall thicknesses are particularly prone to displaying the initial signs of the orange peel effect [Schmid 2015].

Figure 3.23
SLS part with significant sink marks, known as "orange peel" (© Launhardt GmbH)

This thermal post-crosslinking in PA12 can be determined indirectly by physical investigation of the melt viscosity, but more reliably via the chemical solution viscosity number (VZ). As a function of the build-job time, the viscosity number increases asymptotically (Figure 3.24, left; see also [Wudy 2017]). Empirical evidence shows that a viscosity number below 70 mL/g for PA12 exhibits positive properties in the SLS process. This thermal influence can be compensated for by sufficient refreshment with new powder. In Figure 3.24 on the right is a schematic representation of the viscosity number as a function of the virgin powder content.

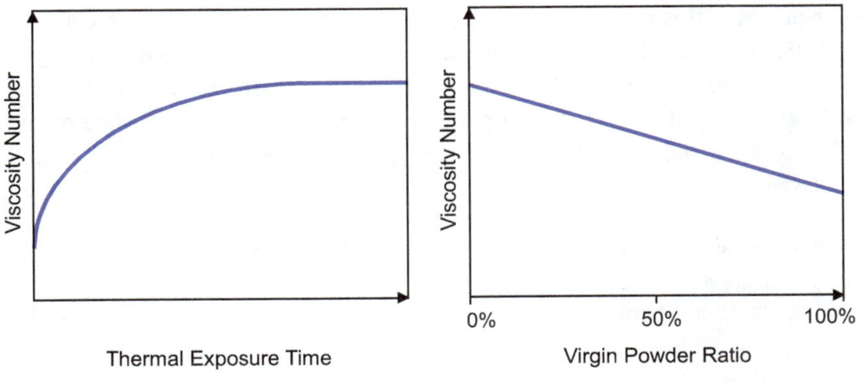

Figure 3.24 Left: Asymptotic increase in viscosity number as a function of thermal load duration. Right: Viscosity number as a function of virgin powder content for PA12 (adapted from [Wudy 2017])

Alternatively, the melt viscosity can be used as an indicator of material quality. If, for example, a melt volume rate (MVR) measurement device is available (as it is in many polymer processing companies), the powder quality can also be indirectly determined with regard to aging (see Section 2.5.5). As a function of the new powder content, the MVR value increases, which corresponds to a reduction in melt viscosity (Figure 3.25).

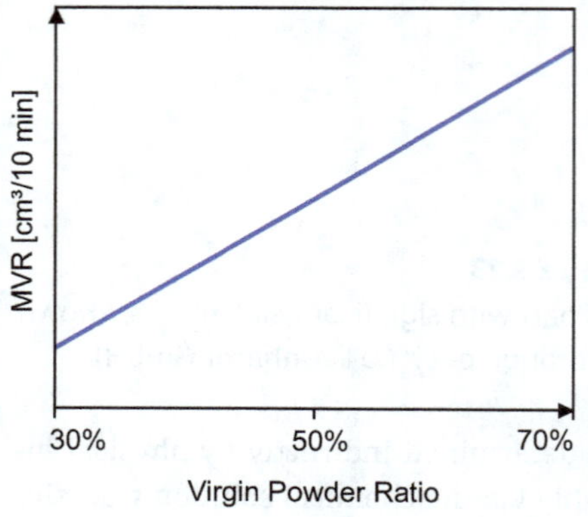

Figure 3.25
Melt viscosity: MVR as a function of virgin powder content for PA12 [Wegner 2015]

In order to reduce the thermal influence on PA12, it is always recommended to refresh the old powder with new powder. Recommended refresh rates are between 30 wt% and 50 wt%, depending on the material supplier. The necessary refresh rate depends on the extent of thermal aging experienced by the powder. This aging primarily hinges on the interplay between process temperature and build time. The build time, in turn, is influenced by factors like packing density and build-job height.

Extreme variations of this refresh rate are a glass-bead-reinforced PA12 material (e.g. PA3200 by EOS, which needs a 70 wt% refresh), and a flame-retardant- and halogen-free PA12 (e.g. PA2210 FR by EOS, which needs a 100 wt% refresh). In contrast to these high refresh rates, PP and TPUs (for example) usually only need to be refreshed by the amount of material withdrawn from the powder pool as manufactured parts, which is around 20 wt%.

Rule of thumb: A PA12 powder is considered "okay" if the following criteria are met:

- Viscosity number ≤ 70 mL/g
- MVR ≥ 15 cm^3/10 min.

Further details on the implementation of SLS powder refreshments are given in Section 3.2.10.

3.2.2 Thermal Process Control

The SLS process heavily relies on temperature control. The thermal processing window is determined by the thermoplastic's properties and spans from the melting temperature T_m to the onset of crystallization temperature (T_c). Within this range, both the solid and the melt coexist at the same temperature in the two-phase mixing region (Figure 3.26).

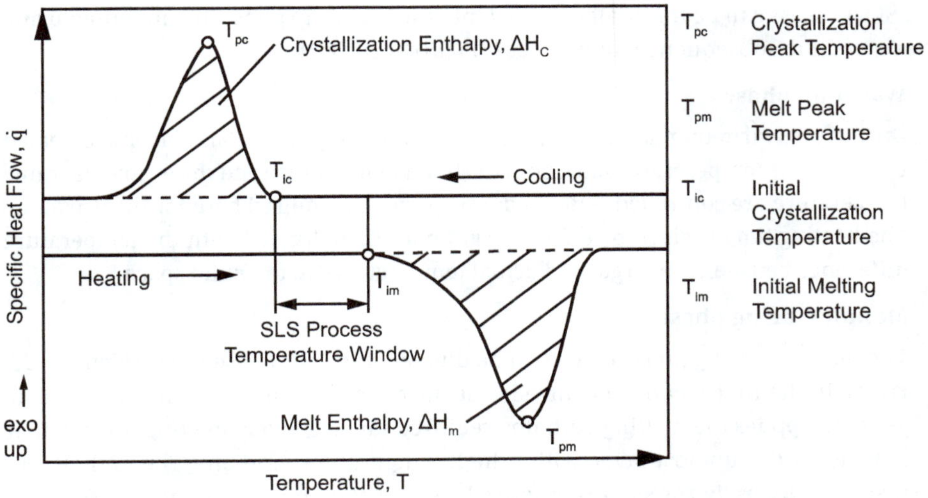

Figure 3.26 Schematic DSC diagram of a semi-crystalline thermoplastic and the resulting thermal process window for SLS (based on [Rietzel 2011, Alscher 2000])

If the build temperature is too high, there is a risk of melting of the powder filling. A build temperature that is too low generally leads to rapid cooling and crystallization of the melt, and thus to high thermally induced distortion (curling effect). This can be so severe that the parts bend upwards at the melt edges during production and are subsequently caught by the recoating system, which subsequently leads to process termination (Figure 3.27). When practically determining the optimum process temperature – representing the controlled powder surface temperature within the pyrometer's measuring range – a common approach involves using test specimens. These specimens, intentionally designed to induce distortion (such as curling crosses), are initially manufactured at low process temperatures. The temperature is then incrementally increased until the noticeable curling effect disappears.

Figure 3.27 Example of the curling effect during the process (© Martin Launhardt)

The SLS process can be thermally divided into the warm-up phase, the manufacturing phase, and the subsequent cooling phase (Figure 3.28, left).

- **Warm-up phase**

  During the warm-up phase, the system parts and the powder base layer are heated up to process temperature. Since both the build temperature and the exchange frame temperature are controlled, a defined time period is required here for the warm-up. The heating time is given by a factor, e.g. 1 min/°C, multiplied with the temperature difference between the target and actual temperatures in the build space.

- **Manufacturing phase**

  The manufacturing phase can again be divided into four subsections (Figure 3.28, right). In the first section, a controlled surface temperature is established. The laser then applies the melting enthalpy required for selectively melting the polymer powder of the upmost layer with a high temperature gradient. The enthalpy of fusion is precisely the energy required to melt the polymer without further increasing the temperature of surrounding material. According to the idealized quasi-isothermal process control, the melt and the surrounding powder should now be at the same temperature [Rechtenwald 2011, Rietzel 2011]. In practice, the polymer melted in this way cools down again only slowly compared to the surrounding powder, since the crystallization process itself releases heat. After lowering the platform, new and significantly colder powder material is applied on top of the last partially melted layer. This is consequently warmed up again to process temperature by controlled heaters [Drexler 2016, Wegner 2015]. Only where areas of melted powder touch (laterally and vertically) do they completely flow together to form a dense part. This complex thermal process is repeated for the subsequent layers until all intended geometries have been exposed to build the desired part [Amado et al. 2012, Amado et al. 2015].

- **Cooling phase**

  At the end of the manufacturing process there is the cooling phase during which the part-cake cools down together with the build parts inside. This causes the polymer to shrink, and clear settlement cracks appear in the part-cake (Figure 3.29 and Figure 3.30). Cooling usually takes place inside the SLS system in a nitrogen atmosphere. In the case of removable exchangeable frames, the cooling phase can also be implemented outside the SLS system. For this purpose, commercially available solutions for external controlled cooling are available, which result in a significant optimization of the manufacturing time in series production. The cooling time depends directly on the build chamber size and specifically the build height, and thus on the build-job time. As a rule of thumb, half of the build time can be planned as cooling time. On the one hand, premature unpacking of the SLS parts leads to a potential hazard for the operator, as the part-cake can be very hot for a very long time, so that there is a risk of burns. On the other hand, the build parts are easily deformed when a hot part-cake is unpacked, and uneven cooling and shrinkage of the parts occurs, resulting in permanent deformation. For other materials, such as when processing elastic materials (TPU, TPE, etc.), unpacking while warm can be beneficial. Here, cleaning the parts from adhering powder is easier in the warm state than in the completely cooled state.

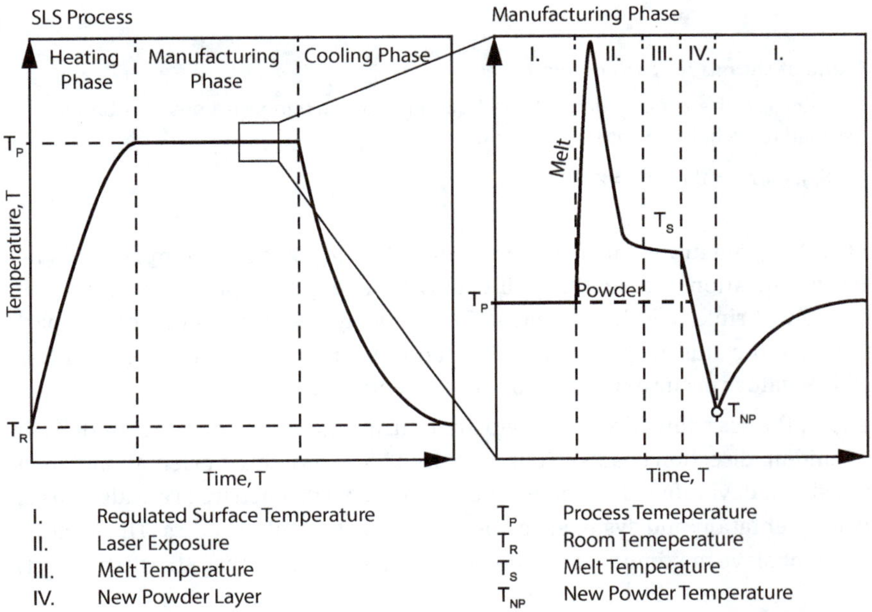

Figure 3.28 Schematic course of the powder or melting temperature at the surface during the SLS process [Launhardt 2021]

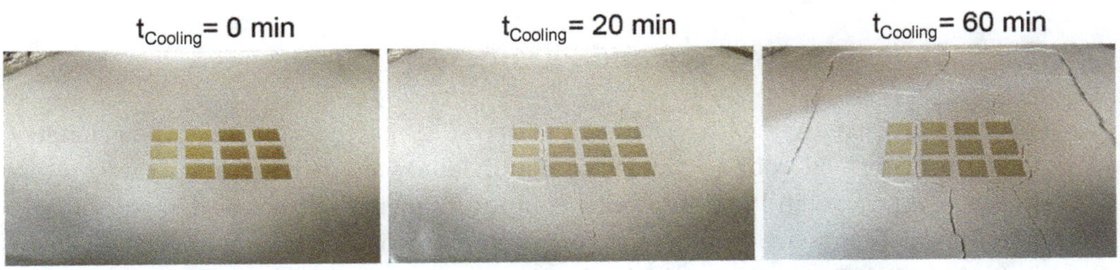

Figure 3.29 Change of the powder-cake surface as a function of the cooling time: Visible settling cracks appear. For test purposes, this was produced without the usual upper powder insulation layer (© Martin Launhardt)

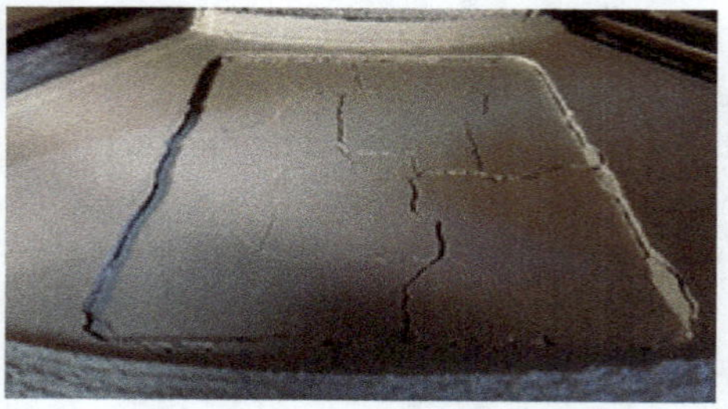

Figure 3.30
Powder cake with characteristic settling cracks at the end of the cooling time (© Launhardt GmbH)

Rule of thumb for the cooling time:

In general, the cooling time of the build-jobs or exchange frames can be assumed to be half the construction time:

$S_{l,z}$ (forz = 0 mm) = 2.6%

During SLS, the geometry is formed by melting polymer powder at temperatures between re-crystallization and melting point, all without applying pressure. As a result, the melt freely shrinks within the part-cake, reducing the dimensions of the part. However, due to the anisotropic nature of the additive manufacturing process, different shrinkage values occur depending on part orientation.

Specifically in the scanning direction (xy-plane), shrinkage values differ from those in the layer build-up direction (z-axis). Consequently, SLS parts exhibit orientation-dependent dimensional deviations. To compensate for this, CAD geometries are scaled larger, considering orientation and system-specific factors of the build process. This ensures that SLS parts achieve maximum dimensional accuracy after shrinkage (Section 3.2.7.1).

Typical scaling factors for the SLS process are as follows:

- Scan direction (xy-plane): 3.2 %
- Layer build-up direction (z-axis): 2.0–2.6 %.

3.2.3 Exposure Strategy and Parameters

To efficiently expose the geometry of a desired CAD part using a laser, the geometry data is transformed into two-dimensional individual scan path vectors within a layer-by-layer exposure strategy. Typically, contour-following vectors and surface-filling parallel hatch lines are employed to represent the two-dimensional image (as shown in Figure 3.31). To accurately depict the geometry's contour, a track offset (beam offset) is defined, shifting the laser spot inward by half the laser focus diameter. This prevents unintentional growth of parts due to adhering powders. For surfaces, the hatching strategy is commonly used, with a defined hatch distance.

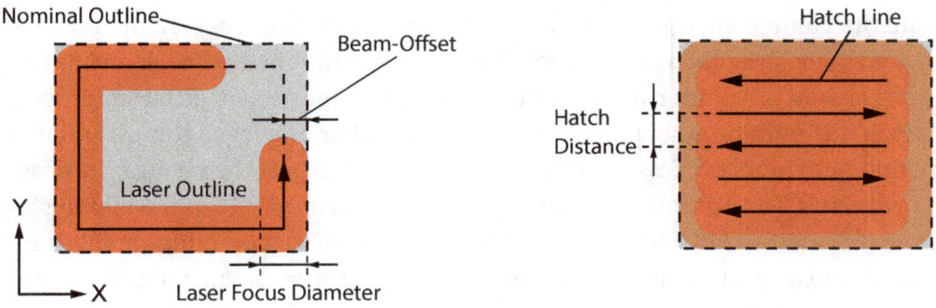

Figure 3.31 Schematic representation of the laser scan paths for producing the contour (left) and for filling a surface using hatched lines (right) [Launhardt 2021]

The laser exposure of the geometries takes place line by line in the xy-plane (scan direction) and layer-by-layer in the z-direction (layer build-up direction). In most cases, the hatch direction is alternated layer by layer in the x- and y-directions (Figure 3.32).

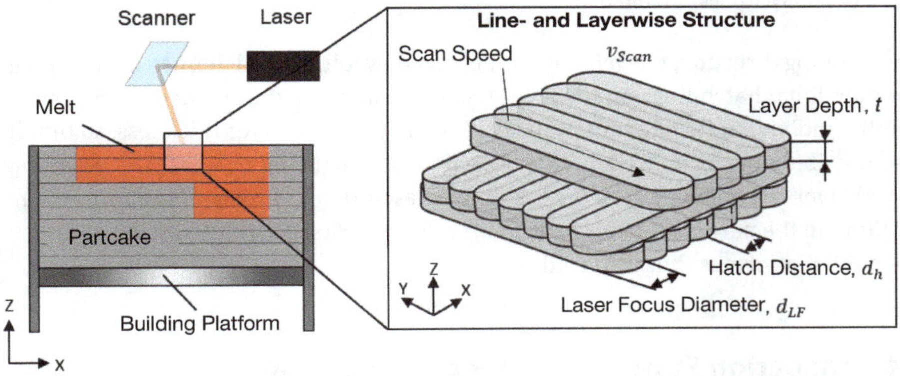

Figure 3.32 Schematic representation of the line-by-line and layer-by-layer build-up of geometries in the SLS process [Launhardt 2021]

The most important process parameters of the laser are the laser power P_L, the scanning speed v_{scan} and the hatch distance d_h, which represents the distance between two hatch vectors. According to Schmid, these parameters can be used to calculate the average area energy input by means of the Andrew number A_Z (Equation 3.1) [Schmid 2015]:

$$A_Z = \frac{P_L}{v_{scan} \times d_h} \tag{3.1}$$

A_Z: Andrew number, surface energy [J/mm^2]

P_L: Laser power [W]

v_{scan}: Scanning speed [mm/s]

d_h: Hatch distance [mm]

To calculate the volume energy density E_{volume} from the surface energy A_Z, the reciprocal applied powder layer thickness, denoted as h_{layer}, is multiplied with the term. The resulting expression is Equation 3.2. This simplifies the comparison of different process parameters of different production systems with each other. However, it is essential to recognize that individual process parameters significantly impact the resulting process and part properties. For instance, scanning speed plays a crucial role in the melting kinetics of the polymer – a key factor in enhancing process efficiency. Additionally, it influences mechanical part properties. When maintaining a constant volume energy density, increasing scanning speed and laser power can accelerate the process, but there are upper limits. This equi-energetic process acceleration affects the dimensional accuracy, density, and mechanical properties of SLS-produced parts [Drexler 2016].

$$E_{\text{volume}} = \frac{P_{\text{L}}}{v_{\text{scan}} \times d_{\text{h}} \times h_{\text{layer}}} \tag{3.2}$$

E_{volume}: Volume energy density [J/mm³]

h_{layer}: Powder layer thickness [mm]

In the scanning direction, the geometry is achieved by joining and defined overlapping of the individual hatch lines. In the layer build-up direction, the geometry is achieved by joining individual two-dimensional layers with a defined layer thickness, so that it is also possible to refer to a "2.5-dimensional" manufacturing process. The resulting practical layer thickness defines the maximum resolution accuracy in the z-direction. In addition to the partially linearly variable scaling factors, a high traversing accuracy of the build platform is thus required.

3.2.4 Production Systems and Process Technology

SLS is a powder-based and layer-by-layer laser-induced melting process. In this process, a powder application system is used to apply thin layers of a thermoplastic powder (e.g. by roller, blade, or double blade). The desired geometry is then melted line by line by a laser and scanner system, depending on the respective layer (Figure 3.33, left). Finally, a homogeneous melt layer is formed during consolidation. The part geometry is positioned in such a way that it is surrounded by powder within the so-called part-cake, which serves as a support structure and improves the thermal insulation. Additional layers of powder are always applied above the last layer to ensure thermal insulation of the melt in all directions. The process then enters the cooling phase. The melt cools and crystallizes in the part-cake without additional pressure (Figure 3.33, center). During this mostly undefined cooling situation, the part has direct interaction with the surrounding powder on all sides. After completion of the cooling phase, the SLS parts can be removed from the part-cake and, if necessary, transferred to further post-processing steps. The parts are characterized by process-typical properties, such as rough surfaces, various shape and dimensional deviations, staircase effects, powder adhesion, powder bonding (depending on the powder), and rounded or sharp part edges (depending on the orientation) (Figure 3.33) [Schmid 2015].

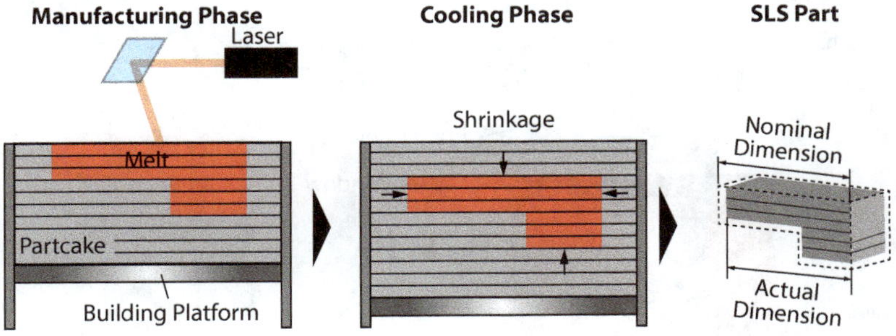

Figure 3.33 Schematic representation of the dimensional formation in the SLS process [Launhardt 2021]

The fundamental principle of SLS remains overall consistent, but there are now various versions of production systems – partly due to patent considerations. Specifically, differences exist in powder feed mechanisms, powder application techniques, build-chamber geometry, and exposure strategies.

Powder feed

While in some production systems the powder is fed into the build-chamber space from below via cylinders (Figure 3.34), in other systems the powder is fed into the build-chamber from above via hoppers and metering rollers (Figure 3.35). There are no clear advantages or disadvantages between the different designs. With poorly flowing powders, bridging can occur with hopper feeding, which can hinder gravimetric powder feeding. In this case, vibration systems are to be provided, but these come only partially pre-installed by the manufacturer.

Overflow

Systems with an overflow provide the possibility to remove excess powder from the process before the powder application system. Generally, this leads to an additional powder processing step. A distinction is made between the mixed powder, the waste powder from the part-cake, and the overflow powder. This has passed through the build chamber at least once, but is thermally less aged than the part-cake powder. An overflow system is advantageous if the powder is contaminated. If there is contamination in the supplied powder (e.g. brush hairs or partially torn out structures), then these can be pushed into the overflow, provided they are small enough, and the build process can continue undisturbed. In systems without this overflow system (e.g. EOS Formiga P110), the powder is circulated and repeatedly applied in the build area. Any contamination present can thus circulate repeatedly through the build chamber, potentially causing process instability. However, this system has the advantage of maintaining consistent powder quality, composed solely of the part-cake remaining at the end of the build process.

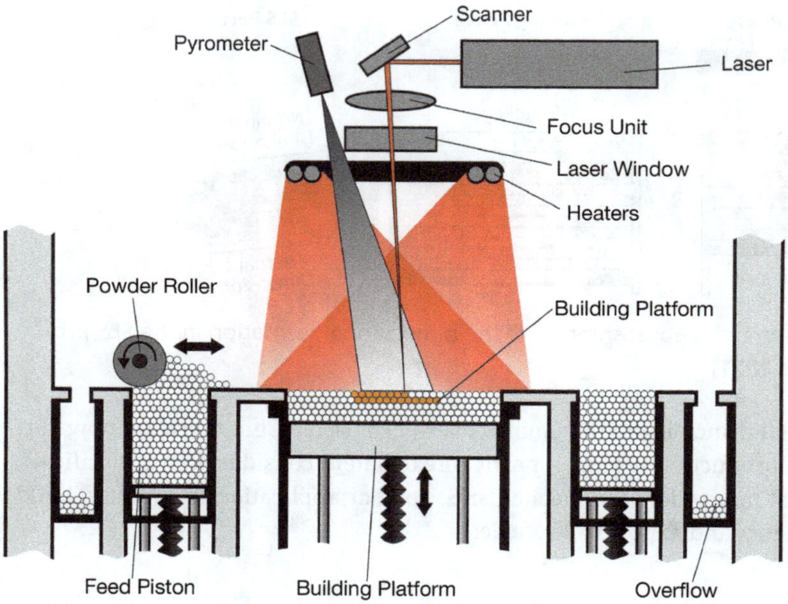

Figure 3.34 Schematic representation of an SLS production system with a powder feed with vertically movable cylinders with overflow containers and with a counter-rotating roller for powder application; this corresponds approximately to the structure of the sintering station system from DTM or Vanguard from 3D Systems [Launhardt et al. 2016]

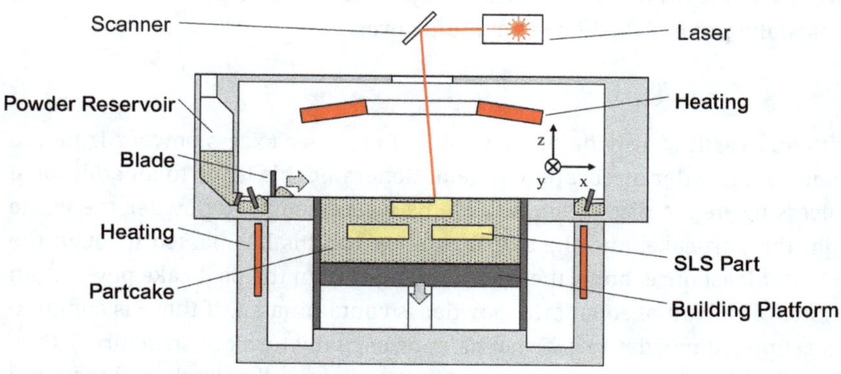

Figure 3.35 Schematic representation of an SLS production system with powder fed in via a hopper without an overflow container and with a blade system and rotating blades, which always re-position the powder in front of the blade; this corresponds approximately to the structure of the P1 system from EOS GmbH [Wolf et al. 2017]

Powder application

For any SLS process, a smooth powder surface is essential to achieve good part results. This smooth powder surface is achieved after the build platform is lowered, either by a blade, a hopper with a double blade, or a counter-clockwise rotating and moving roller. The various application systems have different factors influencing the resulting powder layer in terms of smoothness and powder bulk density. While the smoothness has a decisive influence on the imaging accuracy of the parts to be produced and the process stability, the powder bulk density determines the resulting mechanical part properties. For example, different geometries (radii and angles) can be used with blades and double blades to effect compaction of the powder during powder application. In addition, vibrators can be used to effect a higher compaction of the powder layer. If a powder tends to agglomerate, then a counter-rotating roller can be advantageous for smooth powder application, as this loosens up the powder again before application by breaking up possible agglomerates.

In Table 3.5 currently commercially available equipment types are assigned to the different powder application systems.

Table 3.5 Examples of Different Powder Application Systems Used in Production (Sorted by SLS Equipment)

Roller	Funnel/double-blade	Single-blade
DTM Sintering station 3D Systems Vanguard HS SinterIt Lisa, LisaX, Lisa Pro Prodways P1000	EOS P3 Weirather WLS 3232	EOS FORMIGA P110 Sintratec KIT, S1

Build space geometry

Depending on the system manufacturer, different concepts are pursued for the build space geometry. Some of the build spaces are permanently installed (e.g. DTM Sinterstation 2000) and others are designed as interchangeable frames. The interchangeable frame technology has the clear advantage that the systems can be reloaded with an empty interchangeable frame at the end of the process and production can be resumed promptly after the end of the previous build-job. In addition, the geometry is differentiated between round and rectangular build spaces. Round cavities offer the advantage of simpler homogeneous temperature control on the equipment side, whereas part positioning in the cavity is problematic with mostly square parts and leads to lower packing density. For interchangeable frame systems, a passive or active cooling station is usually required in which the hot part-cake can cool down slowly under defined conditions, e.g. under nitrogen supply.

Most industrial machines today offer rectangular interchangeable frames. Here, the systems often differ in the ratio of the build space dimension in the scan direction (XY) and in the layer build-up direction (Z). While some systems (e.g. EOS P3) have a ratio of the smallest dimension in XY to the extension in Z direction of 1:2, other systems (e.g. EOS P7) offer a rather wide build space in relation to the build space depth. This ultimately has an influence on the positioning of long parts in the build space and thus, depending on the build space geometry, an influence on achievable part strength. The lowest mechanical properties are achieved with SLS in the layer build-up direction (Z). If this is also the largest possible build space dimension, then long parts are positioned by lack of alternative in the less favorable build direction. This means that, depending on the part geometry involved, the installation space geometry of the system to be selected should also be taken into account.

Exceptional in terms of the build space geometry is the HSS (high-speed sintering) machine VX1000 by Voxeljet, as here there is a larger expansion in xy-direction than in the z-direction (build volume 1000 × 600 × 500 mm).

In Table 3.6 currently commercially available SLS production systems are assigned to rectangular and circular build space geometries.

Table 3.6 Examples of Different Build Space Geometries Used in SLS Machines According to Machine Manufacturers

Rectangular	Circular
3D Systems Vanguard HS DTM Sintering station 2500 SinterIt Prodways P1000 Farsoon 403P EOS P1-P7 Weirather WLS 3232	DTM Sintering Station 2000 Sintratec S2

Parameter systems

While in other thermoplastic processing methods, e.g. injection molding, extrusion, and thermoforming, the operator naturally demonstrates their expertise in this field by selecting the necessary process parameters and fitting the intended parts individually, in additive manufacturing, closed parameters are popular on the equipment side. This approach is similar to the razor-blade business model or printer-toner dependency, where the service provider is incapacitated from being able to select its own process parameters for the specific applications. Opening up parameter systems is sometimes possible in return for financial resources. New equipment manufacturers are focusing on more openness here, and offer equipment systems with an open parameter set directly from the factory (Table 3.7).

Table 3.7 Examples of Open and Closed Parameter Systems and Interchangeable Frame Options Used In SLS Machines, According to Machine Manufacturers

Open parameter system	Closed parameter system	Without interchangeable frame
SinterIt Nils Weirather WLS 3232 Prodways Farsoon DTM Sintering Station 2000	EOS P1-P7 Formlabs Fuse 1	DTM Sintering Station 2000 Farsoon 403P Prodways P1000 SinterIt

3.2.5 Build-job Preparation (Pre-processing)

To operate industrial SLS systems effectively, several key parts are essential as well as the manufacturing system itself. These include a stable infrastructure, supporting peripheral equipment, customized software, and well-trained operators.

Even small SLS systems, such as a FORMIGA P110 from EOS GmbH, require an electrical power of approx. 5 kW at maximum utilization, which means that a 16 A plug must be kept ready. Larger systems require correspondingly more power. Likewise, the power grid should be free of voltage fluctuations.

Depending on the production system, the required nitrogen is generated within the production plant as part of a nitrogen generator or can be supplied from outside. In some cases, compressed air is also required for functions such as locking the doors and vibrators for the powder fluidization. Particularly when the compressed air is generated in an internal nitrogen generator, high demands are placed on the supplied compressed air. This should be as constant a pressure as possible, and should be free of oil, water, and impurities. Otherwise, there is a risk that the microfilter of the nitrogen generator will become clogged and have to be changed more frequently. Especially, in small enclosed rooms and when operating many SLS systems in one room, the installation of a nitrogen warning device is mandatory in order to detect the hazardous effect of nitrogen due to oxygen displacement.

An exhaust device should be provided above each SLS system to draw off the outgases generated during the heating of polymers. The exhaust air must not be returned to the work area. In order to transmit the required data to the manufacturing station, most stations require a network connection.

Table 3.8 shows the technical requirements for a FORMIGA P110 SLS system from EOS GmbH.

Table 3.8 Infrastructure Requirements for the Implementation of SLS Production

	Technical requirement
Power	At least 16 A/400 V
Compressed air	Approx. 10 m^3/h at 7 bar Contamination: Class 1 solids (particle size ≤ 0.1 µm, particle density ≤ 0.1 mg/m^3) Water content: Class 4 (pressure dew point ≤ 3 °C) Oil content: Class 1 (oil concentration ≤ 0.01 mg/m^3)
Nitrogen	In some cases, an external nitrogen supply is necessary/possible for some production systems
Safety device	Nitrogen warning system
Fume hoods	Min. 10 m^3/h (e.g. for EOS FORMIGA P110), diameter 100 mm
Network connection	Fast ethernet, TCP/IP

3.2.5.1 CAD Part Design

In order to be able to manufacture parts on SLS systems in the first place, you need suitable software. On the one hand, design software (computer-aided design software) is necessary for the design of a part geometry. Standard engineering software such as SolidWorks, SolidEdge, Fusion 360, or similar can be used for this purpose. However, there are also various design software solutions that appeal to less technical users and allow design-based freeform surfaces. Freeware options of both also exist to design parts accordingly.

Figure 3.36 Thermoplastic polyurethane SLS PowderMonkeys FlipFlop with parametric lattice design. Designed by *Roman Reiner GmbH* and Launhardt GmbH (© Launhardt GmbH 2024)

Especially with the outstanding manufacturing capabilities of additive manufacturing technologies, the importance of a parametric design of parts becomes more relevant. Therefore, the CAD is partially added with programmed similar structures which allow the specific variation of geometries in a larger scale. For example the lattice structure of this bathing flipflop was generated using a parametric approach using Grasshopper. This allows the operator to adjust for example the lattice structure strut thickness, the gap size as well as the overall shape with simple input. This was realized using Rhino in combination with Grasshopper (Figure 3.36).

Table 3.9 lists commercially and freely available software solutions for the construction of technical parts and designs.

Table 3.9 Available CAD and Design Software Solutions (as of 2022)

Software manufacturer	Focus	Costs
Technical CAD software		
SolidWorks, Dassault Systems	Professional product development	From approx. 5,000 €/year
Solid Edge	Good product development	From approx. 4,000 €/year
CATIA, Dassault Systems	Professional product development	From approx. 10,000 €
ProEngineer/Creo	Professional product development	From approx. 4,000 €/year
FreeCAD	Basic CAD capabilities	Free of charge
Inventor, AutoCAD	Professional product development	From approx. 3,000 €/year
Fusion 360, Autodesk	Especially for additive manufacturing	From approx. 500 €/year
NX, Siemens	Professional product development	From approx. 7,000 €
OpenSCAD	Programming knowledge required	Free of charge
Rhino	Designing and modelling of parts	From approx. 1,000 €
Design CAD Software		
SketchUp	Beginner: Architecture, desktop and web-based	From approx. 274 €/year
Blender	Design: Advanced	Free of charge

3.2.5.2 Design Recommendations for SLS parts

Additive manufacturing processes enable the tool-free production of parts directly from vector-based CAD data. For this purpose, the CAD data must first be translated into a set of readable instructions for the respective additive production system. In industrial additive manufacturing settings, the "surface tessellation language" (STL) format has gained prominence.

This necessitates converting proprietary CAD data formats to STL. In STL files, geometry surfaces are represented by varying-sized triangles [Gebhardt 2016, VDI 2014, Breuninger et al. 2013]. Transforming data into triangles can lead to data loss. Insufficient triangle resolution may cause deviations in part shapes and dimensions. These deviations are evident from the characteristic triangles on the surface (Figure 3.37). For the SLS process, EOS GmbH recommends a minimum imaging detail size of 0,01 mm and a maximum triangle angle of 2° [EOS —a].

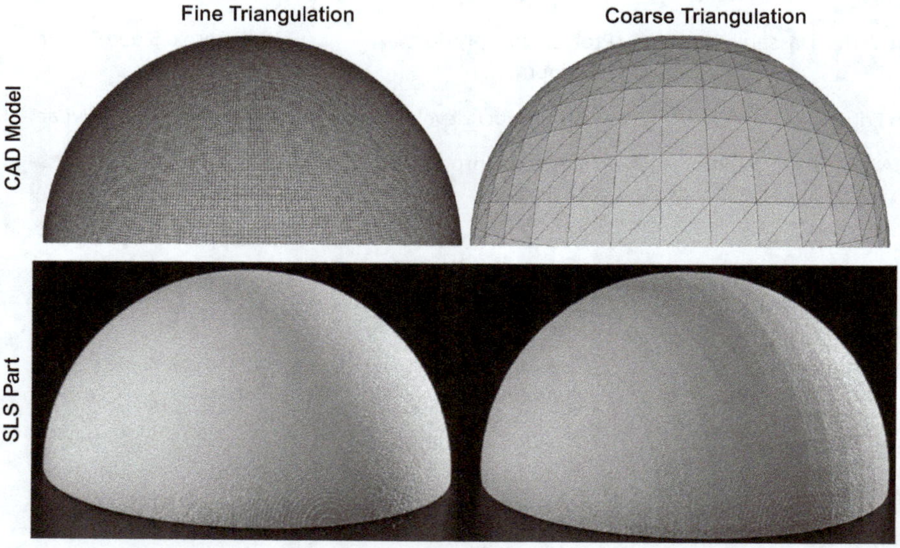

Figure 3.37 CAD geometry with fine triangulation of the surface (top left) and coarse triangulation (top right): Below, the resulting SLS part surfaces with characteristic triangulation are shown [Launhardt 2021]

According to DIN EN ISO 17296-4, various file formats, e.g. container formats with additional manufacturing information for additive manufacturing, are permitted (e.g. STL, 3MF). The STL-based geometries are positioned and oriented in a virtual build space of the SLS system. Here, guidelines, manufacturers, and individual publications give recommendations for design, positioning, and orientation, as well as distance between individual parts, distance to the build space edge, gap dimensions, and smallest possible geometries [VDI 2014, EOS —a, Seepersad et al. 2012, Wegner & Witt 2012]. These outcomes arise from process-specific minimum resolution requirements, which stem from factors such as particle size, laser focus diameter, and layer height. Figure 3.38 provides a summary of crucial guidelines and recommendations for achieving optimal dimensional accuracy and reproducibility in the design of parts and build-job layouts for SLS.

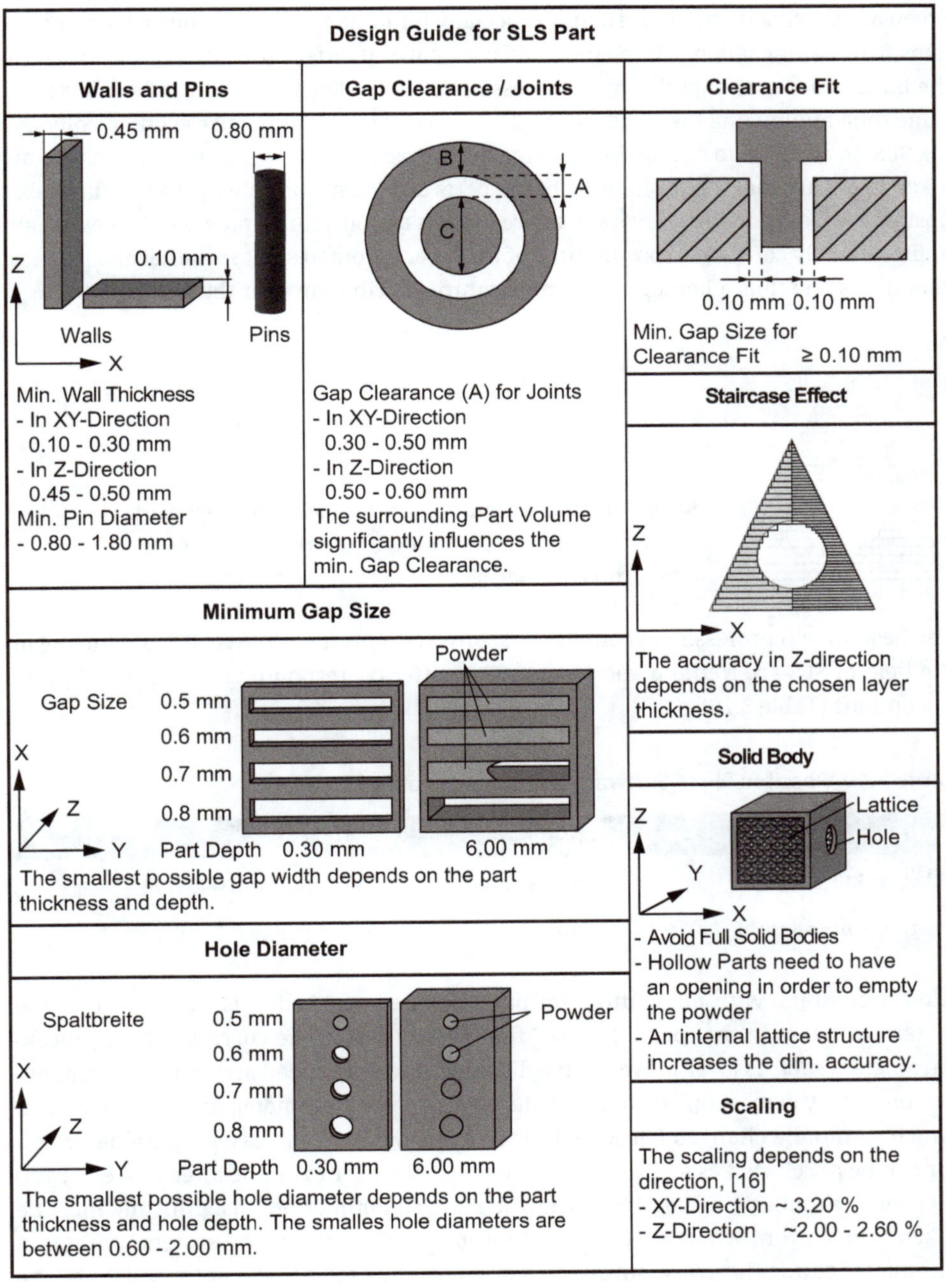

Figure 3.38 General design recommendations for SLS parts [Launhardt 2021]

3.2.5.3 Slicing

In addition to pure design software, every additive manufacturing system requires suitable slicing software. Here, the previously designed part geometries are positioned in a virtual build space, oriented, and then sliced to the desired layer thickness (Figure 3.39). Freeware slicer software is available, especially for desktop systems and consumer systems. The major challenge is to efficiently pack parts of different geometry and size into the build space such that the greatest possible packing density is achieved while at the same time minimizing the build height. This is crucial to manufacture economically using SLS. In addition to optimizing packing density and build height, care should also be taken to ensure that the positioning of the parts is done in such a way that the laser exposure area, of individual layers in z-direction, is as constant as possible. Achieving an approximately constant exposure time of the layers promotes quasi-isothermal process conditions and thus a homogenous temperature distribution over the build volume.

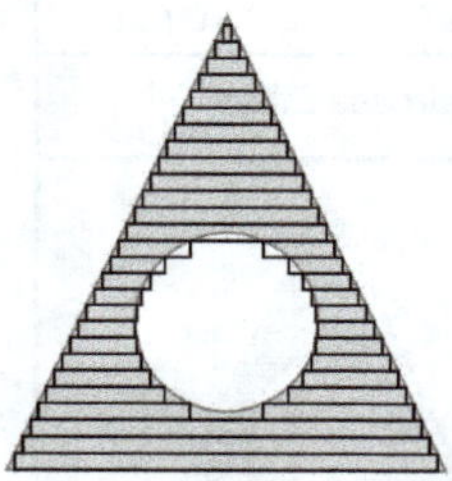

Figure 3.39
Schematic representation of the slicing process and the associated incrementing of the nominal geometry (© Martin Launhardt)

The best-known professional commercial software tools for additive manufacturing in the field of SLS are Netfabb and Magics. They offer correspondingly automated packing options (Table 3.10).

Table 3.10 Possible Slicer Software Solutions (as of 2023)

Software manufacturer	Focus	Costs
Magics, Materialise	Professional	From approx. 6,000 €/year
Netfabb, Autodesk	Professional	From approx. 6,000 €/year

After a build-job with many different parts has been packed, it is transferred to the system software in the form of "slice" files (SLI). Therein the corresponding process parameters such as laser power, hatch distance, and scan speed are assigned. Depending on the system supplier, only predefined process parameter sets are available, which cannot be changed (closed parameter setting). In some cases a parameter editor can be purchased as an upgrade that enables the user to define alternative parameter sets. The additional costs for this, which are sometimes not insignificant, must be taken into account when selecting an appropriate SLS production system. Without a parameter editor, the processing of new or alternative materials – not specified by the system manufacturer – is impossible. Since equipment manufacturers are frequently also material suppliers, some charge additional costs for the possibility of processing alternative materials on their equipment.

3.2.5.4 Recommendations for Build-job Layout

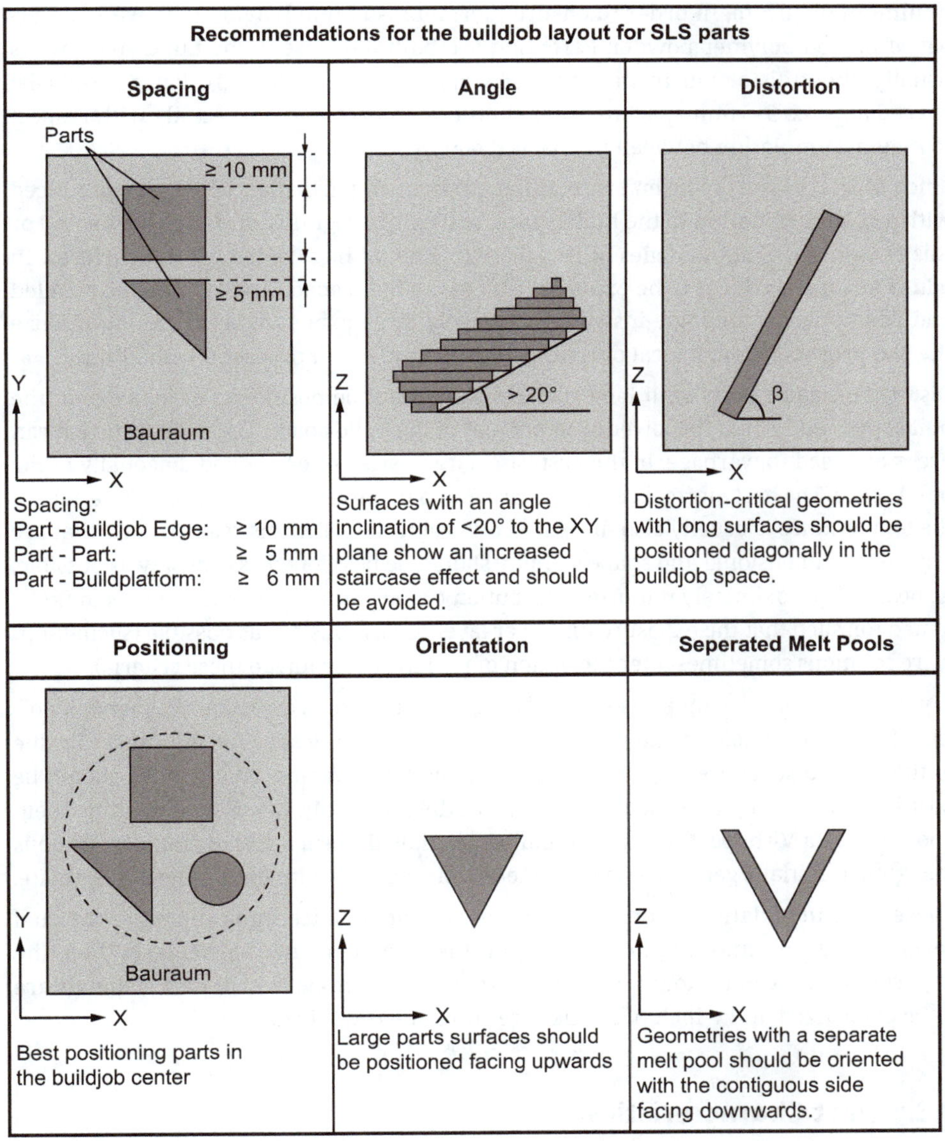

Figure 3.40 Summary of recommendations for the build-job layout for dimensionally stable SLS parts (based on [VDI 2015, EOS —a, EOS 2013, Adam 2015, Adam & Zimmer 2014]

When positioning CAD parts in the virtual build space, a wide variety of aspects must be taken into account in order to avoid part errors, dimensional and shape deviations, or (in the worst case) process terminations. In the following sections these recommendations are discussed, and they are summarized in Figure 3.40.

Predefined distances should be maintained from the edges of the build space and also between the individual parts. Typical distances to the build room walls should be a minimum of 10 mm, in order to ensure thermal insulation between the wall and areas of melted polymer powder. Parts that are built too close to the build space walls usually show distortion in the form of warpage (Section 3.2.7.3). Parts should be spaced approx. 3–5 mm apart to prevent mutual interference, especially with respect to melt-accumulation between the parts (Section 3.2.7.3).

Since SLS is a layer-by-layer process, flat surfaces in particular, which are produced with a slight inclination in the build space, will exhibit significant staircase effects. For this reason, inclination angles of less than 20° should be avoided for flat surfaces. In addition, large surfaces to be exposed (>1/2 of the build space area) should be avoided and positioned at an angle instead. Excessively large exposure areas can destabilize the SLS process through local overheating and may lead to process termination.

Distortion-critical parts with long surfaces should not be positioned at the edge of the build space and not at the lower or upper end of the build space. Thermal distortion can occur and lead to warpage. In the best case, large parts are positioned diagonally in the build space. The best thermal conditions are found in the center of the build space, where the temperature is usually also measured and controlled. For this reason, parts that are critical for dimensional and shape stability should be positioned as centrally as possible. At best an approximately uniform distribution of the parts in the build space should be aimed for, such that the exposure times per layer are as constant as possible (slicing software solutions sometimes offer the option of packing according to these criteria).

Especially shape-relevant parts should be aligned in the scan direction (xy-plane) if possible (e.g. circular holes or fits). In other orientations, elliptical distortion can occur due to the layered structure. If parts are tapered, the larger melt-accumulations should be positioned at the top and filigree elements at the bottom. In the case of dividing melt-pools, e.g. in a V-shape, the parts should be positioned in such a way that a continuous melt-pool is initially generated and the melt-pools separate with increasing z-coordinate.

These recommendations represent a summary from practice, and can be contradicted several times within a part, depending on the part geometry. Particularly then, the experience of the operator is required to know, for example, which part elements are to be prioritized and which part sides are still to be reworked.

3.2.6 Part Characteristics

Due to the layered structure of the geometry within a powder-based additive manufacturing process, parts obtain typical characteristics, such as anisotropic mechanical properties and a characteristic surface roughness. The latter is often a point of discussion compared to the usually smooth surfaces achieved in injection molding. The SLS process, without any mechanical or chemical smoothing, provides a new type of surface structure and haptic experience, which in some design products is already advertised as a positive attribute.

Surface structure

The parts produced by SLS have a very characteristic surface structure. Since the powder serves as its own support structure, regardless of the polymer used, the melted parts are always in contact with a surrounding powder. Consequently, the part surface always features a more-or-less adhering powder layer.

Surface roughness

The roughness of SLS parts is influenced both by the powder adhesion and by the layer transition. In combination with the meniscus effect at the melt edges on the upper side, generally a higher roughness is found on the upper side of the part and a lower roughness on the lower side. The highest roughness is usually with a result of unfavorable build orientations in the build space (see "staircase effect").

> For roughness, a distinction is made between the average roughness R_a and the maximum roughness R_z. The usual roughness of SLS PA12 parts is as follows:
>
> - Average peak-to-valley height: $R_a \approx 10\ \mu m$
> - Maximum roughness depth: $R_z \approx 60\ \mu m$.

Staircase effect

Process-induced defects occur in particular in the build-up direction. At unfavorable angles and radii (see Section 3.2.5.4), there is a certain "staircase effect" (Figure 3.41). As a result of meniscus effects at the melt edges, this is particularly evident on the upper side of the part.

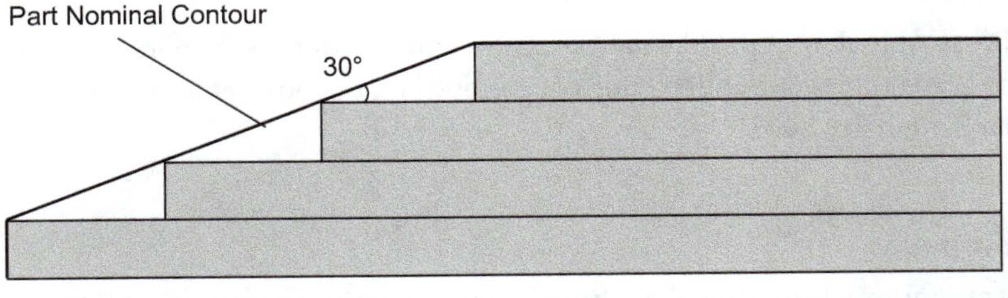

Figure 3.41 Schematic representation of the emergence of extreme manifestations of the staircase effect (© Martin Launhardt)

Meniscus

When the thermoplastic powder is melted by the laser, a meniscus-shaped contact of the melt with the surrounding powder bed results at the edges of the melt pool, which is controlled by surface tension. At the solidified part, this results in sharper upright edges at the top of the part (Figure 3.42b). Subsequently, due to the meniscus effect and the sharper edges at the top, the staircase effect at the top is also more intense. These resulting sharp top edges are always a very clear indication of the part's orientation in the build space. The effect can be reduced by smoothing post-processing procedures (see Section 3.2.9.3).

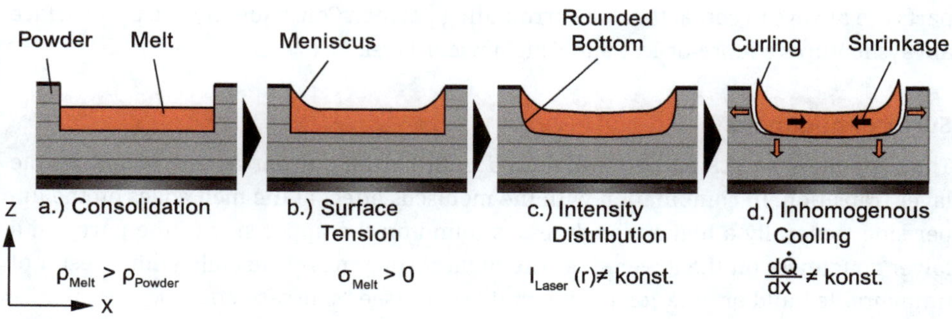

Figure 3.42 Schematic representation of the meniscus effect during melting of the thermoplastic powder [Launhardt 2021]

Since the laser power cannot be applied unambiguously and precisely due to the Gaussian power distribution, the overlapping of hatch lines, and the increased heat conduction effects at edges, the result is a rounded edge at the bottom, in contrast to the sharp edges at the top (Figure 3.42c and Figure 3.43).

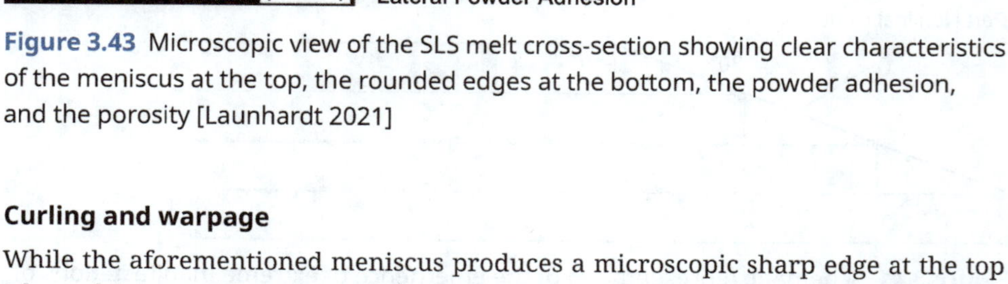

Figure 3.43 Microscopic view of the SLS melt cross-section showing clear characteristics of the meniscus at the top, the rounded edges at the bottom, the powder adhesion, and the porosity [Launhardt 2021]

Curling and warpage

While the aforementioned meniscus produces a microscopic sharp edge at the top edges of the part, if the powder surface is too low in temperature, an upward bending of the melt edges occurs as a result of shrinkage, which leads to upwards curling (Figure 3.41d). In extreme cases, this can lead to the part edges protruding so far upwards during production that they are caught by the powder application system and thus torn out. The consequence is a build-job termination. If the job does not terminate, the effect of curling on parts can be recognized mainly by a strong curvature at the bottom of the part.

However, this should not be confused with thermally induced global part distortion, also known as warpage (Figure 3.44). Warpage occurs when the part-cake experiences strong thermal inhomogeneities and the parts thus cool unevenly. This causes the entire part to warp. Naturally, this effect is mainly visible with large part dimensions. A more detailed explanation of the origin of curling and warpage is given in Section 3.2.7.1.

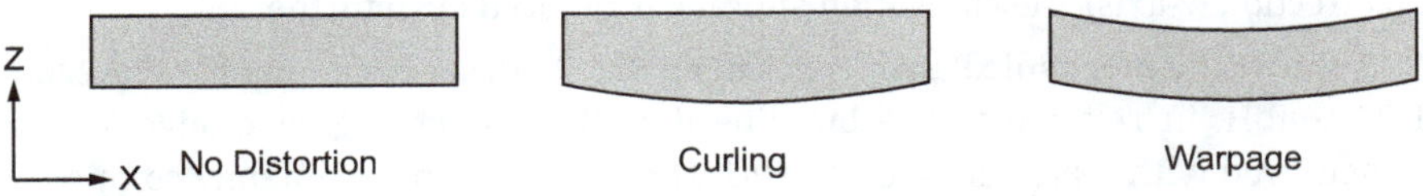

Figure 3.44 Representation of curling and warpage errors, on the basis of a simple rectangular specimen (based on [Schmid 2015])

Porosity

PA12-based parts manufactured by SLS usually have an unevenly distributed porous internal structure. The usual porosity of SLS PA12 parts is 1.5–4.0 % by volume, with pores 30–300 µm in size, predominantly located in the interior of the part. The edge layers, with thicknesses of 0.3 mm, are usually free of pores (Figure 3.45).

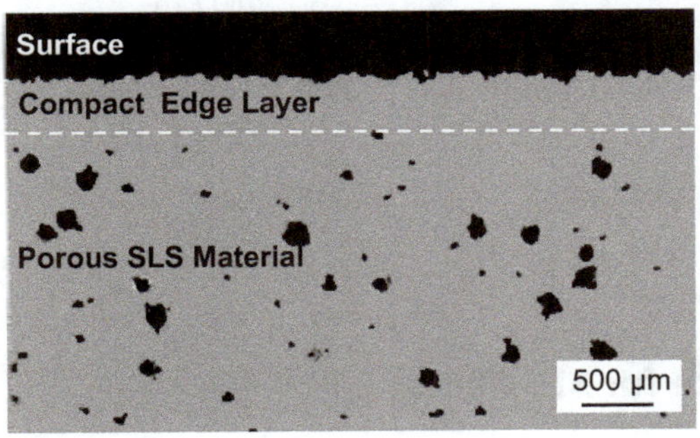

Figure 3.45 Schematic representation of an SLS PA12 part at the edge with a characteristic compact edge layer and a typical pore distribution inside (© Martin Launhardt)

Density

The resulting porosity of SLS parts together with the material's density ($\rho_{PA12} = 1.02\ g/cm^3$) results in a usual SLS PA12 part density of $\rho_{PA12,SLS} \approx 0.97\ g/cm^3$. The density depends on many parameters of the SLS process, e.g. powder deposition speed, powder application system, powder condition, laser power, etc. The density of SLS parts made of TPU, for example, is approximately 1.20 g/cm^3.

Mechanical properties

The mechanical properties of SLS parts are close to those of injection-molded parts made of the same material. However, due to the layer-by-layer build-up process, there is a clear anisotropy, especially in the mechanical properties. For example, it can be seen that in the scan direction (xy-plane), the mechanical properties, such as tensile strength and modulus, are somewhat above those achieved with injection molding, which is probably related to the increased crystallinity of the slowly cooling melt in the SLS process. In the case of elongation at break at 18 %, the SLS part is clearly inferior to the possible elongation at break achieved with injection-molded PA12 parts (>50 %). Particularly in the layer build-up direction (z-axis), the elongation at break is reduced even further.

Table 3.11 shows the mechanical properties of an SLS PA12 specimen made of PA2200 from EOS GmbH, in relation to the building direction. Much higher elongation at break is achieved with materials such as TPU. These exhibit approximately comparable values to injection-molded TPU, especially if they have been chemically smoothed during post-processing, thus reducing microcracks at the surface. Here, an elongation at break can be achieved of up to 700 %.

Table 3.11 Mechanical Properties of SLS Parts Made of PA12 as a Function of the Direction of Installation (Values According to Manufacturer's Data)

	Scan direction (xy-plane)	Layer build-up direction (z-axis)	Injection-molded
Manufacturer and material	EOS, PA2200	EOS, PA2200	EMS Chemistry, Grilamid L16
Elongation at break	18 %	4 %	>50 %
E-modulus	1,650 MPa	1,650 MPa	1,500 MPa
Tensile strength	48 MPa	42 MPa	45 MPa

3.2.7 General Quality Assurance for SLS Parts

There are many different approaches to the quality assurance of SLS parts. The most important distinctive features are dimensional accuracy, visual appearance, and mechanical properties. Dimensional accuracy on SLS parts is usually measured by random sampling. Free-form surfaces present a difficulty, which can be addressed by matching a 3D-scan with the original CAD data. Assessing the appearance provides subjective information about the roughness present, possible defects, and optical blemishes. In practice, the mechanical properties are tested by parallel manufacturing of test specimens next to the parts, e.g. tensile bars. In this way, layer bonding defects can be detected efficiently.

3.2.7.1 Shape Deviation

The accurateness of each part is defined by its coarse and fine shape [Westkämper & Warnecke 2011, Keferstein & Dutschke 2007]. The coarse shape is further divided into the dimensional, shape, and position tolerance. According to DIN 4760, the fine shape is divided further into the waviness and the roughness (Figure 3.46; see also [Jorden & Schütte 2017, DIN 1982]). The roughness and waviness as part of the fine shape also contribute to the part shape. For this reason, the individual factors are explained in more detail below.

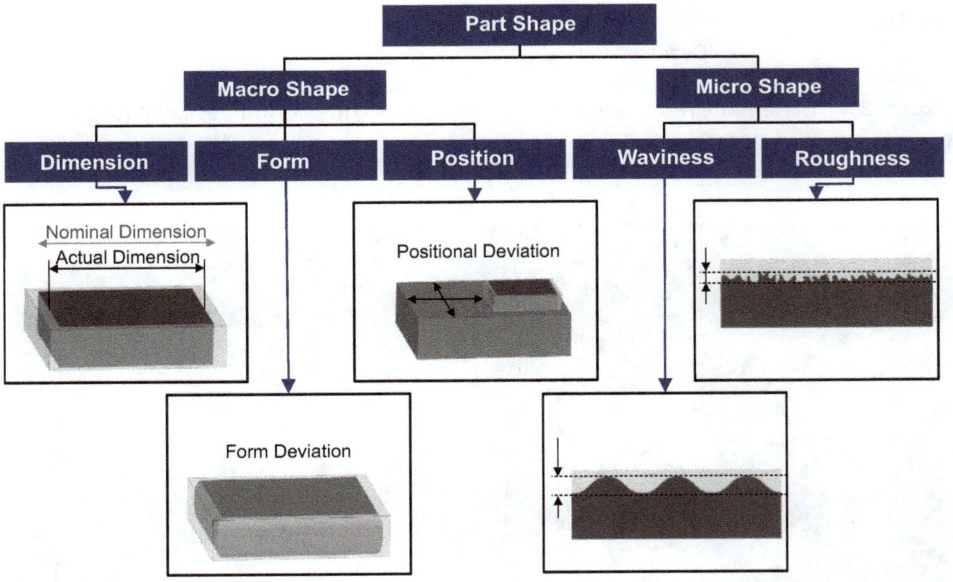

Figure 3.46 Assessment of the part shape (according to [Westkämper & Warnecke 2011, DIN 1982])

Fine shape

The fine shape of SLS parts is determined by the characteristic surface roughness. This is caused by the contact of the melt with the powder during the manufacturing process. In addition, there are layer bonding transitions and staircase effects in the layer build-up direction (z-axis) which additionally impart a waviness. As described in Section 3.2.6 the typical roughness values of SLS parts are $R_a \approx 10$ µm (average roughness) and $R_z \approx 60$ µm (maximum roughness).

In addition, the fine structure may show increased roughness as a result of sink marks ("orange peel skin"). The roughness is usually determined by means of tactile roughness gauges, in which a fine needle traces the surface structure and evaluates the deflection in the direction of the roughness peaks (Figure 3.47). A closer view of these measuring needles on the SLS part surface shows that the actual roughness is typically higher than measured as a result of adhering powder particles and complex undercuts, which can only be represented with moderate accuracy by a large needle (Figure 3.48). The result is a smoothed profile line of the real surface roughness.

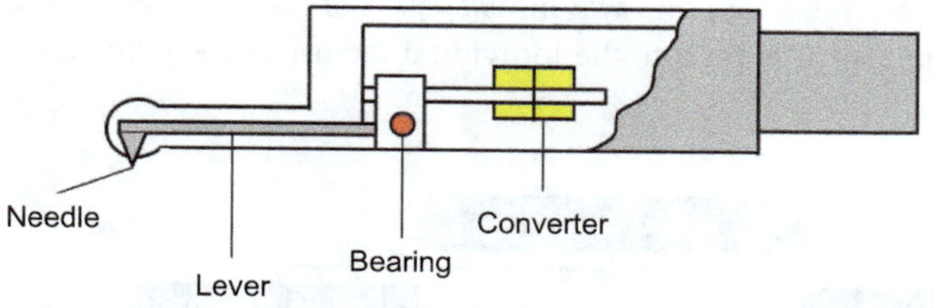

Figure 3.47 Schematic representation of a roughness measuring device (based on [Launhardt et al. 2016])

Figure 3.48 Electron microscope image of a probe tip of a roughness tester on an SLS part surface [Launhardt et al. 2016]

Dimensional deviation

Dimensional deviations are usually linearly measured deviations between the nominal and the actual geometry. Shrinkage of the polymer from the time of laser exposure to the crystallized part at room temperature contributes significantly to the dimensional deviation of SLS parts. Shrinkage is composed of two major sub-shrinkages: thermally induced shrinkage and crystallization shrinkage. The thermal shrinkage is quantified by the thermal expansion coefficient α, and depends on the temperature difference ΔT, and can be considered linear to a good approximation for small temperature ranges (Equation 3.3). Crystallization shrinkage occurs during crystallization and depends only to a small degree on the applied temperature and the cooling rate, the given time, and thus the resulting degree of crystallization.

$$\Delta l_{\text{therm}} = a \times l \times \Delta T \quad (3.3)$$

Δl_{therm}: Thermally induced change in length [mm]

α: Coefficient of linear expansion [1/K]

l: Initial length [mm]

ΔT: Temperature difference [K]

Shrinkage also occurs during the injection molding of polymers, but this can be compensated by applying a controlled holding pressure. Pressure–volume–temperature (p–V–T) diagrams are used to read out the expected shrinkage behavior for the injection-molding process. For injection-molded parts, the processing shrinkage represents the deviation between the cavity dimensions to the part dimensions, 16 hours after demolding at room temperature [Meyer & Falke 2013].

In contrast, there is no holding pressure in SLS, so a different approach is necessary. With a homogeneous melt and homogeneous boundary conditions, isotropic volume shrinkage can be assumed, which should in principle lead to a uniform dimensional deviation of the part. However, due to the line- and layer-wise part structure, a distinction is made in SLS between shrinkage in the scan direction (xy-plane) and shrinkage in the layer build-up direction (z-direction). In addition, the shrinkage in the layer build-up direction decreases with increasing z-positioning, so that a linearly variable shrinkage must be applied in the z-direction [Launhardt et al. 2015, EOS —b, Eschey 2013, EOS 2013, Soe et al. 2013].

The thermally induced volume difference can be calculated by means of Equation 3.4 into the one-dimensional linear dimensional change Δa, and converted by means of Equation 3.5 into the relative volumetric and one-dimensional shrinkage parts.

$$\Delta V = (a + \Delta a)^3 - a^3; \Delta a = \sqrt[3]{\Delta V + a^3} - a \quad (3.4)$$

ΔV: Volume change, $\Delta V = V_2 - V_1$ [mm³]

a: Edge length of the volume increment a [mm]

Δa: Length change [mm]

V_2, V_1: Volume [mm³]

$$S_V = \frac{\Delta V}{V} \times 100\% \quad ; \quad S_1 = \frac{\Delta a}{a} \times 100\% \tag{3.5}$$

S_V: Volumetric shrinkage ratio (scaling factor) [%]

S_l: One-dimensional length shrinkage (scaling factor) [%]

In the field of additive manufacturing, the specification of direction-dependent linear shrinkage in the x-, y-, and z-directions has become established. The shrinkage data can be used directly for adapting the nominal geometry, therefore typically the scaling factors are given in percent.

> Typical scaling factors for SLS are as follows [Launhardt et al. 2015, Eschey 2013, EOS 2013]:
>
> - In scan direction (xy-plane): $S_{l,xy} = 3.2\%$
> - In the layer build-up direction (z-direction):
> - $S_{l,z}$ (for z = 0 mm) = 2.6%; $S_{l,z}$ (for z = 300 mm) = 2.0%

In order to adjust the dimensional accuracy of SLS parts to suit the production system, in practical implementation test specimen are manufactured with two main parameters varied: The scaling factors and the global beam offset. Thus, typical test specimens must feature defined inner and outer surfaces that are subsequently measured. If the inner and outer dimensions are oversized, this can be compensated by adjusting the scaling factors in the respective spatial direction. If, on the other hand, the inner dimension is undersized and the outer dimension is oversized, this can be compensated for by adjusting the beam offset (Figure 3.49). However, this compensation is only possible in the xy-plane and thus allows the dimensions to be adjusted in a limited range independent of the other set process parameters (e.g. laser power, scanning speed, etc.). In the layer build-up direction, only the correction by z-scaling is possible.

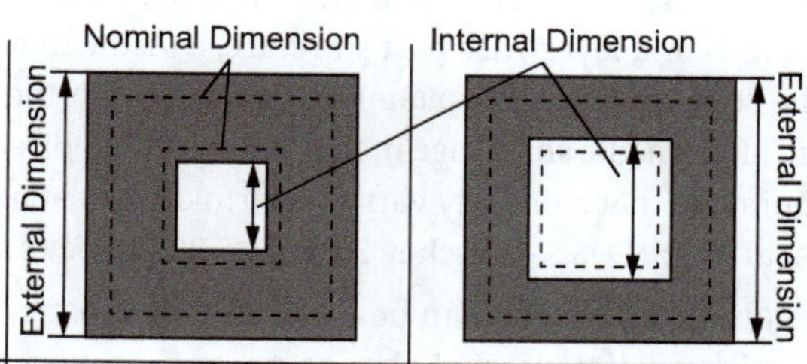

Internal Dimension	Oversize	Undersize	Oversize	Undersize
External Dimension	Undersize	Oversize	Oversize	Undersize
	Beam Offset		Scaling Factor	
➤ Adjust	Reduce	Increase	Reduce	Increase

Figure 3.49 Practical procedure for adjusting over-sizes and under-sizes to inside and outside dimensions using the beam offset and the scaling value [Launhardt 2021]

In order to quantify the dimensional accuracy of a production system, it is necessary to manufacture and measure corresponding dimensional accuracy test specimens with internal and external dimensions in the x-, y-, and z-directions. Additionally, it is advisable to co-produce these test specimens with each build-job and to measure them afterwards to ensure continuous indirect quality assurance. Depending on the build-job pack-density, build-job height, and duration, different dimensions may result. Ultimately, it is advisable, where possible, to measure and document at least one dimension in each spatial direction on each part to ensure the dimensional accuracy.

Some of the typical dimensional stability test specimens used in the literature are depicted in Figure 3.50. In addition to the basic dimensions in the x-, y-, and z-directions, some of these specimens offer angles, free-form surfaces, and graded wall thicknesses in order to assess the manufacturing capabilities. To a certain degree these specimens allow for pre-assessment of the resolution accuracy and dimensional accuracy by visual inspection.

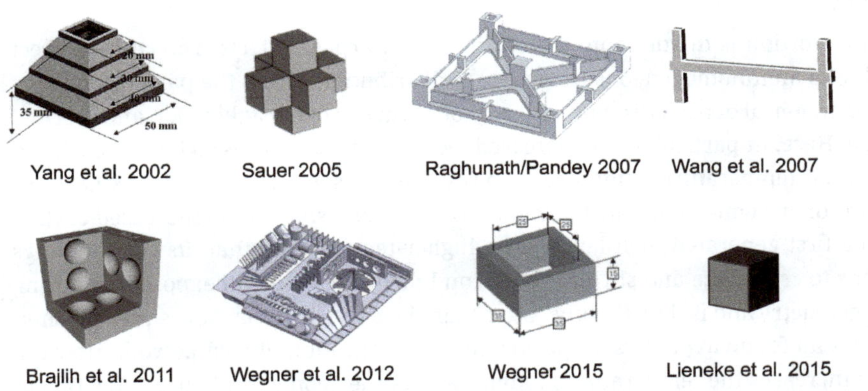

Figure 3.50 Overview of selected specimens from the literature for characterizing the dimensional and shape stability of SLS parts [Yang et al. 2002, Sauer 2005, Raghunath & Pandey 2007, Wang et al. 2007, Brajlih et al. 2011, Wegner & Witt 2012, Wegner 2015, Lieneke et al. 2015]

Shape deviation

With homogeneous temperature distributions, only isotropic shrinkage should occur and thus mainly cause volume-dimensional deviation. In most cases, however, there is a degree of undefined cooling, which leads to locally deviating crystallization and shrinkage. This results in residual stresses in the solidifying part, which ultimately lead to part distortion and are among the most important causes for shape deviations in SLS.

A distinction can be made between system-related and material-related shape deviations. System-related shape deviations include, for example, the resolution accuracy in the scan and build-up direction. In the scan direction, the laser focus diameter represents the minimum exposure increment in the xy-direction. However, in the xy-direction the accuracy of the scanner systems is usually very high. The applicable layer thickness, on the other hand, limits the accuracy in the layer build-up direction (Figure 3.32). In the case of free-form surfaces, this can lead to the previously-mentioned staircase effect if the design and part positioning is unfavorable (Section 3.2.6). The powder particle size also indirectly influences the resolution accuracy. However, since the particle size is usually smaller than the laser focus diameter and the layer thickness, this influence is relatively small.

In some literature sources, four shape-deviating phenomena – shrinkage, growth, warpage, and over-sintering – are mentioned [Sauer 2005]. As described above, shrinkage is a linear dimensional deviation induced by crystallization. The growth in the scan direction causes oversizing due to neighboring melting of powder particles as a result of energy densities that are too high and/or high local temperatures. This dimensional growth effect counteracts part shrinkage, making effect isolation difficult. An example of growth is shown in Figure 3.58. If this effect occurs in the layer build-up direction, it is referred to as over-sintering or over-hardening (which is actually a term derived from stereolithography).

The part distortion is divided into the curling and warpage effects. The curling effect occurs due to an inhomogeneous temperature distribution across the part, resulting in an uneven material density distribution as a consequence of crystallization and thermal shrinkage. Here, in particular, the increased local density change which can be caused by only minor temperature differences as a result of the sometimes complex kinetics of crystallization in semi-crystalline thermoplastics is responsible for an increased curling effect. The first generated melt layer has a higher temperature than its surroundings and begins to crystallize and shrink. The second melt layer is again exposed according to target geometry and is thus slightly larger than the previous layer. Subsequent shrinkage distorts all following layers. In particular, the application of cold powder after the exposure increases the temperature gradient within the geometry (Figure 3.51). Due to the lack of part stiffness, the first melt layers in particular tend to curl intensively. Intensive curling often leads to "pull-out", in which the already molten parts bend upwards at the surface to such an extent that they are either displaced or completely torn out by the powder applications, ultimately leading to early process termination. This effect occurs mainly within the first layers. If the process is not aborted, the subsequent layers generate smooth powder layers again, and thus resulting in a non-curved top surface, potentially on a displaced base part, increasing the loss of powder and production time.

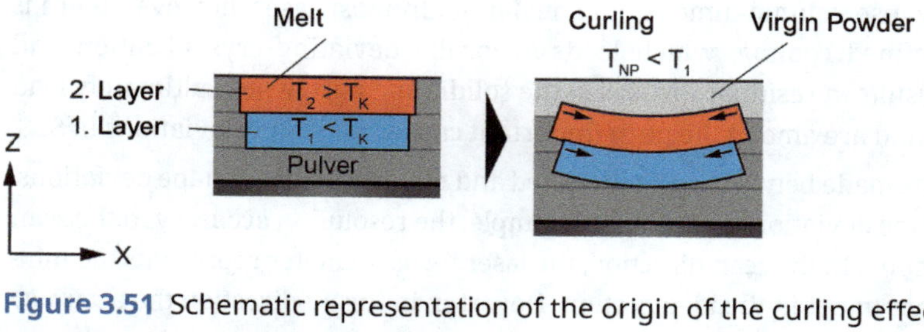

Figure 3.51 Schematic representation of the origin of the curling effect (based on [Soe et al. 2013, Nöken 1997])

While curling is the result of thermal distortion due to the high temperature differences between the melt, part-cake, and new powder, warpage is a distortion caused by uneven thermal cooling throughout the part-cake. Warpage is based on the same thermal distortion as curling, but differs in the way it occurs. Warpage occurs after the end of the exposure process and during the cooling phase in the part-cake. Due to the very high thermal insulating effect of polymer powder, it can be assumed that there is a concentric temperature distribution in the build space and thus also concentric shrinkage lines. Therefore parts positioned in the center experience less warpage compared to those positioned at the edge (Figure 3.52; see also Sauer 2005). An example of a part with significant thermally induced warpage is shown in Section 3.2.7.3 (Figure 3.57).

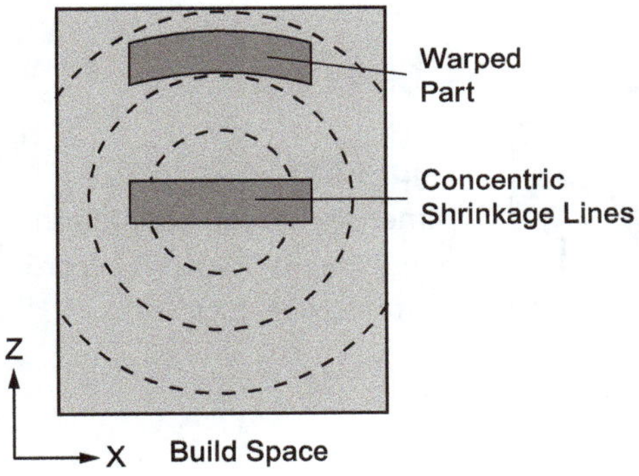

Figure 3.52 Representation of uneven shrinkage distribution in the build space, which leads to warpage outside the center (according to [Sauer 2005])

In addition to the types of warpage mentioned above, the geometry itself offers potential for warpage, comparable to injection-molded parts. At corners, for example, there is a natural accumulation of melt, which offers a larger surface area for heat compensation on the outside and a smaller surface area on the inside. Particularly in geometries with local melt accumulations, this leads to an uneven heat distribution and thus to uneven shrinkage and thus warpage of the parts [Bonten 2016].

Since manual inspection of complex three-dimensional parts in three spatial directions is usually difficult, efforts are currently being made to facilitate quality control with regard to dimensional and shape accuracy by automatically comparing the original CAD geometry with images generated by 3D-scans of the printed part.

3.2.7.2 Mechanical Properties

In order to monitor the mechanical properties of SLS parts, tensile bars are usually manufactured in the corner areas of each build-job and then tested on the mechanical properties. For this purpose, the tension rods are usually oriented vertically (in the z-direction) in order to provoke the minimum strength and the greatest possible probability of defects over many layers. The tensile strength of parts positioned in the corners is usually lower than in the center of the build space due to potential thermal inhomogeneities in the corner areas.

For PA12 or similar high modulus polymers, the test specimen of type 1A is particularly suitable according to DIN EN ISO 527 (Figure 3.53). For parts build from more elastic polymers, the tensile bar geometry of type S2 according to DIN 53504 is better suitable, especially for assessing the elongation at break properties of the SLS-printed parts (Figure 3.54). The reason for providing a shorter tensile bar for elastomeric parts is ultimately the elongation limit of the testing devices. However, the shorter and thinner test specimen may lead to misleading results, especially for elastomeric parts. Different to injection-molded parts, the downscaling of SLS specimens leads to a larger influence of the rough surface structure. This is not scaled, but is the same for all parts, and thus plays a significant role during elongation tests, as the porosity acts as a notched part and thus induces early failure. This can be solved by testing all specimens after chemical smoothing.

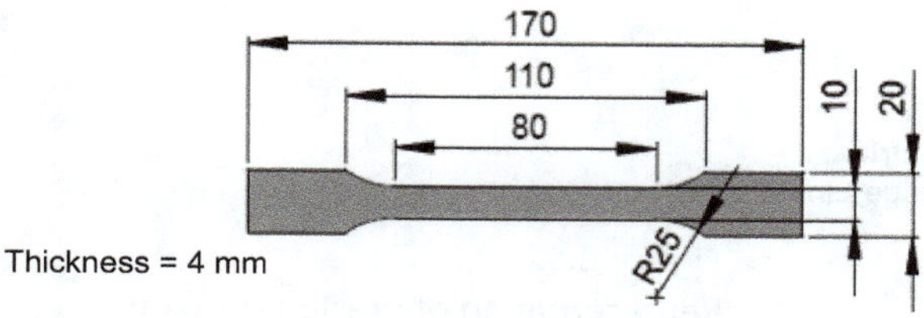

Figure 3.53 Dimensioned representation of tension rod type 1A according to DIN EN ISO 527

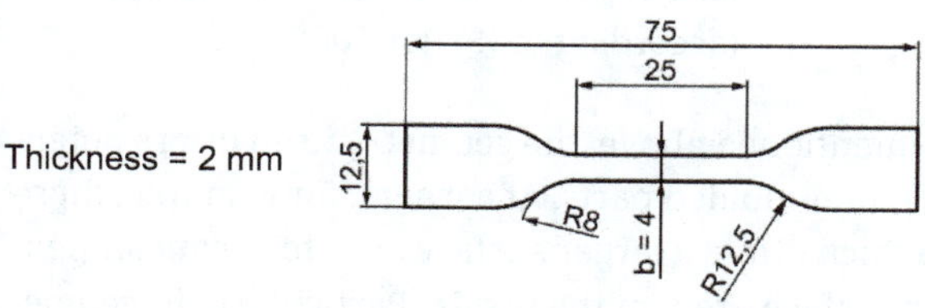

Figure 3.54 Dimensioned representation of tension rod type S2 according to DIN 53504

The build process has several failure modes. Powder impurities in particular often cause process interruptions. In some cases even a single hair of an eyebrow in the powder can prove critical. A fault on the equipment side, e.g. the failure of a heater segment, can cause the part-cake to melt on only one side of the build chamber. Faulty CAD files lead to local build errors, e.g. assemblies with overlaps that can cause several contours to be displayed inside each other with the slicer software being unable to decide where the real volume of the part is and where it is not. For this reason, it is important to carefully check each individual design file and printed part by visual inspection.

Besides the named common failure modes that can occur during an SLS build-job, more typical part and process errors are described in Section 3.2.7.3.

3.2.7.3 Typical Part- and Process Failures

Both process-related and material-related failures can take a wide variety of forms, and a selection of significant resulting part defects are summarized below.

Slicing error

In most cases, the 3D printing service provider receives finished CAD geometries to be processed. They are mostly based on standardized and triangulated STL files. These usually do not contain any information about how a design was generated. Nevertheless, depending on the experience of the designer, there may be awkward designs that cause geometry overlap even in the STL file format, and when slicing the parts, there is an ambiguous assignment of solid material and no material. This can lead the slicing software to omit some layers and not expose them to the laser during the print job (Figure 3.55).

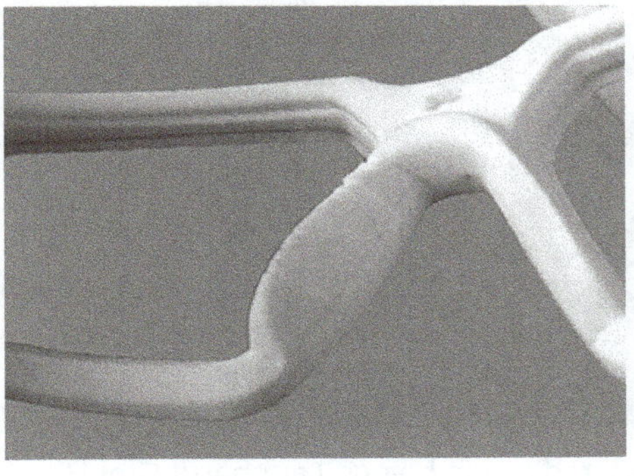

Figure 3.55
Example of a layer build-up error which occurs as a result of faulty slicing process (© Launhardt GmbH)

Contaminated powder material

Particularly in manually operated SLS systems, one of the most common failure characteristics is contamination of the starting material. Since the accuracy of the process depends significantly on a uniformly fine particle size distribution of the powder, impurities of any kind can have a significant influence on process stability and part quality. The obligatory sieving of the powder material before use on SLS systems serves not only to break up powder agglomerates, but also to filter out contamination. However, not all contaminants can be removed safely from the powder. In particular, hairs, impurities from other foreign materials and colored contaminations are frequent causes of visual part defects and process terminations. Brush-hairs, artificial hairs from vacuum-cleaner brushes, and body-hairs (which fall into the open powder due to lack of protective clothing during part-cake unpacking), can “stand upright” in the sieve and fall through. In the best case, these hairs become integrated into the part (Figure 3.56), but in the worst case, they can lead to the last melt layer being torn out and consequent process termination.

Strategies to avoid contamination of the SLS powder:

- "Hair-brushes" should not be used to unpack SLS parts
- When unpacking build-jobs by hand, protective clothing (gloves, gown, and hair net) should be worn at all times.

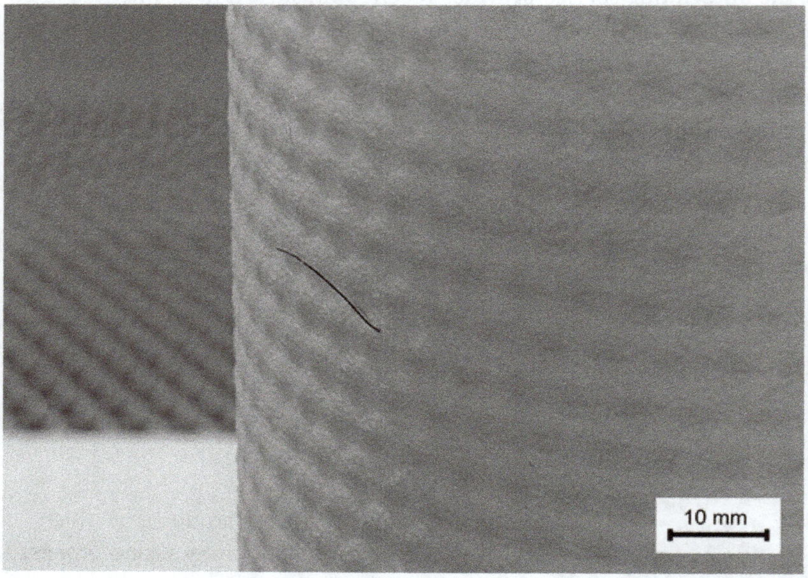

Figure 3.56 Example of an SLS part with a sintered brush-hair (© Launhardt GmbH)

Foreign material poses a particular risk to process safety if the foreign material has a lower melting temperature. If, for example, PP is processed in parallel with PA12 in the same chamber, low-melting PP powder can get into the PA12 powder and, even in small quantities, lead to melting and agglomeration of powders in the process chamber. Here, process termination with considerable rework is usually unavoidable with intensive cleaning and change of sealings.

A common, but less dramatic contamination is the contamination of the powder with foreign particles of the same size. If these do not interfere with the mechanical process of powder application and are not altered by the thermal conditions in the build-space, then these impurities are not critical to the process, but can affect the color of the resulting SLS parts. Colored (mostly dark) dust particles, paint particles, etc. can lead to stains on the SLS-printed parts.

Warpage

Very large parts in relation to the available space tend to suffer from thermal distortion due to an inhomogeneous temperature distribution. In particular, large parts without ribbing and stiffeners tend to show warpage, very similar to the situation in injection molding. Figure 3.57 shows a large flat part which was exposed to a considerable temperature gradient due to a heater defect and thus suffered considerable warpage.

Figure 3.57
Example of a flat part with warpage (© Launhardt GmbH)

Growing

An insufficient part spacing in combination with a high melt accumulation can lead to so-called "growing" in the SLS process (as shown by the example in Figure 3.58). This describes the melting of neighboring powder material, which leads to oversizing and, in extreme cases, to the joining of two parts.

Figure 3.58
Example of two solid cubes with an edge length of 40 mm and a spacing of 3 mm, joined due to growing

Enclosed powder

No support structures are required in the SLS process of e.g. hollow bodies. However, an opening must be provided by the part design to facilitate the removal of potentially enclosed powder. Depending on the purpose, it can be desired for the powder to remain in a completely enclosed hollow body. However, sometimes the uppermost melt layers of an otherwise closed hollow body can bulge during production due to unwanted gas capture and lead to a build-job failure (Figure 3.59).

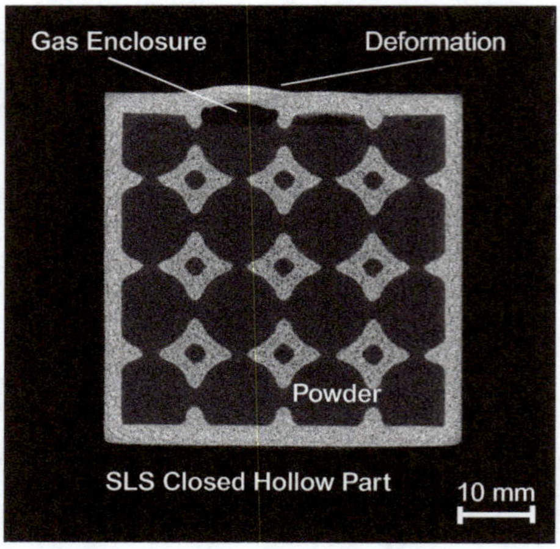

Figure 3.59
Computer tomographic image of a part defect as a result of a fully enclosed hollow body [Launhardt et al. 2015]

3.2.8 Profitability Analysis

Compared to other additive manufacturing processes (e.g. FLM or SLA), SLS is characterized by the fact that a large number of parts can be manufactured simultaneously. Depending on the part size and complexity, SLS offers an economical alternative to injection molding for small batches of up to 10,000 units. With injection molding, it is assumed that the high tooling costs will be amortized within the first 15,000–50,000 pieces. Until then, additive manufacturing processes can be more economical than the classic injection-molding process. For higher quantities, the injection molding process becomes progressively more economical and at the same time offers very high part quality.

Injection-molded parts are designed and manufactured specifically for production in a corresponding mold cavity. Complex parts including subassemblies are either implemented in complex and very cost-intensive multi-part injection molds or by subsequent manual assembly of the individual parts. For example, the housing of a simple kitchen appliance usually consists of more than ten individually injection-molded parts, which subsequently have to be assembled in a complex process. Taking full advantage of the design freedom offered by additive manufacturing processes, functional parts such as hinges, film hinges, joints, snap-on hooks and closures, flexible roller shutters, and any moving parts can be integrated directly into the design. Thus, a part designed using the full design freedom offered by additive manufacturing in SLS can eliminate those ten assembly steps. This saves labor costs and also gets rid of cost-intensive injection molds. In the best case, a part is designed specifically for production in SLS and thus enables the full exploitation of the economic potential of SLS. Using the example of a simple and small part geometry, the break-even point of the two manufacturing processes of SLS and injection molding is approx. 15,000 units and higher (Figure 3.60).

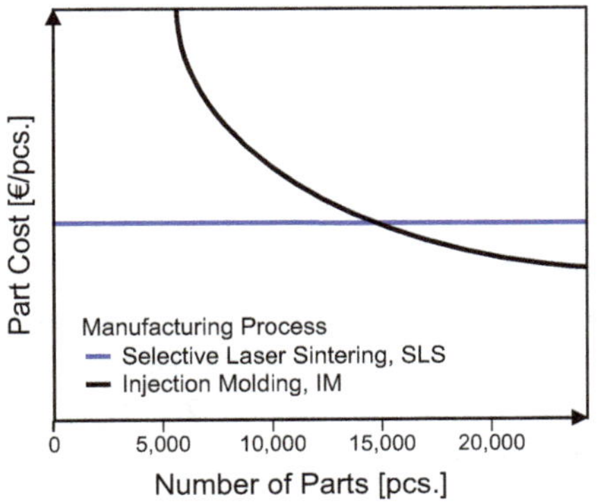

Figure 3.60
Schematic representation of the break-even point for manufacture by SLS and injection molding, using the example of a simple and small part geometry

The more complex the part geometry, and thus the more complex the required mold geometry and mechanics of the injection mold, and the smaller the part geometry, the further the break-even point can shift in favor of higher part numbers for SLS. If, in addition, the possibilities of the assembly steps that may be required are exploited by an appropriate part design, this can have a further positive effect on the cost-effectiveness of SLS.

As a part of a research project, the economic analysis of a simple everyday part that nevertheless has a complex manufacturing process was conducted on the basis of a soft TPU material and a medium-sized SLS machine. This demo part is well-suited for analyzing the profitability of the SLS process.

> **!** A simple rule of thumb is that the smaller the part, the cheaper and more economic it will be in manufacturing in SLS.

Due to the currently complex manufacturing method for spiral elastic hair bands, the production of such "complex" geometries could be simplified using additive manufacturing processes. Traditionally, these common parts are manufactured by spiral extrusion on a rotating rod, then cut and welded to shape a spiral ring. In addition, an effective attribute such as the customizability of a hair band can be developed as a unique selling point.

This specific example shows the increased possible packing density within the build space. With 2560 units per build frame on an SLS system of type P396 by EOS GmbH, the manufacturing price would be at around 0,35 € per unit. This price is directly competitive. Figure 3.61 shows the CAD geometry of the hair band design, and Figure 3.62 shows the packing of these parts in the build frame. Finally, in Figure 3.63, the final thermoplastic elastomeric part is shown.

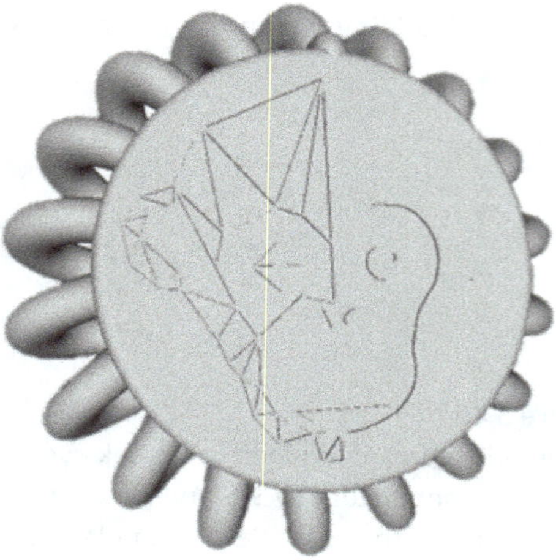

Figure 3.61
CAD of the demo part for the economical analysis. PowderMonkeys spiral elastic hairband with a diameter of 37.5 mm (© Launhardt GmbH)

Figure 3.62 Positioning of the spiral elastic hairband inside a P396 SLS machine by EOS GmbH. Number of pieces per build-job: 2,560 (EOS, P396); Manufacturing price: ~0.35 €/piece (© Launhardt GmbH)

Figure 3.63
Final PowderMonkeys spiral elastic hairband (© Launhardt GmbH)

The production costs of such a part were calculated using the example of a reconstructed hairband with an individual embossing and the current equipment, material, and energy prices.

This proves that the powder-based additive manufacturing process provides high potential for the economic mass production of small parts, and should be evaluated before investing in complex and expensive injection molds.

The economic efficiency of an SLS system is based on a few main factors: the available build volume, the exposure and layer application speed, the packing density, the refresh rate used with new powder, and the cooling time allowed.

The *build volume* depends on the system. The more parts fit into an build space, the more can be manufactured simultaneously, and the less often the system has to be re-equipped and pre-/post-processed. This orientation to the largest possible build space is critical if individual large parts are to be manufactured, because otherwise the volume of "unproductive" powder that is thermally aged as a pure support structure is increased, leading to a need for a larger volume of new powder, which leads to inefficiency.

The *exposure speed* is usually limited by the scanning speed of the galvo scanner. The usual maximum value here is 6000 mm/s but it can reach up to 20,000 mm/s. According to the Andrew number (Equation 3.1), the laser power must also increase as the scanning speed increases. Accordingly, the maximum values available on the equipment side in terms of scanning speed and laser power must be taken into account for economical production. However, even at a constant area or volume energy density, the scanning speed cannot be varied at will without influencing the part properties. Depending on the type of polymer, thermoplastics are slow in absorbing energy, so that the scanning speed, for example, cannot be increased beyond a given threshold [Drexler 2016].

The time-limiting factor in SLS is the layer build-up. Due to the mostly very slow build-up of layers in the z-direction, high process times result for larger build volumes; in some cases up to 48 hours and more. For this reason, the build volume speed (in L/h) is often used by system manufacturers as a comparative value.

The *packing density* is the ratio of the effectively used part volume and the total build space volume (Equation 3.6). The effective volume depends on the height of the part, so that, for example, a job with a low height but closely packed part volume has a higher packing density.

$$D_{\text{packing}} = \frac{V_{\text{parts}}}{V_{\text{build space,eff}}} \tag{3.6}$$

D_{packing}: Part packing density in the build space [%]

V_{parts}: Volume of all parts in the build space [mm^3]

$V_{\text{building space,eff}}$: Effective build space volume as a function of the maximum build height [mm^3]

Typical packing densities in SLS processes are on average 5–12 % by volume.

The refresh rate directly influences the economic efficiency of an SLS system, and is determined by the volumetric quantity of parts removed plus the powder loss due to cleaning and adhering powder and the thermal aging of the material. In running production, it is the goal to maintain a constant powder pool size. For the packing densities mentioned above, and taking into account the ratio of density between the SLS part and the powder bulk of approximately 1:2, the minimum refreshing rate is approximately 10–24 wt%. Since the real thermal load varies with each build-job, and in order to ensure good process reliability, the refreshing rate should be somewhat higher.

The *cooling* of a part-cake with internal SLS parts and surrounding loose polymer powder takes a very long time (according to the rule of thumb mentioned above, about half the build time, see Section 3.2.2). During this time, the SLS system does not operate, but mostly just cools down passively. In order to achieve an increase in efficiency here, it makes sense to use an external cooling chamber, so that the part-cake can cool down slowly and as homogeneously as possible. In the best case, this is done under a nitrogen atmosphere, in order to reduce negative oxidation effects on the powder material. Reloaded with a new build-chamber, the SLS system can start a new build-job directly after the end of the last job and thus making use of the heated machine. This already significantly increases the productivity, as the machines usually need a heat up time of at least 2 hours and more. Studies have shown that even for a relatively small SLS machine, constant heating is achieved not before 7 hours of heating [Launhardt 2021].

Key factors influencing the economic viability of SLS plants:

- Build space volume
- Packing density
- Exposure speed
- Laser power
- Layer application speed
- Refresh rate
- Cooling concept

However, economic series production of SLS parts does not only depend on the SLS system. In fact, additive manufacturing is still very personnel-heavy and depends directly on the usually high material costs. In total, the manufacturing costs per part can be divided into one third each for equipment technology, material costs, and personnel costs (Figure 3.64). This includes:

- The entire production plant technology from SLS equipment to post-processing equipment
- The material costs for the SLS powder
- The personnel costs for job preparation, plant preparation, and post-processing
- The personnel costs for standard post-processing steps, such as unpacking, glass bead blasting, cleaning, measuring, and packaging.

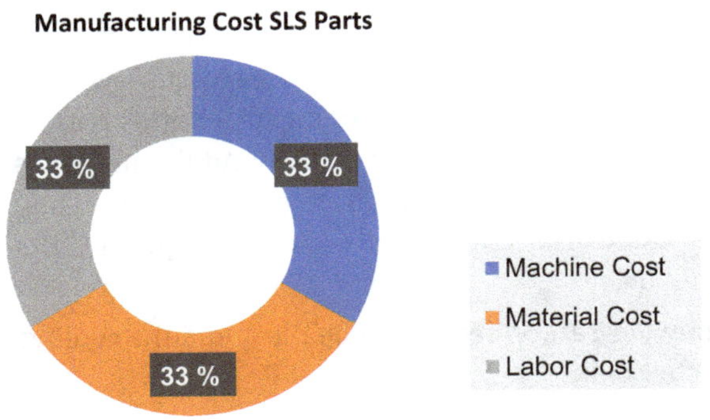

Figure 3.64 Production cost allocation for the SLS process

3.2.9 Post-processing Procedures

After the end of an SLS printing process, post-processing follows. This concerns the SLS system, the manufactured part, and the powder. In this section we will outline the typical steps involved in post-processing and suitable equipment. Further, we consider common post-processing procedures specific to the produced parts (see Section 3.2.9.1). Finally, we provide a more detailed explanation of the powder post-processing and its limitations (see Section 3.2.10).

During the post-processing steps in an SLS production plant, a high degree of cleanliness must be ensured. After the build-job, all powder residues must be removed thoroughly from the SLS printer, since these often show significant thermal ageing and can affect the next build-job.

Process heaters are positioned above the powder surface. Residues or deposits from the powder tend to accumulate on these heaters or above, potentially falling onto the powder surface during the next build-job, resulting in potential disruption of the process or severe part defects. If powder accumulates on the heaters, it can melt and carbonize during operation. Consequently, it is crucial to meticulously clean both the heaters and air circulation shafts. However, cleaning the heaters carries a risk of damage, so all heaters should undergo regular inspection. The heaters are controlled by sensors, typically non-contact thermometers (pyrometers), which are strategically positioned to observe the powder surface. Ensuring the sensor surface of the pyrometer remains clean and debris-free is essential. A contaminated protective window in front of a pyrometer can lead to inaccurate readings, and in extreme cases, cause melting within the entire build chamber.

Polymer powders constantly exposed to high temperatures can lead to deposits and build-up within the SLS system. System manufacturers usually counteract this by installing nitrogen rinsing in front of relevant assemblies, e.g. in front of the pyrometer and laser windows. Nevertheless, deposits may occur and should be carefully removed with lint-free cloths and/or suitable cleaning agents. In the last step, wiping with a lint-free cloth with isopropanol is often recommended. The choice of solvent depends on the powder material used.

All SLS systems require computer units for control, especially of the scanner systems. These are usually not designed for use in powder-rich environments. For this reason, all vents to and from the computer units should be cleaned regularly. Additionally, it is recommended to shut down and restart the computer units at regular intervals. Similar to all computers, long run times can cause "crashes" to occur more quickly, ultimately leading to a build-job termination.

Finally, before each build-job, the powder reservoirs should be filled and the cooling, compressed air or nitrogen supply checked.

In a nutshell: Notes on SLS system post-processing

- Remove waste powder residues removed from internal surfaces
- Clean heaters
- Vacuum out and clean any recirculating air ducts
- Clean laser window or f-theta lens
- Clean powder application system (blade, roller, etc.)
- Clean sensors, especially pyrometers
- Clean computer units from powder and restart if necessary
- Fill up powder
- Check cooling and compressed air or nitrogen supply.

3.2.9.1 Unpacking a Build-job

All SLS parts are supplied to the post-processing station in the form of a part-cake. The parts are completely embedded in powder and must first be freed from the surrounding powder. This can be done either manually or can be partially automated. For manual unpacking, for example, unpacking stations are available from the system manufacturers. Here, a holder for the respective printer-specific exchange frame, an extraction unit, and a sieve unit are usually provided as part of a device. Thus, the residual powder can be sieved and directly refreshed with new powder (Figure 3.65 and Figure 3.66).

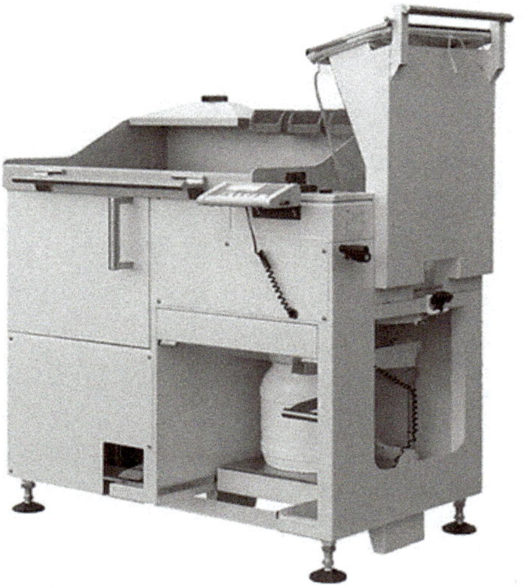

Figure 3.65
Unpacking station of the company EOS GmbH for a FORMIGA P110 line (© EOS GmbH)

Figure 3.66
Unpacking station of a third-party supplier of the company Hasenauer & Hesser GmbH (© Hasenauer & Hesser GmbH)

For fully automated unpacking of parts out of the part-cake, and for the subsequent sieving, refreshing, and mixing of the powder, the company Ossberger GmbH provides an "unpacking system" consisting of two stations with a minimal footprint. This comprises a spinner, which separates the parts from the powder. This powder is sucked into the powder tower, which by means of an airstream, processes the powder and places it in ready-to-use handy powder containers. This solution is provided for frames of the EOS P110 Formiga (Figure 3.67), EOS P3, and Farsoon machines. The extracted parts stay in the exchange frame and can be put into the post-processing blasting cabins for further depowdering.

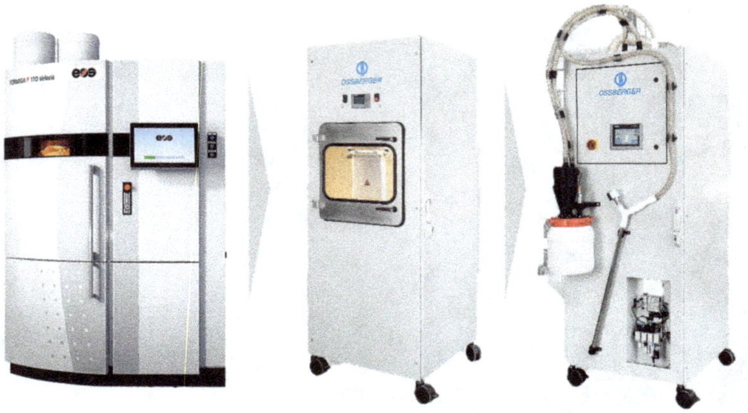

EOS Formiga P110 Ossberger FS Spinner Ossberger FS Powder

Figure 3.67 Unpacking systems for SLS parts prepared on an EOS P110 Formiga (left): An Ossberger FS Spinner (center) and an Ossberger FS Powder (right) (© Launhardt GmbH and © Ossberger GmbH & Co. KG)

3.2.9.2 Part De-powdering

After the first unpacking and cleaning step, there are still powder adhesions on the parts. These are freed from powder residues either manually or by automated blasting processes (mostly with glass beads). The automated systems are mostly rotating drum systems, which guide the parts past blasting and blow-off nozzles and thus free them from residual powder (Figure 3.68 and Figure 3.69). The choice of blasting material (e.g. glass beads, epoxy spheres, walnut shells, etc.) and size in conjunction with the preset air pressure have a considerable influence on the resulting part surface. A blasting material with particle sizes in the same size range as the pore size of SLS parts leads to coarser surfaces, while larger sphere diameters lead to smoother surfaces.

Figure 3.68 DyeMansion Powershot C for automated removal of powder residues on SLS parts (© DyeMansion GmbH)

Figure 3.69
Automatic blast cabinet with rotating drum from MHG, Normfinish (© MHG Strahlanlagen GmbH)

When blasting manually, it must be ensured that the blasting nozzle is always guided in a meandering pattern over the parts, with a minimum distance of approx. 50 mm. Care must be taken to ensure continuous movement, otherwise the high kinetic energy of the blasting material will affect the surface of the part. If the blasting nozzle remains static above a thermoplastic part, it will melt after a short time and finally char the surface of the part. In this case, the part is irreversibly damaged.

3.2.9.3 Smoothing

SLS parts always have characteristic surface features, such as staircase effects, roughness, and sharp edges on the top surface. It can therefore be worthwhile to rework the parts, either by using a surface compacting process such as bead blasting, abrasively using mechanical vibratory finishing, or by chemical smoothing.

Smoothing by blasting

Within the scope of a downstream blasting process with an adapted blasting material, the surface roughness of SLS parts can be significantly reduced (Figure 3.70). Care must be taken to use a blasting material with a large particle diameter to achieve smoothing, with the additional effect that the impact energy causes compaction of the material layers near the surface. The result is a toughened rough-smooth surface with a pleasant feel to touch and with improved performance in subsequent dyeing processes. This process is sometimes referred to as "PolyShot smoothing". In combination with dying the parts, it provides a homogenous surface and thus leads to a more homogenous coloring. A second blasting process after dye-coloring enhances the color saturation, ensures a more robust coloring, and provides a certain glossiness to the parts. Figure 3.71 depicts the characteristic surface structure of a PA12 part that is SLS-manufactured, dye-colored, and PolyShot-smoothed.

Figure 3.70
Automated Powershot S blast cabinet for smoothing and surface compaction of the company DyeMansion (© DyeMansion GmbH)

Figure 3.71
A part assembled from a PA12 white part and a PA12 dye-colored black part. Especially on the dyed part the certain waviness yet glossy appearance of smoothing by blasting is visible (© Launhardt GmbH)

Mechanical smoothing

Furthermore, the surface of SLS parts can be mechanically smoothed in rotating drums or vibratory finishing systems by means of abrasive vibratory bodies. Here, either the pure gravitational force of the parts is used, as e.g. in rotating drums, or additionally induced energy through vibration is used for the smoothing process (e.g. trough vibrator, see Figure 3.72 and Figure 3.73). Gravitational-based smoothing through media tumbling is softer for the parts but may take a longer time, whereas the trough vibrators allow faster processing. Continuous liquid circulation of the so-called compound (e.g. water-based) carries off abrasion and powder residues, and preserves the abrasiveness of the vibratory finishing tools. In this context, a good concept for water recycling must always be provided.

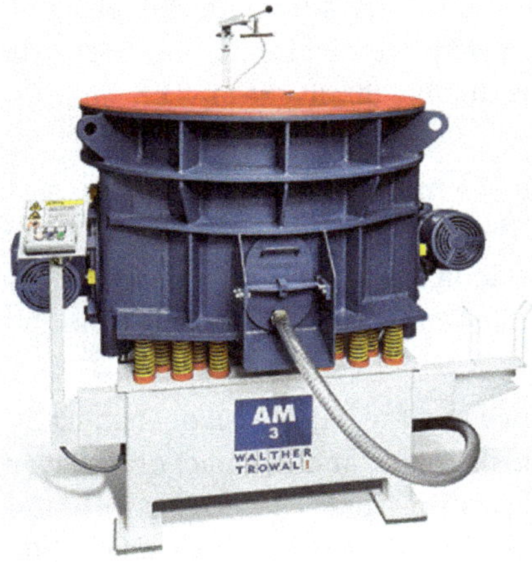

Figure 3.72
Compact trough vibrator system from Trowal for mechanical smoothing of SLS parts (© Walther Trowal GmbH & Co. KG)

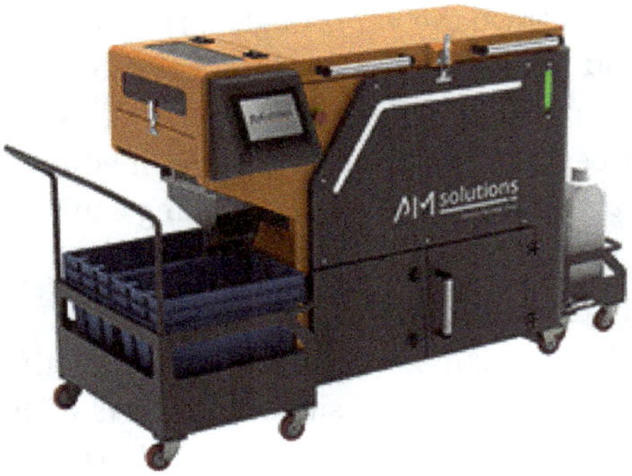

Figure 3.73
Smoothing and polishing 3D-printed parts made of metal or polymer: vibratory grinding with the M1 system from AM Solutions (© AM Solutions S.r.l.)

Chemical smoothing

In chemical smoothing, sometimes also referred to as "vapor smoothing", the SLS parts are exposed to a solvent that is capable of dissolving the material's surface. For uniform exposure, the solvent is evaporated and the part is thus exposed to a saturated solvent atmosphere. In contacting with the part the solvent vapor dissolves the surface and due to the liquid character and the surface tension this leads to a smoothing effect. The exposure of the part and the defined retraction of the solvent is crucial for the resulting smoothing effect. However, by dissolving the areas close to the surface, detailed geometric features are smoothed out, which can lead to a "washed-out" impression, and filigree structures may completely dissolve. In addition, the part must be fixed at least at one point (standing or hanging), causing there to be a visible impairment of the surface. Important process parameters are the solvent volume, the exposure time, the part and the solvent temperature, and the curing and drying times. In order to simplify and partially automate this process, some manufacturers offer machines exclusively for chemical smoothing of SLS parts. This is especially crucial for achieving reproducible quality results on smoothing.

The choice of the solvent very much depends on the material of the SLS parts, and should fulfill the local standards on toxicity. This is especially important for the parts themselves if, for example, they are to be used for food and drug purposes, or for applications close to the body. Therefore, the complete removal of the solvent from the part is essential.

Besides the positive effect on the optical appearance and the haptic feel of chemically smoothed SLS parts, chemical smoothing generates a closed and compact outer layer on the whole part. This unifies the otherwise typical anisotropic characteristics of the additive manufacturing process. Characteristic artefacts such as sharp edges, the staircase effect, and surface roughness disappear. Moreover, the shaping of a compact outer layer and thus the removal of grooves and crazing on the surface leads to significantly enhanced mechanical properties (in contrast, the very rough surface of unsmoothed SLS parts usually leads to cracks and failures under mechanical loads). If continued, chemical smoothing allows a structure to be fully sealed, allowing especially thin structures or elastomeric TPUs to become impermeable to the medium. This particularly enhances elastomeric TPU parts intended for use as sealings, tubes, or air flow bellows.

There are currently three companies offering chemical smoothing systems for SLS parts: Luxyours (Germany), AMT (UK), and ASM (Germany), with small sized smoothing machines (Figure 3.74 and Figure 3.75). The basic process of smoothing parts with solvent deposited on the part surface is similar for all three systems, but there are significant differences in the solvent used, the reuse of solvent, and the way in which solvents are delivered. In Figure 3.76 the difference between an untreated and vapor-smoothed SLS TPU part is shown.

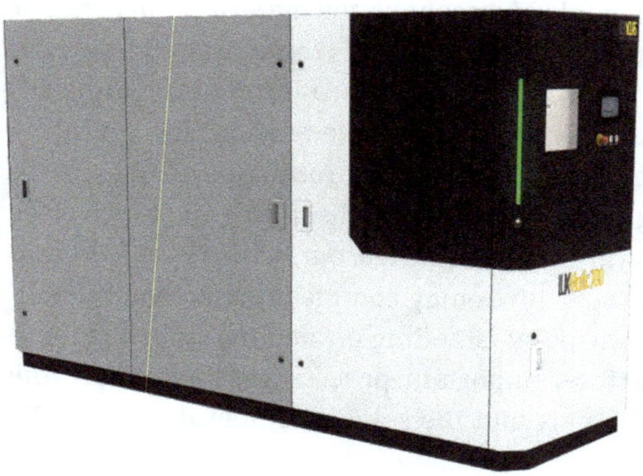

Figure 3.74
System for smoothing SLS parts: the LUXMatic 700 from LuxYours (© LuxYours GmbH)

Figure 3.75
System for smoothing SLS parts: the SF100 by AMT (© Additive Manufacturing Technologies Ltd)

Figure 3.76
Example of the above-mentioned PowderMonkeys TPU spiral elastic hairband, untreated (left) and vapor-smoothed (right) (© Launhardt GmbH)

3.2.9.4 Dye Coloring

The expectation of direct manufacturing of parts using powder bed fusion processes also brings with it the necessity of colored and differently colored parts. Since the coloring of the initial powder materials is often not practical, except in the case of low-laser-power SLS systems, most parts come out of the SLS systems either white or natural colored (color of the polymer), and the surfaces must be dye-colored during post-processing. For this purpose, there are manual and also partially automated systems that simplify the coloring process as much as possible (Figure 3.77). In most cases, dyes known from the textile industry are used, using (for example) so-called dip-dyeing. In this process, the parts are immersed in a dye bath at a specified temperature and for a specified time so that the dye penetrates up to 200 µm into the surface. In some cases, additional processes are used here, e.g. to improve dyeing under pressure and/or with ultrasonic technology. Nevertheless, the characteristically anisotropic surface structure of SLS parts usually leads to an inhomogeneous dyeing result, so that pretreatment e.g. smoothing of the part surfaces, is crucial to achieve a uniform color appearance. Since the colors used have usually already been tested for textiles, skin compatibility and abrasion resistance are mostly known. In this way, SLS parts made of e.g. PA12 or TPU can be dyed very well. Dyeing a RAL tone is also possible, but requires much more trial and error to match the tone exactly (Figure 3.78).

Figure 3.77
Automated dyeing system for SLS and MJF parts: the DeepDye DM60 from DyeMansion (© DyeMansion GmbH)

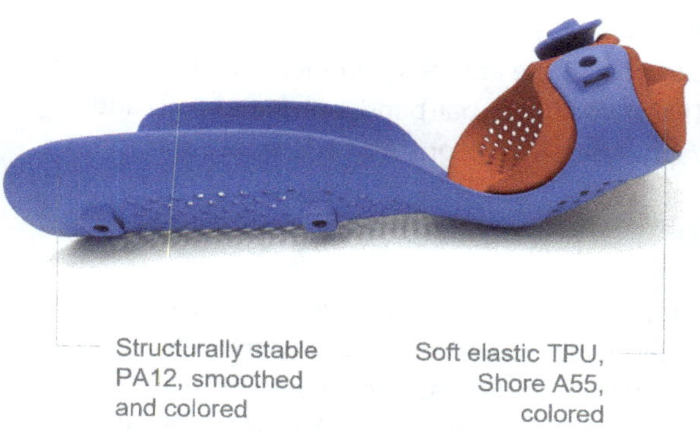

Figure 3.78
Orthoses made by SLS PA12 (blue) and TPU (red). Both are dip-dye-colored SLS parts (© Launhardt GmbH)

Many base powder materials are already colored white, since white parts are generally easier to over-color subsequently. The coloring of a base powder material can have several advantages. For example, completely through-colored blue polymer parts are used in the food industry, since blue is a rare food color, allowing broken-off parts to be more easily detected. Furthermore, the coloring of light-colored and low-absorption polymer powders offers the possibility of processing them on SLS systems with low laser power. Fully colored SLS parts also offer security against discoloration, fading, and/or light scratch marks in the application. Obviously the disadvantage of pre-colored base powders is their lack of flexibility, since only parts of the same color can be manufactured.

In addition to these essential post-processing methods, there are many other conceivable processes that can be used to increase 3D printed part quality and functionality. Below is a brief list of some of them:

- Metallizing to generate a metallic surface
- Infiltration, e.g. with epoxy resin, to guarantee media tightness for thin parts. In addition, infiltration increases the part strength and enables the production of smooth surfaces
- Painting to overcolor dark parts or to generate class A surfaces
- Flocking to create a soft surface
- Functionalization through inserts, e.g. through hot-inserted or screwed-in thread inserts, improving the robustness of selected functions (Figure 3.79).

Figure 3.79
Colored SLS part made of PA12 with threaded insert for functionalization (© Launhardt GmbH)

Many of the post-processing methods mentioned are processes that are not yet fully suitable for series production. The initial costs for the acquisition of special treatment technology are often considerable, so that many service companies have developed independent solutions in this area.

3.2.10 Materials Management

Depending on the production system, different powder circuits are necessary for the operation of an SLS machine. A general distinction can be made between three powder types:

- Virgin powder

- Used powder
- Mixed powder.

The virgin powder should always be free of contamination and agglomerates, so that it can be added to the used powder without sifting. Used powder is the powder that remains as residual powder after the end of the build-job. The mixed powder is the ready-to-use material that has already been refreshed.

The usual powder circuit, using the FORMIGA P110 system from EOS GmbH as an example, is shown in Figure 3.80 and includes the following essential steps:

1. **Quality control of the new powder**

   Before using virgin powder, the powder quality should be examined according to appropriate standards and with suitable test methods (Section 3.2.1.2). At the very least, however, the flowability and the presence of visible contamination should be checked.

2. **Refreshing process – mixing**

   Subsequently, the sifted used material is mixed with the defined refresh rate of virgin powder. For this purpose, a mixing machine is necessary. As a minimum, the powder can be mixed by rotating or tumbling in rotating containers. The mixing process should be implemented carefully, as some materials, especially TPEs, can be destroyed by excessive mixing energy.

3. **Build-job unpacking**

   After the SLS process, the build-job is unpacked and the used powder of the part-cake is separated from the manufactured parts. Here, it is important that the powder residues adhering directly to the parts are not added to the powder pool, as these are particularly thermally stressed. The unpacking of parts is often still done manually, so that a clean environment must be ensured here in particular. Brushes should not be used for unpacking, as these could lose brush-hairs, which may enter the powder pool and also will not be completely removed by sieving.

4. **Sifting of used powder**

   The separated used powder is now sieved in a fine sieve, to sieve contaminants out of the powder and to break up powder agglomerations. Usually, sieves with a mesh size of 200 µm are used for standard PA12 powder types.

The cycle then starts all over again.

If, for example, a SLS system operates with overflow containers, the overflow material is usually treated as used powder. These four powder process steps result in a circuit with a powder pool, which should have at least ten times the volume of the build space. Depending on the selection of the refresh rate, the powder pool quantity should be kept approximately constant. With high refresh rates, the quantity of the powder pool increases steadily. In this case, used material must be regularly removed from the powder circuit, in order to prevent the powder pool size from increasing too much.

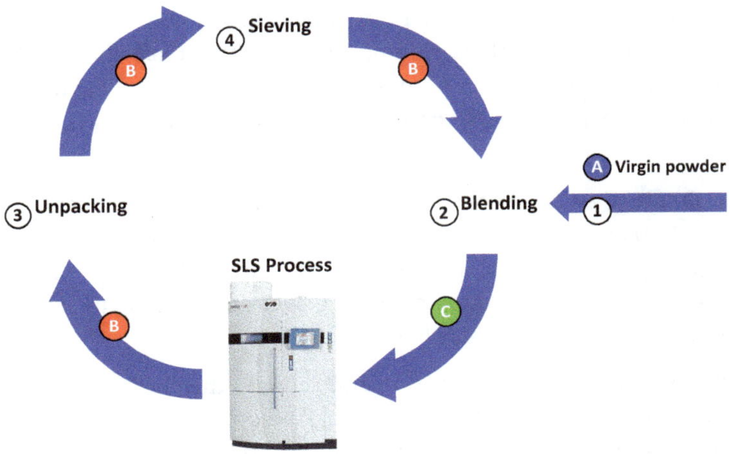

Figure 3.80 Example of a PA12 powder circuit diagram using the example of the FORMIGA P110 SLS system from EOS GmbH (© Launhardt GmbH)

Even with thermally less sensitive materials than PA12, constant refreshing of the powder makes sense. Since usual packing densities are in the range of 5–10 %, to maintain a constant powder pool, refreshing rates should be at least 10–20 wt%.

Depending on the temperature and humidity of the environment, it may be necessary to slightly moisten PA12 powder with distilled water. In very dry conditions, this reduces the electrostatic charge of the powder particles, and thus increases the flowability and processability in the powder sieve.

3.2.11 Industrial Safety

When processing fine polymer powders in SLS systems, a wide variety of safety measures must be provided, with regard to both the powder handling and the manufacturing process itself. Operator safety is the top priority, while simultaneously ensuring that the powder material remains protected from contamination by the operators. Since fine polymer powder is used in SLS, it is essential to minimize dust exposure during processing and handling. Typical SLS powders usually consist of particles that are not respirable (>10 µm). Nevertheless, due to the manufacturing processes, such particles may be present, if mostly agglomerated with larger particles. The polymer particles, but also the mineral fillers partly incorporated as a dry-blend, can be stirred up and inhaled when handling the powders. In particular, titanium dioxide (TiO_2), which is mostly used as a white coloring pigment, is suspected of being carcinogenic. For this reason, the following safety measures should be taken when processing powders in the open:

1. A **fine particle mask** according to FFP2 or FFP3 should always be used, in order to avoid the direct inhalation of swirled-up powder dust. Care should always be taken to ensure that the mask fits correctly, especially around the nose. Complete face protection mask is not necessary in the polymer area but is beneficial.

2. **Protective gloves** should always be used to avoid skin damage in direct contact with the powders. PA12 powder is not toxic, but prolonged contact can lead to dry skin because of the hydrophilicity of the powder. Latex or nitrile gloves, for example, are suitable protective clothing. Cotton or work gloves should be avoided, in order not to contaminate the powder with fibers.

3. Wearing **safety shoes** ensures a secure footing and reduces the risk of slipping on powder-contaminated floors, as well as reducing the danger of powder explosion by static discharge. Due to the spherical or nearly spherical shape of the powder particles, powder residues on smooth floors lead to increased risk of slipping.

4. The wearing of **protective goggles** is recommended when handling polymer powders, to avoid contamination of the eyes with powder. If powder nevertheless gets into the eyes, they should be carefully rinsed out with water.

5. A **fixed cleaning schedule** should be established (e.g. twice a week) to reduce the spread of polymer powder in the working environment. In the case of direct powder handling, basic cleaning should be carried out regularly according to a fixed plan, in addition to all existing safety and automated cleaning equipment.

The following permanently installed safety devices should be considered when operating SLS production equipment:

1. An **exhaust air vent** with point extraction is absolutely necessary, in order to extract locally swirled up powder dust to achieve an explosion-proof design. Either a mobile version or permanently installed suction nozzles are suitable for this. The following parameters have proven to be helpful for the extraction:
   - Air flow rate: approx. 3000 m^3/h
   - Filter area: approx. 20 m^2
   - Filter for dust class M.

2. An **explosion-proof vacuum cleaner** (ATEX) with a fine filter of dust class M should be used. Depending on the local conditions, both a vacuum cleaner powered by compressed air and an electric vacuum cleaner can be suitable.

3. **Eye wash** should be available to wash out the eyes in case of possible contamination with powder.

4. A **non-slip floor** is necessary. The fine polymer powders mostly consist of spherical powder particles, which in fine concentration on a smooth floor lead to an increased risk of slipping. For this reason, some companies use carpeted floors or special rough floors.

5. **Permanently installed exhaust systems** should be provided above each SLS system to ensure the extraction of outgassing and vaporization of various polymers under the influence of heat. Most industrial SLS printers are equipped with appropriate exhaust connections at the top of the equipment.

6. A **nitrogen warning system** is particularly necessary for small rooms, as industrial SLS production systems use nitrogen as an inert gas to flood the build chamber. The human body has no sensory system for detecting nitrogen, which can lead to asphyxiation by displacing oxygen from the room air.

7. **Additional exhaust systems** are used to increase air exchange and air purity.

8. **Smoke detection systems** directly above each SLS system increase fire protection. SLS systems are basically baking ovens, which often run overnight and on weekends due to the long build-job times. For this reason, SLS systems in particular should be additionally equipped with smoke detection units.

All SLS systems use a laser. In functioning systems without modifications, no access to the laser system is possible, so that no special laser protection device is required outside the systems. Only system maintenance may require the use of laser protection measures. A temporary blocking of the corresponding areas for the employees may be necessary here.

3.2.12 Sample Applications

Model making

In model making, many complex parts are required in small quantities. These are used in model flying, model railroads, or even remote-controlled drones. Here, additive manufacturing using SLS and PA12 enables the manufacture of mechanically resilient and rigid parts with a high level of detail and at the same time low weight, as is required, for example, for model airplanes capable of flight. In addition, post-processing methods enable the realization of visually appealing colored and smooth parts. SLS has a special application in the model construction of 1:1 replica helicopter cockpits, which are used as simulators both in the hobby and for training purposes (Figure 3.81). Here, the original parts are reconstructed in elaborate designs and prepared for SLS in a manner suitable for production. In this way, complex individual parts can be manufactured without the need for costly injection molds.

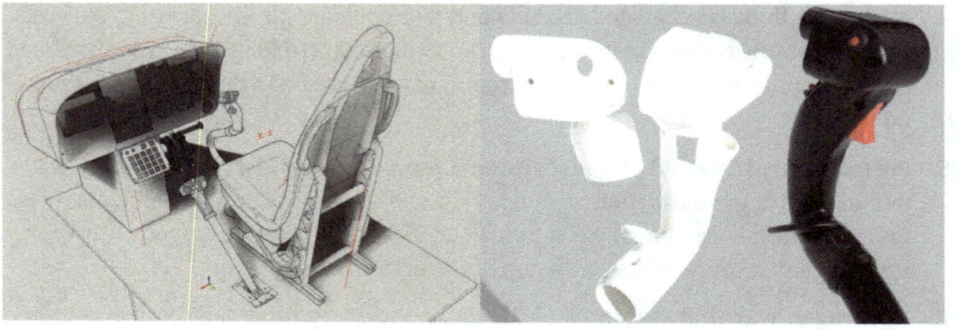

Figure 3.81 A 1:1 helicopter cockpit (left) and a control handle (right) manufactured and post-processed using the SLS process (PA12) (© microHELIS.de)

Shoes and orthotics

In orthotics in particular, supportive and soft parts with geometries individually adapted to the patient are required in low quantities. Additive manufacturing processes with resilient PA12 and soft parts made of TPU are ideal here. Both materials come into direct contact with the skin, so that skin compatibility must be certified. In this way, orthoses (Figure 3.78), insoles, thumb splints, or even entire shoes can be produced individually adapted to the patient (Figure 3.82 and Figure 3.83).

Figure 3.82 An insole made of TPU produced by SLS enables graded degrees of hardness (© Covestro AG)

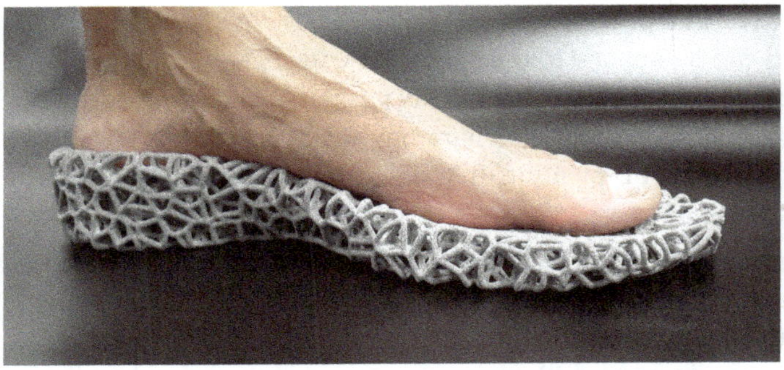

Figure 3.83 Midsole lattice structure with a soft thermoplastic polymer and a linearly graded shore hardness (© Launhardt GmbH).

However, SLS can also be used for high-priced novel shoe models with fine lattice structures in medium series. The fine structures enable high ventilation while exploiting the spatially resolved damping properties of TPU (Figure 3.84 and Figure 3.85).

Figure 3.84
Demonstrator of a complete flipflop shoe made of soft, black TPU material (Shore A hardness 65), untreated (left) and sequentially vapor-smoothed (right) (© Powder-Monkeys, Launhardt GmbH)

Figure 3.85
Shoe sole with fine lattice structure manufactured from TPU using SLS and then smoothed (© Covestro AG)

Depending on the chosen TPU and process method (SLS or HSS (high-speed sintering)), various shore hardnesses can be applied to the shoes. This, for example, allows for a selective harder area in the heel and toe area. This way it is possible to create orthopedic shoes with maximum comfort without downgrading the visual appearance. Furthermore, due to the character of 3D-printing, individualization of the shoe, both in shape and in marking, is possible.

Work support systems represent a large area of application for the combination of rigid and elastic materials. For example, hand splints or assembly aids for the hand can be individually adapted to the activity and to the employees, thus simplifying strenuous activities. This can prevent health impairments and make everyday assembly work easier.

Wearables and fashion

Elastic SLS materials are ideal for manufacturing wearables as functional prototypes and in small series. These include, for example, elastic wristbands or digital readout devices with adapted wear straps (Figure 3.86). As in the case of orthotics, an elastic and skin-compatible polymer is also required here.

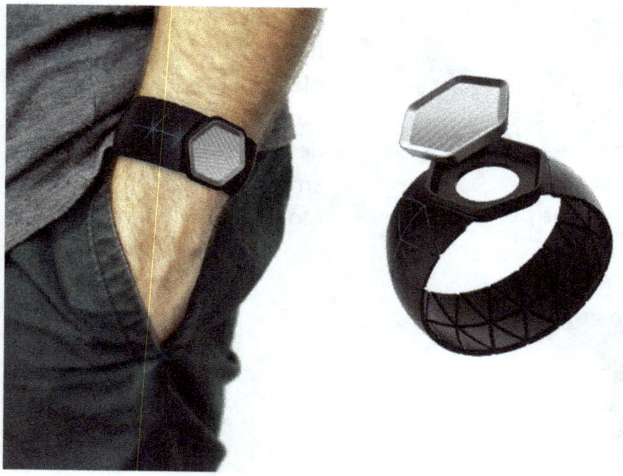

Figure 3.86 Design study of an additively manufactured elastic wristband for wearables (© Covestro AG)

In addition, movable structures such as a chainmail structure made of TPU can also take on a textile-like property in SLS and find use in the fashion sector (Figure 3.87). The movable and elastically adaptable properties of TPU together with the movable geometry enable a comfortable wearing sensation. Solid structures can be implemented in the fashion sector by means of SLS and PA12, and thus even highly filigreed structures can be achieved. This can be illustrated by the example of a designer bowtie (Figure 3.88).

Figure 3.87 A textile-like chainmail structure (made from TPU using the SLS process) enables applications in the fashion sector (© Launhardt GmbH)

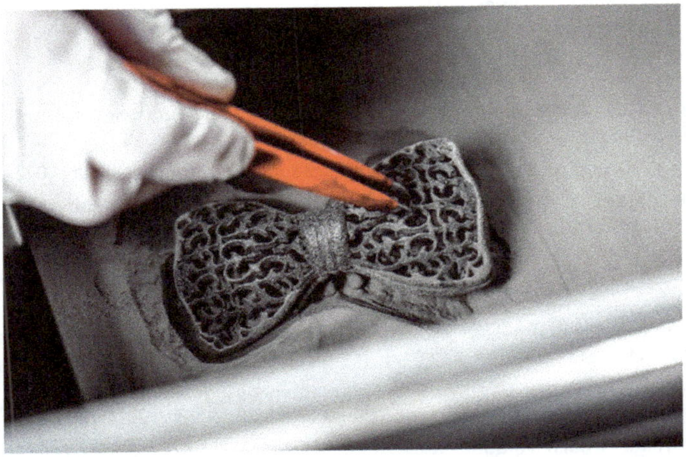

Figure 3.88 Designer bow-tie made from PA12 using the SLS process (© PowderMonkeys, Launhardt GmbH)

Robotics

In the field of robotics, aspects of artificial intelligences, but also the direct interaction between man and machine, are becoming increasingly important. Especially in everyday work, it is more efficient and economical if robot systems can act in direct contact with workers. Doing this, however, requires robot structures that are elastic so as not to cause injuries in direct contact with humans. In addition to fine sensor technology and sensitively controlled actuators, elastic outer structures and adapted gripper systems are relevant here. Adapted gripper systems can also be used to handle bulky objects, and enable largely harmless direct human–machine contact (Figure 3.89).

Figure 3.89 Adaptive gripper (made of polypropylene using the SLS process) that adapts to the geometry of the target object

References

3Druck 2017: What's the point of a 3D printer with more than 3 axes? 3Druck, 2017. *https://3druck.com/blog/was-bringt-ein-3d-drucker-mit-mehr-als-3-achsen-0755445* (retrieved 27 February 2022)

3Druck 2022: 3D Printer Software. 3Druck. *https://3druck.com/3d-drucker-software* (retrieved 03 May 2022)

Achten et al. 2021: D. Achten, F.-S. Stern, C. Tomczyk, R. Wagner, B. Mettmann, T. Büsgen, N. Degiorgio, J. Künzel & M. Wolf, Method for producing an additively manufactured and treated object. Patent specification: EP3880441A1. Patent proprietor: Covestro Intellectual Property GmbH & Co KG. European Patent Office, 2021

Adam 2015: G. A. O. Adam, Systematische Erarbeitung von Konstruktionsregeln für die additiven Fertigungsverfahren Lasersintern, Laserschmelzen und Fused Deposition Modeling. Shaker Verlag, Aachen 2015 (ISBN 978-3-8440-3474-5)

Adam & Zimmer 2015: G. A. O. Adam & D. Zimmer, Design for Additive Manufacturing – Element transitions and aggregated structures. In: CIRP Journal of Manufacturing Science and Technology 7, 2014, pp. 20–28

Aiden 2022: E. Aiden, The Fourth Industrial Revolution. *https://www.academia.edu/38203483/The_Fourth_Industrial_Revolution_pdf?* (retrieved 27 February 2022)

Alscher 2000: G. Alscher, The behavior of semi-crystalline thermoplastics during laser sintering. Dissertation. University-GH Essen, Mechanical Engineering, 2000

Amado et al. 2012: A. Amado, K. Wegener, M. Schmid & G. Levy, Characterization and modeling of non-isothermal crystallization of polyamide 12 and co-polypropylene during the SLS process. In: Proceedings of the 5th International Polymers & Moulds Innovations Conference, Ghent, 2012

Amado et al. 2015: A. Amado, K. Wegener & M. Schmid, Manfred, Simulation of warpage induced by non-isothermal crystallization of co-polypropylene during the SLS process. In: AIP Conference Proceedings, 2015, 160002, pp. 1–5

Azimi 2016: P. Azimi, D. Zhao, C. Pouzet, N. E. Crain & B. Stephens, Emissions of ultrafine particles and volatile organic compounds from commercially available desktop three-dimensional printers with multiple filaments. In: Environmental Science & Technology 50, 2016, pp. 1260–1268

BG BAU 2019: (i) Technical rules for hazardous substances. TRGS 500 – Protective measures. Berufsgenossenschaft der Bauwirtschaft, 2019 (GMBl No. 66/67, 2019, p. 1330; last corrected GMBl No. 4, 2020, p. 88); (ii) Priority order of protective measures – "STOP principle". *https://www.bgbau-medien.de/handlungshilfen_gb/daten/tr/trgs500/5.htm* (retrieved 27 February 2022)

BG BAU 2022: Explosion limits. Berufsgenossenschaft der Bauwirtschaft, https://www.*bgbau.de/themen/sicherheit-und-gesundheit/gefahrstoffe/sicherheitsdatenblatt/explosionsgrenzen* (retrieved 27 February 2022)

Bonten 2016: C. Bonten, Plastics Technology. Introduction and Fundamentals. 2nd ed. Carl Hanser Verlag, Munich, 2016 (ISBN 978-3-446-44674-8)

Brajlih et al. 2011: T. Brajlih, B. Valentan, J. Balic & I. Drstvensek, Speed and accuracy evaluation of additive manufacturing machines. In: Rapid Prototyping Journal 17, 2011, pp. 64–75

Breuninger et al. 2013: J. Breuninger, R. Becker, A. Wolf, S. Rommel & A. Verl, Generative Fertigung mit Kunststoffen. Concept and design for Selective Laser Sintering. Springer Verlag, Berlin/Heidelberg, 2013 (ISBN 978-3-642-24324-0)

BAUA 2022a: GS mark. Federal Institute for Occupational Safety and Health (BAUA). *https://www.baua.de/DE/Aufgaben/Gesetzliche-und-hoheitliche-Aufgaben/Produktsicherheitsgesetz/GS-Zeichen/GS-Zeichen_node.html* (retrieved 27 February 2022)

BAUA 2022b: 3-D printers. What to consider when using 3-D printers. Federal Institute for Occupational Safety and Health (BAUA). *https://www.baua.de/DE/Themen/Anwendungssichere-Chemikalien-und-Produkte/Produktsicherheit/3D-Drucker.html* (retrieved 27 February 2022)

CAMPUS 2022: CAMPUS® – a material information system for the plastics industry. *https://www.campusplastics.com* (retrieved 24 February 2022)

Covestro 2017: Method for the mechanical testing of a structure formed as one part on the basis of test pieces generated by a 3D printing process. Covestro Deutschland AG. Patent specification, Publication number: 20200309656. June 20, 2017. *https://patents.justia.com/patent/20200309656* (retrieved 27 February 2022)

Deckard 1997: C. R. Deckard, Apparatus for producing parts by selective sintering. IPC US Patent 5597589. patent specification: US 5597589 A. Patent owner: University of Texas System. 28 January 1997

Deckard et al. 1992: C. R. Deckard, J. J. Beaman, J. F. Darrah, Method for Selective Laser Sintering with layerwise cross-scanning. Application no. US 07/611,025. patent specification: US 5155324 A. 13 October 1992

DGUV 2022: (i) Safe working in the plastics industry. DGUV Rule 113-011. Deutsche Gesetzliche Unfallversicherung e.V. *https://publikationen.dguv.de/regelwerk/dguv-regeln/986/sicheres-arbeiten-in-der-kunststoffindustrie* (retrieved 27 February 2022); (ii) Annex 8. safe working in the plastics industry (previously: BGR 223). Deutsche Gesetzliche Unfallversicherung e.V. *https://www.arbeitssicherheit.de/schriften/dokument/0:4989067,18.html* (retrieved 27 February 2022)

DIN 1982: DIN 4760: Shape deviation; terms; classification system. German Institute for Standardization e.V. 1982

DIN 2019: Publication of DIN SPEC 17071 – First guide for industrial additive manufacturing. DIN e.V. 18 November 2019. *https://www.din.de/de/din-und-seine-partner/presse/mitteilungen/veroeffentlichung-der-din-spec-17071-erster-leitfaden-fuer-die-industrielle-additive-fertigung-689584* (retrieved 27 February 2022)

DIN 2022a: DIN – Briefly explained. DIN e.V. *https://www.din.de/de/ueber-normen-und-standards/basiswissen* (retrieved 27 February 2022)

DIN 2022b: Additive manufacturing. DIN e.V. *https://www.beuth.de/de/themenseiten/additive-fertigungsverfahren* (retrieved 27 February 2022)

Drexler 2016: M. Drexler, On laser beam melting of polyamide 12: Analysis of time-dependent influences in process control. Dissertation. Friedrich-Alexander-Universität Erlangen-Nürnberg, Faculty of Technology, Chair of Plastics Engineering, 2016

Dyze 2022: Typhoon™ Extruder. Dyze Design. *https://dyzedesign.com/typhoon-extruder* (retrieved 24 February 2022)

EOS —a: Design Guidelines Additive Manufacturing. Presentation. EOS GmbH

EOS —b: Machine, software parameters. EOS GmbH

EOS 2013: EOS FORMIGA P110. Training manual. EOS GmbH, 2013

Erhard 2008: G. Erhard, Designing with Plastics. 4th edition. Carl Hanser Verlag, Munich, 2008, p. 277

Eschey 2013: C. Eschey, Maschinenspezifische Erhöhung der Prozessfähigkeit in der additiven Fertigung. Utz Verlag, Munich, 2013 (ISBN 978-3-8316-4270-0)

Fernandez-Vicente et al. 2016: M. Fernandez-Vicente, W. Calle, S. Ferrandiz & A. Conejero, Effect of infill parameters on tensile mechanical behavior in desktop 3D printing. In: 3D Printing and Additive Manufacturing, 1 September 2016, pp. 183–192

Fitzharris et al. 2018: E. R. Fitzharris, N. Watanabe, D. W. Rosen & M. L. Shofner, Effects of material properties on warpage in fused deposition modeling parts. In: The International Journal of Advanced Manufacturing Technology 95, 2018, pp. 2059–2070. *https://doi.org/10.1007/s00170-017-1340-8*

Gebhardt 2016: A. Gebhardt, Additive Manufacturing. Additive Manufacturing and 3D Printing for Prototyping – Tooling – Production. Carl Hanser Verlag, Munich 2016 (ISBN: 978-3-446-44401-0)

Gebhardt et al. 2016: A. Gebhardt, J. Kessler & L. Thurn, 3D Printing. Fundamentals and applications of additive manufacturing (AM). 2nd ed. Carl Hanser Verlag, Munich, 2016

Glamsch et al. 2019: J. Glamsch, K. Deese & F. Rieg, Methods for increased efficiency of FEM-based topology optimization. In: International Journal of Simulation Modelling 18, 2019, pp. 453–463

Grund 2015: M. Grund, Implementation of layer-additive manufacturing processes. In: Light Engineering for Practice 4. Springer Verlag, Berlin/Heidelberg, 2015

Gregurić 2019: L. Gregurić, Printer Plantations. 3D Printing Farm – 5 Great Showcases. All3DP, 2019. *https://all3dp.com/2/3d-printing-farm-great-showcases* (retrieved 27 February 2022)

GSR 2022: The difference between through hole and blind hole. GSR Gustav Stursberg GmbH. *https://gewindeaufschneider.de/blog/unterschied-durchgangsloch-sackloch* (retrieved 27 February 2022)

Heikkinen et al. 2018: I. T. S. Heikkinen, C. Kauppinen, Z. Liu, S. M. Asikainen, S. Spoljaric, J. V. Seppälä, H. Savin & J. M. Pearce, Chemical compatibility of fused filament fabrication-based 3-D printed parts with solutions commonly used in semiconductor wet processing. In: Additive Manufacturing 23, 2018, pp. 99–107 *https://doi.org/10.1016/j.addma.2018.07.015*

Hobbs et al. 2002: C. Hobbs, D. Ferriss, S. Roberts, S. Gnaniah & C. Brown, Thermal properties of polymer melts at elevated pressure from compression heating and differential scanning calorimetry. NPL Report MATC(A)25, 2002. *https://eprintspublications.npl.co.uk/2100/1/MATC25.pdf* (retrieved 24 February 2022).

Hofstaetter et al. 2015: T. Hofstaetter, R. Pimentel, D. B. Pedersen, M. Mischkot, H. N. Hansen, Simulation of a downsized FDM nozzle. Excerpt from the Proceedings of the 2015 COMSOL Conference in Grenoble. *https://www.comsol.de/paper/download/288221/hofstaetter_paper.pdf* (retrieved 24 February 2022)

IPH 2023: Sensor- and app-based validation of process and product quality for effort-reduced approval of personalized medical devices, Institute for Integrated Production Hannover, 2023 *https://www.iph-hannover.de/de/forschung/forschungsprojekte/?we_objectID=5843* (retrieved 27 February 2022).

Jiang et al. 2020: P. Jiang, Z. Ji, X. Wang, F. Zhoua, Surface functionalization – A new functional dimension added to 3D printing. In: Journal of Materials Chemistry C 8, 2020, p. 12380

Johnson 2019: K. Johnson, How to easily calibrate retraction in 3D printers. CNX Software Limited, 2019. *https://www.cnx-software.com/2019/09/05/how-to-easily-calibrate-retraction-in-3d-printers* (retrieved 24 February 2022)

Jorden & Schütte 2017: W. Jorden & W. Schütte, Form and Position Tolerances. Handbook for Study and Practice. 9th ed. Carl Hanser Verlag, Munich, 2017 (ISBN 978-3-446-44626-7)

Keferstein & Dutschke 2007: C. P. Keferstein & W. Dutschke, Fertigungsmesstechnik. Practice-oriented Fundamentals, Modern Measuring Methods. 6th edition. B. G. Teubner Verlag, Wiesbaden, 2007 (ISBN 978-3835101500)

Kelly 2022: R. Kelly, Multi-material 3D Printer: Types & Printing Guide. All3DP, 2022. *https://all3dp.com/2/multi-material-3d-printing-an-overview* (retrieved 27 February 2022)

Koch 2017: M. Koch, Variable filament paths for efficient and fast 3D printing with FDM. Nuremberg University of Technology, 2017. *https://www.th-nuernberg.de/fileadmin/thn_forschung-innovation/FORSCHUNGSDATENBLAETTER/FlexFilament/FlexFilament.pdf* (retrieved 24 February 2022).

KUZ 2022: Research project: ReUp-3D-PrintwasteURL. Kunststoff-Zentrum Leipzig. *https://www.kuz-leipzig.de/aktuelles/news/forschungsprojekt-reup-3d-printwaste* (retrieved 27 February 2022).

Launhardt 2021: M. Launhardt, Zur Entstehung von Dimensionen und Maßabweichungen beim Laserstrahlschmelzen von Kunststoffen. Dissertation. Friedrich-Alexander-Universität Erlangen-Nürnberg, 2021

Launhardt et al. 2011: M. Launhardt, W. Aquite, N. Rudolph & T. A. Osswald, Manufacturing of micropellets using rayleigh disturbances. Proceedings of the 68th Annual Technical Conference of the Society of Plastics Engineers 2010 (ANTEC 2011), Boston, 2011

Launhardt et al. 2015: M. Launhardt, C. Fischer & D. Drummer, Research on the influence of geometry and positioning on laser sintered parts. In: Applied Mechanics and Materials Journal 805, 2015, pp. 105–114

Launhardt et al. 2016: M. Launhardt, A. Wörz, A. Loderer, T. Laumer, D. Drummer, T. Hausotte & M. Schmidt, Detecting surface roughness on SLS parts with various measuring techniques. In: Polymer Testing 53, 2016, pp. 217–226

Lieneke et al. 2015: T. Lieneke, G. A. O. Adam, S. Leuders, F. Knoop, S. Josupeit, P. Delfs, N. Funke & D. Zimmer, Systematical determination of tolerances for additive manufacturing by measuring linear dimensions. In: Solid Freeform Fabrication Proceedings, 2015, pp. 371–384

Menges et al. 2011: G. Menges, E. Haberstroh, W. Michaeli & E. Schmachtenberg, Menges Werkstoffkunde Kunststoffe. 6th edition. Carl Hanser Verlag, Munich, 2011, p. 79 ff

Meyer & Falke 2013: B.-R. Meyer & D. Falke, Dimensionally accurate plastic moldings. Tolerances and molding engineering. Carl Hanser Verlag, Munich, 2013 (ISBN 978-3-446-43687-9)

Nienhaus 2020: V. T. Nienhaus, Investigation and modeling of plastic extrusion systems in fused filament fabrication. Dissertation. Darmstadt University of Technology, 2020. *https://tuprints.ulb.tu-darmstadt.de/17756/1/2020_Diss_VN_Untersuchung_Modellierung_Kunststoffextrusionssysteme_FFF.pdf*

Nöken 1997: S. Nöken, Technology of selective laser sintering of thermoplastics. Dissertation. RWTH Aachen, 1997

Nycz et al. 2020: A. Nycz, V. Kishore, J. Lindahl, C. Duty, C. Carnal & V. Kunc, Controlling substrate temperature with infrared heating to improve mechanical properties of large-scale printed parts. In: Additive Manufacturing 33, 2020. https://www.*osti.gov/servlets/purl/1606965* (retrieved 24 February 2022)

O'Connell 2022: J. O'Connell, Inspecting the path. The best G-code viewers/simulators of 2022. All3DP, 2022. *https://all3dp.com/2/gcode-viewer-3d-printer-simulator-best-tools* (retrieved 5 February 2022)

Ostermann & Ostermann 2006: H.-J. Ostermann & B. Ostermann, Machinery Directive. Directive 2006/42/EC of the European Parliament and of the Council of 17 May 2006 on machinery and amending Directive 95/16/EC (recast) (Text with EEA relevance) OJ L 157 p. 24 ff. *http://www.maschinenrichtlinie.de/fileadmin/dokumente/2006-42-EG_maschinenrichtlinie_de.pdf* (retrieved 27 February 2022)

Parra-Cabrera et al. 2018: C. Parra-Cabrera, C. Achille, S. Kuhn & R. Ameloot: 3D printing in chemical engineering and catalytic technology: Structured catalysts, mixers and reactors. In: Chemical Society Reviews 47, 2018, pp. 209–230

Philips 2022: 3D printed lighting – Tailored lighting to suit your needs. Philips *https://www.tailored.lighting.philips.com/en/default.aspx* (retrieved 27 February 2022)

Polymer Database 2022: Heat capacity of polymers. *http://polymerdatabase.com/polymer%20physics/HeatCapacity.html* (retrieved 24 February 2022)

Precious Plastic 2022: The Precious Plastic universe. *https://preciousplastic.com* (retrieved 3 May 2022)

Raghunath & Pandey 2007: N. Raghunath, P. M. Pandey, Improving accuracy through shrinkage modeling by using Taguchi method in Selective Laser Sintering. In: International Journal of Machine Tools & Manufacture 47, 2007, pp. 985–995

Rechtenwald 2011: T. Rechtenwald, Quasi-isothermal laser sintering of high-temperature thermoplastics. A consideration of material- and process-specific aspects using the example of PEEK. Dissertation. Friedrich-Alexander University Erlangen-Nuremberg, 2011

Richter & Quitter 2015: F. Richter & D. Quitter, Available plastics for 3D printing. Konstruktionspraxis, 2015. *https://www.konstruktionspraxis.vogel.de/verfuegbare-kunststoffe-fuer-den-3d-druck-a-514289* (retrieved 24 February 2022)

Rietzel 2011: D. Rietzel, Material behavior and process analysis in laser-sintering of thermoplastics. Dissertation. Friedrich-Alexander-Universität Erlangen-Nürnberg, 2011

Samhaber 2011: M. P. Samhaber, Development of a multiextrusion head for an FFF 3D printer. Diploma thesis. Vienna University of Technology, 2021. *https://repositum.tuwien.at/handle/20.500.12708/17035* (retrieved 27 February 2022)

Sauer 2005: A. Sauer, Optimization of part properties in selective laser sintering of thermoplastics. Dissertation. University of Duisburg-Essen, 2005

Schäfer et al. 2018: C. Schäfer, S. P. Meyer & T. A. Osswald, A novel extrusion process for the production of polymer micropellets. In: Polymer Engineering & Science 58, 2018, pp. 2264–2275

Schmid 2015: M. Schmid, Selective Laser Sintering (SLS) with Plastics. Technology, Processes and Materials. Carl Hanser Verlag, Munich, 2015 (ISBN 978-3-446-44562-8)

Seepersad et al. 2012: C. C. Seepersad, T. Govett, K. Kim, M. Lundin & D. Pinero, A designer's guide for dimensioning and tolerancing SLS parts, vol. 23. In: 23rd Annual International Solid Freeform Fabrication Symposium: An Additive Manufacturing Conference, Proceedings, 2012, pp. 921–931

Seppala & Migler 2016: J. E. Seppala, K. D. Migler, Infrared thermography of welding zones produced by polymer extrusion additive manufacturing. In: Additive Manufacturing 12, 2016, pp. 71–76 *https://doi.org/10.1016/j.addma.2016.06.007*

Shaik & Schuster 2021: Y. P. Shaik & J. Schuster, A scientific review on various pellet extruders used in 3D. In: Open Access Library Journal 8, 2021, e7698. *https://doi.org/10.4236/oalib.1107698*

Skrzypczak et al. 2020: N. G. Skrzypczak, N. G. Tanikella & J. M. Pearce, Open source high-temperature RepRap for 3-D printing heat-sterilizable PPE and other applications. In: HardwareX 8, 2020, e00130

Soe et al. 2013: S. P. Soe, D. Eyers & R. Setchi, Assessment of non-uniform shrinkage in the laser sintering of polymer materials. In: The International Journal of Advanced Manufacturing Technology 68, 2013, pp. 111–125

Steinhausler 2017: T. Steinhausler, Better temperature tower 240-180. Thingiverse 2017. *https://makerware.thingiverse.com/thing:2625999/files* (retrieved 27 February 2022)

Swanson et al. 2004: W. J. Swanson, P. W. Turley, P. J. Leavitt, P. J. Karwoski, J. E. LaBossiere & R. L. Skubic, High-temperature modeling method. Patent specification: US20040104515A1 (03.06.2004) and US7297304B2 (20.11.2007)/High temperature modeling apparatus. Patent specification: US6722872B1 (04/20/2004). United States Patent Office

Trpkoš 2018: M. Trpkoš, *MINI* All In One 3D printer test. Thingiverse, 2018. *https://www.thingiverse.com/thing:2806295/apps* (retrieved 27 February 2022)

VDI 2014: Guideline VDI 340: Additive manufacturing processes, rapid manufacturing – Basics, definitions, processes. Association of German Engineers (VDI), 2014

VDI 2015: Guideline VDI 3405, Sheet 3: Additive Manufacturing Processes – Design Recommendations for Part Manufacturing with Laser-Sintering and Laser Beam Melting. Association of German Engineers (VDI), 2015

Wang et al. 2007: R.-J. Wang, L. Wang, L. Zhao & Z. Liu, Influence of process parameters on part shrinkage in SLS. In: The International Journal of Advanced Manufacturing Technology 33, 2007, pp. 498–504

Wang et al. 2019: M. Wang, H. Zhang, Q. Hu, D. Liu & H. Lammer, Research and implementation of a non-supporting 3D printing method based on 5-axis dynamic slice algorithm. In: Robotics and Computer-Integrated Manufacturing 57, 2019, pp. 496–505

Wegner 2015: A. Wegner, Theory on the continuation of melting processes as a basic requirement for robust process control in laser sintering of thermoplastics. Dissertation. University of Duisburg-Essen, 2015

Wegner & Witt 2012: A. Wegner & G. Witt, Design rules for laser-sintering. In: Journal of Plastics Technology 8, 2012

Westkämper & Warnecke 2011: E. Westkämper, H.-J. Warnecke, Einführung in die Fertigungstechnik. 8th edition. Vieweg + Teubner, Wiesbaden, 2010 (ISBN 978-3-8348-0835-6)

Wikipedia 2022a: List of CAD programs. *https://de.wikipedia.org/wiki/Liste_von_CAD-Programmen* (retrieved 3 May 2022)

Wikipedia 2022b: G-code. *https://en.wikipedia.org/wiki/G-code* (retrieved 03 May 2022)

Wipperfürth 2018: J. Wipperfürth: Automated load-path-compliant design and optimization of 3D infill structures. 29th International Colloquium on Plastics Technology, 2018, p. 259

Wohlers 2020: History of additive manufacturing. Wohlers Associates, 2020

Wolf et al. 2017: M. Wolf, M. Launhardt & D. Drummer, Investigation into the welding of SLS parts. In: Joining Plastics, 2017

Wudy 2017: K. Wudy, Aging behavior of polyamide 12 during Selective Laser Sintering. Dissertation. Friedrich-Alexander-Universität Erlangen-Nürnberg, 2017

Yang et al. 2002: H.-J. Yang, P.-J. Hwang & S.-H. Lee: A study on shrinkage compensation of the SLS process by using the Taguchi method. In: International Journal of Machine Tools & Manufacture 42, 2002, pp. 1203–1212

YouTube 2022: Remelting 3D printed thermoplastics (the salt method). User: free spirit 1. YouTube. *https://www.youtube.com/watch?v=nRLJ4ylGTFc* (retrieved 27 February 2022)

4 Additive Building Materials and Applications in the Thermoplastic Sector

In the industrial environment, there is still intense debate about the extent to which 3D printing can be used in a way that makes economic and social sense. Are the often-cited advantages of 3D printing in technical applications already sufficient to establish it on a larger scale as a new standard manufacturing technology? These advantages include the following:

- Simple integration of different functions in just one component, for example a completely printed gearbox [Asche 2020]
- High availability and production on demand, for example for the fast and location-flexible production of spare parts [Umweltbundesamt 2019]
- Economic production of small batches, for example for prototypes, reproductions, or products with a high degree of individualization [Feldmann 2016]
- Exploiting new design and layout options, for example to produce better and lighter products.

Crucial for the success of 3D printing is that the sum of advantages and disadvantages at least balance each other out. The most important advantages are production on demand – even in small quantities – and the "cost-free complexity" for manufacturing new products or products with high functional integration. These are offset by the disadvantages of the still limited availability of relevant materials, the tendency toward lower production efficiency (parts or volume per hour) and, in particular, the partially still unsolved problem of testing and quality certification of 3D-printed products.

While for resin-based 3D printing the production formulations have to be developed from scratch, the processes based on thermoplastic materials can at least draw on a number of known and tested materials. The main issue here is to make more and more known thermoplastic materials accessible to 3D printing by slightly modifying the materials and/or printing processes, as well as to understand in detail the consequences of the printing process on the component properties and to take them into account for component design.

Section 4.1 provides an up-to-date overview of thermoplastic printable materials, their manufacturers and suppliers, as well as the corresponding AM processes and the associated printers and their manufacturers. A distinction is made between free and proprietary/certified materials and processes for both extrusion-based processes and powder-based processes. In Section 4.2 selected products and applications generated by the thermoplastic additive processes (which continue to dominate in volume) are presented as examples.

4.1 Additive Building Materials and Value Chains

4.1.1 Extrusion-based Processes (MEX)

4.1.1.1 Building Materials

The number of materials available for extrusion-based processes (material extrusion, MEX) appears very large at first glance. In fact, the extrudability of a thermoplastic processable material (which can almost always be assumed) is only one of the requirements for the material to be readily processable by MEX (e.g. FLM/FFF/FDM or FGF/PAM). Again, a clear distinction must be made between use as a non-functional prototype component for visualizing products and true functional components that can be used in a technical/medical application.

Many materials can be processed quite well in extrusion-based processes, but as components they often exhibit only some of the properties known from classical extrusion or injection molding of the same material formulation. One reason for this is the principle of the layer-based structure in combination with the very untypical temperature profile (cooling rate) to which the components are subjected in MEX processes compared to the classic processes.

The temperature of the receiving layer, in combination with the temperature of the applied layer, is decisive for the adhesion between the layers. For each material there are theoretical ideal process parameters, which are determined by the intrinsic material properties of the thermoplastic (fast or slow crystallization, high or low viscosity, high or low glass transition temperature, material properties depending on the cooling rate such as degree of crystallization, crystal sizes, free volume, etc.).

These ideal process parameters usually do not perfectly match the real process conditions (build space and extrusion temperature), since these themselves depend on the build-job or the number of products and the individual product geometry (layer thickness, wall thickness, high or low layer speed), the conversion to the G-code to be executed (print path), and the printer capabilities (nozzle geometry and temperature, local build space temperature, traverse speed, etc.).

Ideally, both the process and the material to be processed are perfectly matched. Many additive production processes for manufacturing functional components are therefore based on proprietary solutions that provide clear specifications for the material and the printing process, as well as the permitted printing geometries and print path design.

Another problem with the introduction of AM as a reliable production process lies in the currently still widespread "one size fits all" mentality of many production devices and input materials. Similar to modern thermoplastic processing, it will be necessary to develop dedicated materials and AM processes for specific products in the future, in order for production applications in AM to be taken seriously. How flexible these material and process solutions will then still be in order to handle a wide variety of problems and production tasks will have to be evaluated on a case-by-case basis.

4.1.1.2 Value Chains (Thermoplastic, Compounding, Filament)

In this section, the value chain of AM with MEX processes is from the material to the sales product. For this purpose, the material chains from the plastics manufacturer to the sales product (e.g. filament) are listed for the most important extrusion-based materials. The supply of blended (compounded) pellets for pellet extruders is not dealt with separately, as the classic supply chains of the plastics industry continue to apply here. Proprietary solutions, where the materials are only sold in combination with selected printers, are specially marked.

Many suppliers offer both pure polymers and blended (compounded) variants. Particularly for the higher-grade materials (high-performance/structural plastics), fiber-filled variants (glass, carbon, and aramid fibers) are often available, but these are not listed individually. Table 4.1 to Table 4.19 are intended to provide guidance as to which materials are freely available on the market (and from which source), and which are linked to specific printers.

The property pyramid of polymers (Section 2.6.1) provides a good initial overview of material properties and costs, as well as use volumes of plastics. In this context, the material price usually correlates very well with the usage volume. The pyramid has developed over the last few years in such a way that currently the inexpensive thermoplastic standard plastics such as PE, PP, PVC, PET, and PS have by far the largest annual volumes. With prices falling in real terms compared to alternative materials such as metal, wood, and inorganic materials such as ceramics, an increasing number of suitable processing technologies and applications for polymers have been found over the years. However, in the case of AM processes that process thermoplastics (extrusion/sintering), as discussed in Section 3.1, there are new added variables that resort the product price, product properties, and consequently the application quantities of the known polymers. The most important are the following:

- Component properties
- Possible processing speed
- Processing risks (temperature/decomposition)
- Printer/operating costs
- Geometry options/limitations (bridging/warping etc.)
- Interlayer adhesion
- Building platform adhesion.

AM processes are much more process-intensive and show a much higher cost rate (€/kg material processed) in terms of volume throughput. It is not the classic and cheapest materials that basically have the largest sales volume, but the materials that can be processed well and in a process-safe manner.

Apart from the granule-based extrusion processes, the manufacturing costs of the raw materials in the extrusion-based processes are still dominated by the costs of filament production, especially for engineering and standard plastics. Plastics with raw material costs of a few euros are consistently traded as filament in the double-digit euro range. AM also repeatedly uses materials that appear to be "over-engineered" in terms of their basic property level for some of the applications. This choice of materials reflects not only the need for higher safety margins, but also the different processability of the respective materials. Indeed, particularly favorable materials such as PP, PE, PVC, and PET are not always easy to process and require special adjustments in polymerization, formulation, or process.

The new polymer pyramid in AM can be visually defined, at least in a first approach, by the number of suppliers for the respective products. Starting from a manageable number of raw material suppliers and compounders, a whole new layer of suppliers for formulated plastic filaments and also powders is developing. As usual, the product and supplier overviews are divided into standard, engineering, and high-performance plastics. Here, the market volume of the respective plastics corresponds amazingly well with the number of commercial suppliers for the corresponding filaments.

4.1.1.2.1 Value Chain of High-performance Plastics

Table 4.1 Polyvinylidene Fluoride (PVDF)

Core properties:	Chemical-, oxidation-, fire-, and UV-resistant, tough
Price:	100–200 €/kg
Material manufacturer:	Arkema, Daikin Industries, Dyneon, Kureha Corporation, Shanghai 3F New Materials Company, Shanghai Ofluorine Chemical Technology, Solvay, Zhejiang Fotech International, Zhuzhou Hongda Polymer

Compounders/Filament Manufacturers/Product Brands:	3DXTech FluorX (PVDF), AddNorth (Koltron G1), Ensinger Tecafil (PVDF), Fillamentum (Fluorodur), Nanovia (PVDF), Solvay Solef (PVDF)
Product variation:	Molecular weight, copolymer composition
Features (max. values):	Tensile modulus: 2200 MPa Elongation: >30 % Tensile strength: 60 MPa Heat resistance: <150 °C
Compounding:	Colorants, fillers, fibers, stabilizers, flame retardants
Processing:	Nozzle: 240–270 °C Platform: 90–110 °C Installation space: 20 °C Speed: 25–40 mm/s

Table 4.2 Polyetherimide (PEI)

Core properties:	Temperature-resistant, flame-resistant, high strength
Price:	150–300 €/kg
Material manufacturer:	Kuraray, Mitsubishi Chemical, RTP Company, SABIC, Solvay SA, Toray Industries
Compounders/Filament Manufacturers/Product Brands:	3D4Makers (PEI), 3DXTech (ULTEM), Arevo Labs (ULTEM), BASF (Ultrafuse PEI), electrafil 3dp Techmer PM (PEI), Ensinger Tecafil (PEI), Essentium (PEI), Evo-Tech (ULTEM), FIL-A-Gehr (ULTEM), Filament PM (ULTEM), iSANMATE (PEI), Itamsys (ULTEM), Kimya (PEI), MCPP (ULTEM), Perlon (PEI), Prima Select (ULTEM), Push-Plastic (PEI), Sabic (Ultem), Thermax (PEI)
Product variation:	Molecular weight
Features (max. values):	Tensile modulus: 3000 MPa Elongation: 40 % Tensile strength: 90 MPa Heat resistance: <175 °C
Compounding:	Colorants, fillers, fibers, stabilizers, flame retardants
Processing:	Nozzle: 350–400 °C Platform: 180–220 °C Installation space: 180–220 °C Speed: 10–30 mm/s

Table 4.3 Polyarylether Ketones (PAEK, PEEK, PEKK)

Core properties:	Temperature-resistant, high strength, implantable
Price:	200–500 €/kg
Material manufacturer:	Cendres+Métaux SA, Evonik, Panjin Zhongrun Polymers Co. Ltd, Solvay S.A., Victrex plc
Compounders/Filament Manufacturers/Product Brands:	3D4Makers (PEEK), 3DXTech (PEEK), 3DXTech (PEKK), Apium (PEEK), Arevo Labs (PEEK), Arevo Labs (PAEK), Arkema (PEKK), Electrafil TechmerPM 3dp (PEEK), Ensinger Tecafil (PEEK), Ensinger Tecafil (PEKK), Essentium (PEEK), Essentium (PEKK), Evonik INFINAM (PEEK), FIL-A-Gehr (PEEK), Intamsys (PEEK), iSANMATE (PEEK), Kimya (PEKK), Luvocom 3F (PEEK), Luvocom 3F (PEKK), MSNJ (PEEK), MIA (PEEK), Nanovia (PEKK), Perlon (PEEK), Treed (PEEK), Treed (PAKK), Solvay KetaSpire (PEEK), Thermax (PEEK), Thermax (PPEK), Victrex AM (PAEK)
Product variation:	Molecular weight, copolymer composition
Features (max. values):	Tensile modulus: 3000 MPa Elongation: 20 % Tensile strength: 100 MPa Heat resistance: <250 °C
Compounding:	Colorants, fillers, fibers, stabilizers
Processing:	Nozzle: 350–450 °C Platform: 160–250 °C Installation space: 160–250 °C Speed: 10–30 mm/s

Table 4.4 Polyphenyl Sulfone (PPSU, PES, PPS, PSU, PAS, etc.)

Core properties:	Temperature- and chemical-resistant, high strength
Price:	200–600 €/kg
Material manufacturer:	BASF, Mitsubishi Chemical Advanced Materials, Sabic, UJU New Materials
Compounders/Filament Manufacturers/Product Brands:	3D4Makers (PPSU), 3DXTech (PPSU), 3DXTech (PSU), electrafil 3dp TechmerPM (PPS), electrafil TechmerPM 3dp (PPSU), Ensinger Tecafil (PES), Ensinger Tecafil (PPSU), Evo-Tech (PPS), Fabrix (PPS), Fiberforce (PPS), FIL-A-Gehr (PPSU), iSANMATE (PPSU), Kimya (PEKK), Kimya (PPSU), Luvocom (PPS), MCPP (PPSU), MSNJ (PPSU), Nanovia (PSU), REC (PSU), Sabic (PPE), Solvay Radel NT1 (PPSU), Thermax (PES), Thermax (PPS), Thermax (PPSU), Thermax (PSU), Thermax (TPI), Treed (PPS)

Product variation:	Molecular weight, copolymer composition
Features (max. values):	Tensile modulus: 2400 MPa Elongation: >50 % Tensile strength: >70 MPa Heat resistance: <180 °C
Compounding:	Colorants, fillers, fibers, stabilizers, flame retardants
Processing:	Nozzle: 320–400 °C Platform: 160–220 °C Installation space: 160–220 °C Speed: 10–30 mm/s

4.1.1.2.2 Value Chain of Engineering Plastics

Table 4.5 Polyvinyl Alcohol (PVA); Polybutenediol Vinyl Alcohol Copolymer (BVOH); Starch

Core properties:	Support material, water-soluble and/or brittle
Price:	30–60 €/kg
Material manufacturer:	Anhui Wanwei Group, Chang Chun Group, China Petroleum & Chemical, Denki, Eastman, Japan Vam & Poval, Kuraray, Nippon Gohsei, Nippon Synthetic Chemical Industry, OCI Company, Sekisui Chemical, Sinopec, Synthomer, Wacker
Compounders/Filament Manufacturers/Product Brands:	3DFuel (Hydro), 3DJake (nicePVA), AzureFilm (PVA), BASF (Ultrafuse BVOH), BestFilament (PVA), Bigrep (PVA), Dynamism Breakaway, Dynamism (PVA), BCN3D (BVOH), BCN3D (PVA), BY (PVA), E Motion Tech (Soluble), EasyPrint PVB, Essentium (Support), eSun (PVA), Evo-Tech (SUPP), Fiberlogy (BVOH), Flashforge (PVA), Formfutura (AquaSolve), Formfutura (Atlas Support), Formfutura (BVOH), Formfutura (Helios Support), Formfutura (PVA), Herz (PVA), ICE (PVA), Infinite Material Solution (AquaSys), Junshan (PVA), Kuraray (Mowiflex 3D 2000), L.J.J (PVA), MatterHackers (PVA), Maertz (PVA), MCPP (PVA), MSNJ (PVA), Neofil3d (PVA), Phaetus (aeSupport), Polymaker (Polydissolve PVA), Polymaker (PolySupport), Prima Select (PVA), PushPlastic (Water soluble), Redline Filament (PVA), Raise3D (Premium PVA+), Raise3D (Support), R3D (PVA), RS PRO (PVA), REED (PVA), Sabic AMS31F, SaintSmart (Water soluble), Smartfil (PVA), Super-Filament (PVA), Taulman3D (SAC Support), ThermaX (HTS Support), TiZYX (PVA), Ultimaker (Breakaway), Ultimaker (PVA), Verbatim (BVOH), Village Plastics, Volumic Ultra (PVA), XYZprinting (PVA), Yimihua (PVA), Z3D (PVA), Zmorph (BVOH), Zoretrax-z-Support

Table 4.5 Polyvinyl Alcohol (PVA); Polybutenediol Vinyl Alcohol Copolymer (BVOH); Starch (*continued*)

Product variation:	Molecular weight, degree of saponification
Features (max. values):	Various
Compounding:	Fillers
Processing:	Nozzle: 190–240 °C Platform: 25–70 °C Installation space: 20–80 °C Speed: 20–45 mm/s

Table 4.6 Polyvinyl Butural (PVB)

Core properties:	Hard, tough, good reworkability, good printability
Price:	30–50 €/kg
Material manufacturer:	Eastman Kuraray, Sekisui, ShinEtsu, ChangChun Group
Compounders/Filament Manufacturers/Product Brands:	Polymaker (Polycast PVB) Polymaker (Polysmooth) Polymaker (PolySmooth PVB) Prusa (Prusament PVB)
Product variation:	Molecular weight, copolymer composition
Features (max. values):	Tensile modulus: 2300 MPa Elongation: 70 % Tensile strength: 50 MPa Heat resistance: <60 °C
Compounding:	Paint, fillers
Processing:	Nozzle: 180–220 °C Platform: 60 °C Installation space: 20 °C Speed: 30–60 mm/s

Table 4.7 Polyethylene Glycol Terephthalate (PETG)

Core properties:	Hard, transparent, good printability
Price:	15–50 €/kg
Material manufacturer:	EASTMAN, Liaoyang Petrochemical, LSB, Shandong Shenghe Plastic Ltd, Simona Plaskolite, SK Chemicals

Compounders/Filament Manufacturers/Product Brands:	3D4Makers (PETG), 3DFuel (PETG), 3DJake(PETG), 3DPlatform (PETG), 3DXTech (PETG), AddNorth (PETG), Airwolf 3D (PETG), Amazon Basic (PETG), AMOLEN (PETG), Anisopring (PETG), AnKun (PETG), AzureFilm (PETG), BASF (Ultrafuse PET), Basicfil (PETG), BCN3D (PETG), Bigrep (PETG), CERPRiSe (PETG), Color-Fabb (PETG), Creality (PETG), DASFilament (PETG), Devil Design (PETG), Digitalrise (PETG), Dremel (PETG), Duramic 3D (PETG), Dynamism (PETG), EasyPrint (PETG), E-DA (PETG), eMotionTech (PETG), Enotepad (PETG), Eolas (PETG), Eono (PETG), Eryone (PETG), Essentium (PETG), eSun (PETG+), Evo-Tech (PETG), Extrudr (XPETG), FDPlast (PETG), Fiberforce (PETG), Fiberlogy (PETG), FIL-A-Gehr (PETG), Filament P (PETG), Filament PM (PETG), Filamentworld (PETG), FilaTex3D (PETG), Filkemp (PETG), Fillamentum (CPE), Flashforge (PETG), Formfutura (PETG), Fron-tierFila (PETG), GEEETECH (PETG), Giantarm (PETG), HATCHBOX (PETG), Hello3D (PETG), Herz (PETG), IC3D (PETG), ICE (PETG), I-Filament (PETG), iSANMATE (PETG), Jabil (PETG), Jayo (PETG), KEHUASHINA (PETG), Kimya (PETG), MAKEASY (PETG), Material-4Print (PETG), MatterHackers (PETG), MCPP (PETG), MIA (PETG), Midori (PETG), Minadax (PETG), Nanovia (PETG), Nexeo (PETG), Noulei (PETG), NuNus (PETG), Orbi-Tech (PETG), Overture (PETG), OWL (PETG), Perlon (PETG), Polymaker (PETG), POPESQ (PETG), PrimaCreator (PETG), Prinsfil (PETG), Print-Me (PETG), Procatec (PETG), Proto Pasta (PETG), Prusa (Prusament PETG), PushPlastic (PETG), R3D (PETG), Raise3D (Premium PETG), Recreus (PETG), REDD (PETG), Redline Filament (PETG), Re-pet3D (rPETG), Reprapper (PETG), Roffelsen (PETG), Rosa3D (PETG), SaintSmart (Pro-3 PETG), Sienoc (PETG), Spectrum (PETG), Stronghero3D (PETG), Sunlu (PETG), Super-Filament (PETG), Taulman3D (PETT), TECBEARS (PETG), TEQStone (PETG), Tianse (PETG), TinMorry (PETG), TiZYX (PETG), Treed (PETG), TronHoo (PETG), Ultimaker (PETG), Verbatim (PETG), Village Plastics (PETG), Winkle (PETG), Xioneer (PETG), XVICO (PETG), XYZprinting (PETG), Yorking (PETG), Yoyi (PETG), Z3D (PETG), Zmorph (PETG), Zortrax (PETG)
Product variation:	Molecular weight, copolymers with different monomers
Compounding:	Colorants, fillers, fibers, stabilizers, plasticizers, flame retardants
Features (max. values):	Tensile modulus: 2200 MPa Elongation: 50 % Tensile strength: 50 MPa Heat resistance: <80 °C
Processing:	Nozzle: 220–260 °C Building platform: 70–110 °C Installation space: 20 °C Speed: 40–100 mm/s

Table 4.8 Polylactic Acid Esters (PLA)

Core properties:	Hard, transparent, very good printability
Price:	15–50 €/kg
Material manufacturer:	COFCO, eSun, Futerro, Jiangsu Supla, JiangxiAcademy, NatureWorks, Synbra Technology, Tong-jie-liang, Total Corbion, Zhejiang Haizheng
Compounders/Filament Manufacturers/Product Brands:	3D4Makers (PLA), 3DJake (PLA), 3D Platform (PLA), 3DXTECH (PLA), AddNorth (Textura PLA), Airwolf 3D (PLA), Amazon Basic (PLA), AMOLEN (PLA), Anycubic (PLA), BASF (Ultrafuse PLA), Basicfil (PLA), BCN3D (PLABIQU PLA), Bigrep (PLA), BIQU (PLA), CCTREE (PLA), ColorFabb (PLA), Copper 3D (PLA), Copymaster (PLA), Creality Ender (PLA), DASFilament (PLA), Dynamism (PLAtough), E-DA (PLA), eMotionTech (PLA), Enotepad (PLA), Eolas (PLA), Eono (PLA), Eryone (PLA), eSun (PLA+), Evo-Tech (PLA), Extrudr (PLA NX-2), Fabbrix (PLA), Fiberforce (PLA), FIL-A-Gehr (PLA), Filament PM (PLA), Filamentworld (PLA), FilaTex3D (PLA), Filkemp (PLA), Fillamentum (PLA), Flashforge (PLA), Formfutura (PLA), GEEETECH (PLA), Giantarm (PLA), Gizmo Dorks (PLA), Hatchbox (PLA), Hello3D (PLA), Herz (PLA), IC3D (PLA), ICE (PLA), I-Filament (PLA), Janbex (PLA), Jayo (PLA), Kimya (PLA-HI), KEHUASHINA (PLA), Labists (PLA), Material4Print (PLA), MatterHackers (PRO PLA), MCPP (PLA), MIA (PLA), MSNJ (PLA), Nanovia (PLA), Nexeo (PLA), Noulei (PLA), Orbi-Tech (PLA), Overture (PLA), OWL (PLA), Patona (PLA), Perlon (PLA), Polymaker (PLA), Prima Filaments (PLA), Prinsfil (PLA), Procatec (PLA), Proto Pasta (HAT-PLA), Prusa (Prusament PLA), PushPlastic (PLA), Pxmalion (PLA), Raise3D (Premium PLA), REC (PLA), Recreus (PLA), Redline Filament (PLA), ReForm (PLA), Roffelsen (PLA), Rosa3D (PLA), SaintSmart (PLA), Sienoc (PLA), Sunlu (PLAZ/PLA), Suntop (PLA), Super-Filament (PLA), Taulman3D (PLA), Tianse (PLA), TinMorry (PLA), Treed (PLA), TronHoo (PLA), Ultimaker (PLA), Verbatim (PLA), Xioneer (PLA), XVICO (PLA), Yousu (PLA), Yoyi (PLA), Zmorph (PLA), Zortrax (PLA)
Product variation:	Molecular weight, copolymers with other esters
Features (max. values):	Tensile modulus: 3500 MPa Elongation: 5 % Tensile strength: 50 MPa Heat resistance: <70 °C
Compounding:	Colorants, fillers, fibers, stabilizers, plasticizers, flame retardants
Processing:	Nozzle: 190–230 °C Building platform: 20–70 °C Installation space: 20 °C Speed: 40–80 mm/s

Table 4.9 Thermoplastic Elastomers (TPEs)

Core properties:	Elastic, tough, chemical-resistant
Price:	20–50 €/kg
Material manufacturer:	ApexPolyone, Dynasol, Evonik, Kraiburg, Kraton, Mitsui, Solvay, Star Thermoplastics, Teknor, UBE
Compounders/Filament Manufacturers/Product Brands:	3D4Makers (TPC), 3DFuel (DOW OBC), 3DXTech (TPC), BASF (Ultrafuse TPC), BASF (Ultrafuse TPS), Copper3D (MDflex), ColorFabb (nGen_Flex), Covestro (Arnitel ID 2045), E Motion Tech (Flex), E Motion Tech (SBS), Essentium (CoPE, CoPA), eSun (TPE), Evo-Tech (TPE), ezPrint (TPE), Fiberlogy (FiberFlex), Filaflex Stretch (TPE), Filament PM (TPE), Fillamentum (Flexfill TPE), Formfutura (Arnitel ID), Formfutura (Flexifil), FrontierFila (Flex), Gembird (TPE), ICE (Flex), Jabil (TPE), Kimya (PEBA-S), Kimya (TPC), MCPP (TPC), Nanovia (TPE), Perlon (TPS), Prima (Flex), Recreus (Filaflex), Redline Filament (Flexible), REC (SEBS), REC (TPEE), Selfenbein Flex (TPE), SpiderMaker (SpiderFlex TPE), Taulman3D (PCTPE), Treed (TPA), Treed (TPE), Verbatim (Tefabloc TPE), Verbatim (TPE Primalloy), Zortrax (z-Flex)
Product variation:	Molecular weight, copolymer composition
Features (max. values):	Tensile modulus: 50–1000 MPa Elongation: 50–1000 % Tensile strength: 5–50 MPa Heat resistance: <80 °C
Compounding:	Colorants, fillers, plasticizers, stabilizers, flame retardants
Processing:	Nozzle: 180–260 °C Platform: 20–100 °C Installation space: 20 °C Speed: 10–40 mm/s

Table 4.10 Thermoplastic Polyurethane (TPU)

Core properties:	Elastic, tough, abrasion-resistant
Price:	20–50 €/kg
Material manufacturer:	BASF, Covestro, DOW, Huafon, Huntsman, Kuraray, Lubrizol, Miracll, Sumei Chemical, Wanhua

Table 4.10 Thermoplastic Polyurethane (TPU) (*continued*)

Compounders/Filament Manufacturers/Product Brands:	3D4Makers (TPU), 3DJake (TPU), 3D Platform (TPU), AddNorth (Easyflex TPU), Airwolf 3D (TPU), Amazon Basic (TPU), BASF (Ultrafuse), Bigrep (TPU), ColorFabb (Varioshore TPU), Copper 3d (TPU), Covestro (Addigy TPU), Dynamism (TPU), E-DA (TPU), eMotionTech (TPU), Enotepad (TPU), Eolas (TPU), Eryone (TPU), Essentium (TPU), eSun (eFlex), Extrudr (TPU), Fiberlogy (ABS), FilaTex3D (TPU), Fillamentum (Flexfill), Flashforge (Flexible), Formfutura (Phyton Flex), Formfutura (TPU), GEEETECH (TPU), Giantarm (TPU), Huntsmen Isoprint (TPU), IC3D (TPU), Jabil (TPU), Jayo (TPU), Kehuashina (TPU), Kimya (TPU), MatterHackers (PRO Series TPU), MCPP (TPU), MIA (TPU), Nanovia (TPU), NinjaTek (Ninjaflex, Cheetah, Armadillo/TPU), Noulei (TPU), NuNus (TPU), Orbi-Tech (TPU), Overture (TPU), OWL (TPU), Perlon (TPU), Polymaker (Polyflex TPU), PushPlastic (TPU), Pxmalion (TPU), Raise 3D (TPU), REC (TPU), REC (PPT), Recreus (Filaflex TPU), Redline (TPU), Rep-Flex (TPU), SaintSmart (TPU), Sienoc (TPU), Spectrum (TPU), Sunlu (TPU), Super-Filament (TPU), Tecnikoa (Filafresh), Tianse (TPU), TinMorry (TPU), Treed (Flexmark TPU), TronHoo (TPU), Yoyi (TPU), Zmorph (TPU), Zortrax (TPU)
Product variation:	Molecular weight, polyester polyurethane, polyether polyurethane, polyether polyester polyurethane, aromatic polyurethane, aliphatic polyurethane, hard block content
Features (max. values):	Tensile modulus: 50–1000 MPa Elongation: 50–1000 % Tensile strength: 5–50 MPa Heat resistance: <100 °C
Compounding:	Colorants, fillers, plasticizers, stabilizers, flame retardants
Processing:	Nozzle: 180–260 °C Platform: 20–60 °C Installation space: 20 °C Speed: 10–60 mm/s

Table 4.11 Polyoxymethylene (POM)

Core properties:	Hard, abrasion-resistant, tough, low water absorption
Price:	70–150 €/kg
Material manufacturer:	Asahi Kasai Plastics, BASF, Celanese, DuPont, Kolon Plastics, Polyplastics, RTP Company, Sabic, Techmer PM

Compounders/Filament Manufacturers/Product Brands:	Ensinger Tecafil (POM), Evo-Tech (Iglidur), EzPrint (POM), Frontier-Fila (POM), Gizmo Dorks (Acetal Delrin), NLLeZ (POM), Tarfuse (POM), ZDZQDD (POM)
Product variation:	Molecular weight, copolymer composition
Features (max. values):	Tensile modulus: 2500 MPa Elongation: 50 % Tensile strength: 70 MPa Heat resistance: <100 °C
Compounding:	Colorants, fillers, plasticizers, stabilizers, flame retardants
Processing:	Nozzle: 190–210 °C Platform: 100–130 °C Installation space: 50–120 °C Speed: 10–30 mm/s

Table 4.12 Polycarbonate (PC)

Core properties:	Hard, tough, transparent, heat-resistant, flame-retardant
Price:	50–100 €/kg
Material manufacturer:	Covestro, Formosa, LG Chem, Mitsubishi, Sabic, Samsung, Teijjin, Trinseo
Compounders/Filament Manufacturers/Product Brands:	3DXMAX (PC-ABS), 3DXMAX (PC), 3DXTech (PC), 3DZ (PC), Airwolf 3D (PC), Amazon Basic (PETG), BASF (Ultrafuse PC/ABS), BestFilament (PC-ABS), Clariant (PC-ABS), Covestro (PC-ABS), eMotion-Tech (PC), Ensinger Tecafil (PC), eSun (ePC), Evo-Tech (PC), FIL-A-Gehr (PC), Filament PM (PC-ABS), Fillamentum (PC-ABS), Hello3D (PC), Herz (PC), ICE (PC ABS), ICE (PC Pro), Inovatefil (PC), iSANMATE (PETG), Jabil (PC), Kimya (PC-S), Longsell (PC), Material4Print (PC-ABS), MatterHackers (PC), MCPP (PC-ABS), MIA (PC), MSNJ (PC), Nanovia (PC), Neofil3D (PC-ABS), NLLeZ (PC), NoLogo (PC), Orbi-Tech (PC), Polymaker (PC-ABS), Polymaker (PC-PBT), Polymaker (PolyLite PC), Polymaker (PolyMax PC), Owen Corning, Priline (PC), PrimaCreator (PC-EPETG), PrimaSelect (PC), Proto Pasta (PC-ABS), Prusa (Prusament PC), PushPlastic (PC), PushPlastic (PC-PBT), Redline Filament (PC), Raise3D (Premium PC), Sabic (PC), Sabic (PC-PBT), TechmerPM 3dp (PC), Treed (PC-ABS), Treed (PC), Triton 3D (PC), Ultimaker (PC), Uniqstore (PC), Verbatim (Durabio), Volumic (PC), W2 Filaments (PC), Walmeck (PC), XFC-3D (PC), XSTRAND (PC), Z3D (PC), Zortrax (z-PCABS)
Product variation:	Molecular weight, copolymer composition

Table 4.12 Polycarbonate (PC) (*continued*)

Features (max. values):	Tensile modulus: 2400 MPa Elongation: 80 % Tensile strength: 50 MPa Heat resistance: <130 °C
Compounding:	Colorants, fillers, fibers, plasticizers/polymer blends, stabilizers, flame retardants
Processing:	Nozzle: 250–330 °C Platform: 80–130 °C Installation space: 50–130 °C Speed: 30–50 mm/s

Table 4.13 Polyamide (PA)

Core properties:	Hard, tough, heat-resistant
Price:	50–200 €/kg
Material manufacturer:	Acend Performance, Amco, Arkema, Asahi Kasai, ASF, DSM, DuPont, Evonik, Formosa, Honeywell, Invista, Lealea, Lanxess, Radici, Rhodia, Shifeng, Solvay, UBE
Compounders/Filament Manufacturers/Product Brands:	3D-Prima (Nylon), 3DXTech (PA), 3DXTech (PA12), AddNorth (Adura), Airwolf 3D (Nylon), AMIDEX (Nylon), AMOLEN (Nylon), AprintaPro (PA12 cleaning), AzureFilm (Nylon), BASF (Ultrafuse PA), BCN3D (PA), BestFilament (BFNylon), BigRep (PA 6/66), Clariant (PA 6/66), ColorFabb (PA), Covestro (Novamid ID), Devil Design (PA12), Dremel (Nylon3DFuel, Nylon), Dynamism (nylon), eMotion-Tech (nylon), Ensinger Tecafil (PA), Ensinger Tecafil (PA12), eSun (nylon), Essentium (PA), FDPlast (nylon), FIL-A-Gehr (PA12), Fiberforce (PA), Fiberforce (Nylforce), Fiberlogy Nylon (PA12), Fiberthree (PA), Filkemp (Nylon), Filament PM (Nylon), Fillamentum (Nylon), Flashforge (PA), Formfutura (STYX-12), Formfutura (Nylon), Frontier Fila (PA), Hello3D (PA), Herz (PA), Jabil (PA), Lehvoss Luvocom (PAHT), MatterHackers (Nylon), MCPP (Nylon), MeltFox (PA12), MIA (PA), MSNJ (Nylon, PA12), Nanovia (PA), Orbi-Tech (nylon), Overture (nylon), Owens Corning, Perlon (PA12), Polymaker (nylon), PolyMide (PA 6 and PA12), PrimaCreator (nylon), PrimaSelect (nylon), PushPlastic (nylon), MIA (PA nylon), Phaetus (aeCarbon Ultra PA), Raise3D (PA12), Rosa3D (PA), SaintSmart (ePA), Sienoc (PA), Spectrum (PA), Super-Filament (nylon), R3D (nylon), REC (PA), REDD (nylon), Taulman3D (nylon), TechmerPM electrafil 3dp (PA), Treed (PA 6), Treed (PA12), Ultimaker (nylon), Village Plastics (nylon), Xstrand (PA 6), Z3D (nylon), Zmorph (PA), Zortrax (z-Nylon)

Product variation:	Molecular weight, copolymer composition (PA 6, PA 66, PA 46, PA 410, PA 11, PA12)
Features (max. values):	Tensile modulus: >1000 MPa Elongation: >50 % Tensile strength: >50 MPa Heat resistance: approx. 100–160 °C
Compounding:	Colorants, fillers, fibers, plasticizers, stabilizers, flame retardants
Processing:	Nozzle: 230–320 °C Platform: 90–120 °C Installation space: 50–120 °C Speed: 30–60 mm/s

4.1.1.2.3 Value Chains of Standard Plastics

Table 4.14 Polyolefins (PE/PP)

Core properties:	Chemical-resistant
Price:	20–50 €/kg
Material manufacturer:	**PE:** Braskem, DowChemical, ExxonMobilCorporation, Lyondell-Basell, SABIC, Sinopec Corporation **PP:** Braskem, China Petrochemical, ExxonMobilCorporation, INEOS Group Holding, LyondellBasell
Compounders/Filament Manufacturers/Product Brands:	3DXTech (OBC), AddNorth (OBC), Airwolf3D (PP), Apium (PP), BASF (UltrafusePP), BCN3D (PP), Braskem (PP), Caverna (PP), eMotionTech (PP), Essentium (PP), Evolv3D (OBC PE), Fiberlogy (PP), Fillamentum (PP), Formfutura (Centaur PP), Formfutura (Pegasus PP), Hyperlite (PP), IGUS (Iglidur), Infinite Material Solutions, Love lamp (POM), MatterHackers (PP), MCPP (PP), Owens Corning, PM (PP), Prima Select (PP), Purefil (HDPE), Raise3D (Premium PP), REC (PP Filament), Recreus (PP), Recreus (PP3D), Spectrum (S-Flex), Super-Filament (PP), Treed (PP), Ultimaker (PP), Verbatim (PP), XSTRAND (PP), XSTRAND GF30-PP
Product variation:	Molecular weight, branching, composition
Features (max. values):	**PE:** Tensile modulus: 400–1000 MPa Elongation: 100–1000 % Tensile strength: 20–30 MPa Heat resistance: <80 °C

Table 4.14 Polyolefins (PE/PP) (*continued*)

	PP: Tensile modulus: 1000–1400 MPa Elongation: 150–300 % Tensile strength: 25–35 MPa Heat resistance <100 °C
Compounding:	Colorants, fillers, fibers, stabilizers, plasticizers, flame retardants
Processing:	**PE:** Nozzle: 240–280 °C Platform: 100–120 °C Installation space: 20–80 °C Speed: 20–50 mm/s **PP:** Nozzle: 220–250 °C Platform: 80–100 °C Installation space: 20–80 °C Speed: 20–50 mm/s

Table 4.15 Polyvinyl Chloride (PVC)

Core properties:	Weather-resistant, chemical-resistant, color-resistant, hard and tough
Price:	40–50 €/kg
Material manufacturer:	Axiall, CNSG Anhui Hong Sifang Co. Ltd, Formosa Plastics Corp, Inovyn ChlorVinyls, LG Chem Ltd, ChemChina, Orbia, OxyVinyls (Occidental Corp.), Shin-Etsu Chemical Co. Ltd, Westlake Chemical Corporation.
Compounders/Filament Manufacturers/Product Brands:	Fillamentum (PVC Vinyl)
Product variation:	Molecular weight, manufacturing process, composition
Features (max. values):	Tensile modulus: 2000–4000 MPa Elongation: 15 % Tensile strength: 34–62 MPa Heat resistance: <80 °C
Compounding:	Colorants, fillers, fibers, stabilizers, plasticizers, flame retardants
Processing:	Nozzle: <220 °C Platform: 80–100 °C Installation space: 60–80 °C Speed: 30–50 mm/s

Table 4.16 High-impact Polystyrene (HIPS)

Core properties:	Hard, scratch-resistant, reworkable, good printability
Price:	10–40 €/kg
Material manufacturer:	BASF SE, ChiMei, INEOS Styrolution, LG Chem, SABIC, Total
Compounders/Filament Manufacturers/Product Brands:	3D Platform (HIPS), 3DXMAX (HIPS), 3DZ (HIPS), BASF (Ultrafuse HIPS), BestFilament (HIPS), Devil Design (HIPS), eSun (HIPS), FDPlast (HIPS), Fiberlogy (HIPS), Fillamentum (HIPS), Flashforge (HIPS), Formfutura (EasyFil HIPS), FrontierFila (HIPS), Hello3D (HIPS), Herz (HIPS), ICE (HIPS), Kimya (HIPS), MCPP (HIPS), MIA (HIPS), Minadax (HIPS), Nanovia (HIPS), NuNus (HIPS), Orbi-Tech (HIPS), Prima Filaments (PS-HIPS), Prima Select (HIPS), Procatec (HIPS), REC (HIPS), Sienoc (HIPS), Smartfil (HIPS), Spectrum (HIPS), Super-Filament (HIPS), TAG3D (HIPS), Treed (HIPS), Village Plastics (HIPS), Velleman (HIPS), Zortrax (z-HIPS)
Product variation:	Molecular weight, copolymer composition
Features (max. values):	Tensile modulus: 2600 MPa Elongation: 85 % Tensile strength: 56 MPa Heat resistance: <95 °C
Compounding:	Colorants, fillers, fibers, stabilizers, plasticizers, flame retardants
Processing:	Nozzle: 230–250 °C Platform: 90–110 °C Installation space: 20 °C Speed: 30–50 mm/s

Table 4.17 Acrylonitrile Butadiene Styrene (ABS);

Core properties:	Hard, tough, scratch-resistant, reworkable, good printability
Price:	10–40 €/kg
Material manufacturer:	Celanese, ChiMei, ELIX Polymers, INEOS, Kumho, LG Chem, Polimeri, SABIC, Trinseo, Toray
Compounders/Filament Manufacturers/Product Brands:	3D4Makers (ABS), 3DJake (niceABS), 3D Platform (ABS), 3DPrima (value ABS), 3DXTECH (ABS), Add North (ABS), Airwolf 3D (ABS), Amazon Basic (ABS), AzureFilm (ABS), Basicfil (ABS), BASF (Ultrafuse ABS), Bigrep (ABS), Dremel (ABS), Dynamism (ABS), E-DA (ABS), eMotionTech (ABS), Enotpad (ABS), eSun (ABS+), Evo-Tech (ABS), Extrudr (DuraPro), Fabbrix (ABS), Fiberforce (ABS), Fiberlogy (ABS), FIL-A-Gehr (ABS), Filamentworld (ABS), FilaTex3D (ABS), Fillamentum (ABS), Flashforge (ABS), Formfutura (ABS),

Table 4.17 Acrylonitrile Butadiene Styrene (ABS); (*continued*)

	Giantarm (ABS), GEETECH (ABS), Hello3D (ABS), Herz (ABS), IC3D (ABS), ICE (ABS), Jabil (ABS), Jayo (ABS), Kimya (ABS), Material-4Print (ABS), MatterHackers (Build Series ABS), MCPP (ABS), MIA (ABS), MSNJ (ABS), Nanovia (ABS), Noulei (ABS), NuNus (ABS), Orbi-Tech (ABS), Overture (ABS), OWL (ABS), Patona (ABS), Perlon (ABS), Polymaker (Polylite ABS), Priline (ABS), Prima Filaments (value ABS), Procatec (ABS), Prusa (Prusament ABS), PushPlastic (ABS), Pxmalion (ABS), R3D (PLA), Raise3D (ABS), REC (ABS), Red-line Filament (ABS), ABIC (ABS), SaintSmart (ABS), Sienoc (ABS), Spectrum (ABS), Sunlu (ABS), Super Filament (ABS), TechmerPM 3dp (ABS), TEQStone (ABS), TinMorry (ABS), Treed (ABS), Triton 3D (ABS), TronHoo (ABS), Ultimaker (ABS), Verbatim (ABS), Work-day (ABS), XVICO (PLA), Yoyi (ABS), Zmorph (ABS), Zortrax (ABS)
Product variation:	Molecular weight/branching, composition/ratio of nitrile, butadiene and styrene, transparency, plasticizer content, surface gloss, reaction/mixtures with other polymers (PC, styrene, MMA, TPU)
Features (max. values):	Tensile modulus: 3300 MPa Elongation: 20 % Tensile strength: 40 MPa Heat resistance: <85 °C
Compounding:	Colorants, fillers, fibers, stabilizers, plasticizers, flame retardants
Processing:	Nozzle: 230–270 °C Building platform: 80–110 °C Installation space: 80–100 °C Speed: 40–60 mm/s

Table 4.18 Acrylonitrile Styrene Acrylic Ester (ASA)

Core properties:	Hard, impact-resistant, weather-resistant, good printability
Price:	10–40 €/kg
Material manufacturer:	ChiMei, Formosa Chemicals, INEOS Styrolution, Kumho, LG Chem, Nippon A&I, Sabic, Techno-UMG
Compounders/Filament Manufacturers/Product Brands:	3D4Makers (ASA), 3DJake (ASA), 3DXTEch (ASA), Ampertec (ASA), AzureFilm (ASA), BASF (Ultrafuse ASA), BigRep (ASA), ColorFabb (ASA), C-Tech (ASA), DevilDesign (ASA), Dynamism (ASA), eMotion-Tech (ASA), eSun (ASA), Evo-Tech (ASA), Extrudr (DuroPro ASA), Fabbrix (ASA), Fiberlogy (ASA), Filament PM (ASA), FilaTex3D (ASA), Fillamentum (ASA), Flashforge (ASA), Formfutura (APOX ASA),

	Frontier (ASA), Hello3D (ASA), Heart (ASA), Innofil3d (ASA), KEHU-ASHINA (ASA), Kimya (ASA), Kleverfil (ASA), MatterHackers (ASA), MIA (ASA), MSNJ (ASA), NuNus (ASA), Material4Print (ASA), Natur Cuber 3D (ASA), NEXBERG (ASA), Orbi-Tech (ASA), Overture (ASA), Perlon (ASA), Polymaker (Polylite ASA), Prima Filaments (ASA), Prinsfil (ASA), Print with Smile (ASA), Prusa/ Prusament (ASA), PushPlastic (ASA), Raise3D (Premium ASA), REC (ASA), Redline Filament (ASA), Rosa3D (ASA), RS pro (ASA), Smartfil (ASA), Spectrum (ASA), Sunlu (ASA), Super-Filament (ASA), Swissfil (ASA), TinMorry (ASA), Treed (ASA), Triton 3D (ASA), YourDroid (ASA), Winkle (ASA), Zmorph (ASA), Zortrax (ASA)
Product variation:	Molecular weight/branching, copolymer composition
Features (max. values):	Tensile modulus: 2600 MPa Elongation: 8 % Tensile strength: 50 MPa Heat resistance: <95 °C
Compounding:	Colorants, fillers, fibers, stabilizers, plasticizers, flame retardants
Processing:	Nozzle: 230–250 °C Building platform: 90–110 °C Installation space: 20 °C Speed: 30–50 mm/s

Table 4.19 Polymethyl Methacrylate (PMMA)

Core properties:	Weather-resistant, transparent, scratch-resistant
Price:	30–50 €/kg
Material manufacturer:	Arkema, Chi Mei Corporation, Evonik Industries, KURARAY, Mitsubishi Chemical, Sumitomo Chemical
Compounders/Filament Manufacturers/Product Brands:	Devil Design (PMMA), Material4Print (PMMA), Mitsubishi Chemical (PMMA), PushPlastic (PMMA), REC (PMMA), Treed (PMMA)
Product variation:	Molecular weight
Features (max. values):	Tensile modulus: 3200 MPa Elongation: 3.5 % Tensile strength: 73 MPa Heat resistance: <80 °C

Table 4.19 Polymethyl Methacrylate (PMMA) (*continued*)

Compounding:	Dyes
Processing:	Nozzle: 210–240 °C Platform: 80–90 °C Installation space: 60–80 °C Speed: 30–50 mm/s

4.1.1.2.4 Closed Value Chains From Printers and (Certified) Materials

To ensure that the properties of components meet specifications, the software (product design program, slicer), hardware (printer), and material (filament/granules) must be ideally matched. To achieve this, various printer manufacturers have developed slicers for their printers and sell certified materials with optimized slicer settings together with their printers. This concept has two manifestations:

- In the first case, the entire print chain is closed. This means that foreign materials cannot be processed on the given printers.
- Alternatively, the closed print chain can be opened up by certifying materials for the printer and slicer. This allows printer manufacturers to address different customer strata, from DIY use to professional users for the production of "functional" prototypes, and ultimately to extend the use to production applications.

In closed systems, the input material, e.g. a filament, may be controlled by chipped material cartridges, as in home office inkjet printing applications. This enables the printer manufacturer to ensure timely replenishment in addition to controlling the material qualities used, as the printing systems are often connected online with the print manufacturers. In addition to the material stock, timely updates and improvements can thus be applied, and upcoming user needs can be inferred from the print history and the processed print geometries/product applications. Possession of data on print logic, print material consumption, number of pieces, and product application is becoming increasingly important here.

Currently, there are only a few printing systems that have been developed around specific applications. However, this idea is gathering interest, as printers are increasingly used in production applications. Initial approaches in this direction are already underway [3D-Systems].

Figure 4.1 shows an overview of currently available closed systems and those that prefer to work with certified filaments. In general, the frequency of closed systems is is particularly high among 3D printing pioneers such as 3D Systems (powder, liquid systems) and Stratasys (all materials). This is due in part to the original "razorblade" business model historically, which assumed that the technology was well protected by a strong patent base [Moore 2013]. With the expiration of key patents over the past decade, this business model has been gradually replaced by open systems. The exception remains highly complex processes with integrated pre- and post-processing of products, or value chains that must be supplied from a single source for quality or insurance reasons (safety components, critical medical technology applications, etc.).

Certified materials

System provider	Printer environment		Standard plastic						Engineering plastic					High performance plastic				Short fiber reinforced			
	Closed	*Software*	*ABS*	*ASA*	*HIPS*	*PETG*	*PE/PP*	*PLA*	*TPU/TPE**	*PVA/PVBOH*	*PC/PC-ABS**	*PA**	*POM*	*PVDF*	*PEI*	*PAEK**	*PPSU/PSU*	*Glass*	*Carbon*	*Aramid*	*Other*
3DGence		X	X						X		X	X			X	X			X		
3ntr	X	X	X	X		X			X		X	X	X		X			X	X		
BCN3D		X	X			X	X	X	X	X		X						X	X		
Apium			X				X							X	X	X		X	X		
bigrep		X	X	X		X		X		X		X							X		
CEAD		X	X			X	X	X			X	X				X	X	X	X		
Cosine		X	X	X	X	X	X	X	X	X	X	X						X	X		
Essentium		X	X			X			X	X	X	X				X		X	X		
Evo tech			X	X		X	X	X	X	X	X	X	X		X		X	X	X		
IEMAI3D		X	X	X		X	X	X	X		X	X			X	X	X	X	X		
Intamsys		X	X	X	X			X	X	X	X	X			X	X			X		
Kumovis	X	X													X	X	X				
Makerbot	X	X	X	X		X		X	X	X	X	X						X	X	X	X
Multec						X		X	X	X											X
Orbital composites	nein	X				X		X				X				X			X		
Prusa		X		X		X		X		X	X										
Raise3D		X	X	X		X	X	X	X	X	X	X						X	X		
Roboze			X				X		X		X				X	X			X		
Reprap / Innovatiq			X	X			X	X	X	X	X	X				X		X	X		
Sharebot			X			X	X	X	X			X						X	X	X	
Stratasys	X	X	X					X	X	X	X	X			X		X		X		
Triditive		X	X	X	X	X	X		X		X	X	X						X	X	
Ultimaker		X	X	X	X	X	X	X	X	X	X	X	X	X				X	X		
Vshaper	X	X	X	X		X			X	X	X	X				X		X	X		

* Material class

Figure 4.1 Closed and certified printer systems/material combinations

4.1.1.2.5 Continuous Fiber Printers

As the list of companies marketing continuous fiber 3D printing continues to grow, so does the range of available materials, processes, part sizes, production volumes, applications, and end markets. There are several concepts for direct fiber integration, which are enumerated and briefly explained below [Matschinski 2020] (Figure 4.2):

- **Impregnation on-site:** Dry fibers are fed into the die, while a matrix material is injected through one or more feeders by co-extrusion. The matrix is melted and impregnates the fiber on-site before it is deposited.
- **Co-extrusion with towpreg (pre-impregnated fiber):** Instead of dry fibers, a pre-impregnated fiber (towpreg) is fed into the die, heated and co-extruded with additional matrix material. Typically, the matrix in the towpreg is identical to that used in co-extrusion. An exception is the anisoprinting process, where the towpreg matrix is thermoset and the co-extrusion is thermoplastic.
- **Towpreg extrusion:** The towpreg fiber is heated and extruded without additional material.
- **Consolidation on-site:** Essentially, in-place consolidation describes a scaled-down version of thermoplastic automated fiber placement (AFP), in which the fed thermoplastic towpreg/prepreg tape is consolidated in place as it is deposited. During the feeding process, the towpreg is first heated to the application temperature at the nozzle by an external energy source, and then placed and consolidated or cooled by a pressure roller during application.

With continuous fiber printers, there are two fundamentally different printing philosophies. The continuous fiber can be embedded in a filament before printing, or the fiber (if necessary already plastic-coated) is only specifically used and embedded in the build matrix independently of the build matrix where reinforcement is to take place.

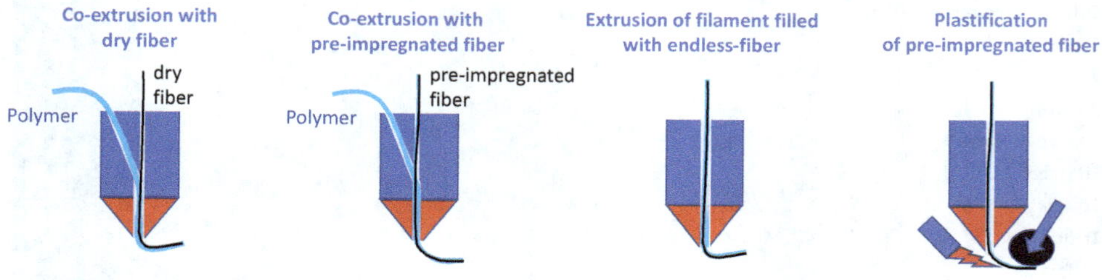

Figure 4.2 Continuous fiber processing

If appropriate print heads are mounted on five- or six-axis robots, continuous fiber printing can be combined with pellet printing. This is often done on the basis of short-fiber-filled thermoplastic blends. The great advantage of multi-axis printing robots in combination with fiber printing is that load paths can now, for the first time, also be designed and specifically reinforced in space or in the z-axis. This makes component geometries with high strength and low component weight accessible, which were previously inconceivable with classic thermoplastic-based manufacturing processes, and are in direct competition with components manufactured from metal (for example in applications from the fields of renewable energies, vehicle/aircraft construction, defense, but also sports and medicine).

The use of continuous reinforcement fibers requires high-quality materials and corresponding precision in process planning and control (Figure 4.3). In order to fully exploit the advantages of the technology, a very specific line management of the printing process must be implemented on the software side and adapted to the printer and material capabilities. This results in a typical candidate for a closed printer-material combination, as found almost exclusively in this technology.

	Certified materials																								
System provider	Printer environment		Standard plastic						Engineering plastic					High performance plastic				Short fiber reinforced				Endless fiber reinforced			
	Closed	Software	ABS	ASA	HIPS	PETG	PE/PP	PLA	TPU/TPE*	PVA/PVBOH	PC/PC-ABS*	PA*	POM	PVDF	PEI	PAEK*	PPSU/PSU	Glass	Carbon	Aramid	Other	Glass	Carbon	Aramid	Other
APS Tech Solutions		X										X										X	X	X	X
Anisoprint	X	X				X					X	X			X	X	X						X		X
AREVO	X	X										X											X		
Desktop metal	X	X										X				X						X	X		
Electroimpacts	X	X										X			X	X						X	X		X
Ingersoll		X										X										X	X		
Mantis Composites	X	X										X											X		
MarkForged	X	X										X			X			X	X			X	X	X	
Markone	X	X		X		X	X	X	X	X					X	X		X	X				X		

* Material class

Figure 4.3 Continuous fiber printer

4.1.1.2.6 Granule Printers

Fused granulate fabrication (FGF; pellet additive manufacturing, PAM) is a variant of the MEX processes that uses granules instead of filaments as the starting material. As with the FLM process, parts are built up layer by layer by extruding molten material from a computer-controlled positioned nozzle. Granule 3D printing is particularly common in large format printing, where it can exploit its raw material cost advantages due to the high material throughput possible.

In order to use pellets as the starting material, an extruder is required which, like the filament extruder, deposits molten material line by line or drop by drop (Figure 4.4). In the case of granulate extruders, which are considerably heavier than filament extruders, it may be advisable to move the build platform or even both the extruder and the build platform instead of the extruder. In this way, undesirable vibrations can be avoided.

In pellet printers, the raw material is fed to the die mainly via a single-screw-cylinder system, which can be coupled with a melt pump to further improve discharge uniformity. Basically, pellets are conveyed by gravity from a hopper into a cooled receiver to be subsequently melted and transported by a screw to a die.

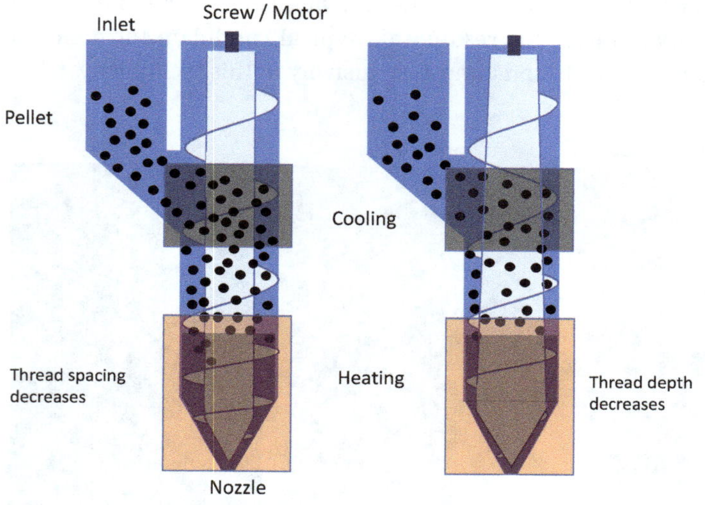

Figure 4.4 Pellet extruder

The design of the screw extruder used can be adapted very flexibly to the needs of part production. With screw diameters smaller than 1 cm combined with die sizes of 0.4–2 mm, throughput rates of 50–150 g/h can be achieved with small pellet extruders [Shaik et al. 2021]. The output volume can be scaled up to throughput rates of >10 kg/h with screw diameters larger than 3 cm and die sizes larger than 10 mm in so-called large-scale additive manufacturing (LSAM) plants.

Granule 3D printing requires either a dedicated granule printer or a portable granule extruder that can be attached to an existing multi-axis positioning machine – such as a robot arm. Granulate extrusion with its screw-barrel system allows up to 100 times higher material extrusion rates, depending on the design, due to the good heat transfer in the screw, in combination with large die diameters. However, this is then associated with sacrifices in accuracy and resolution.

Granulate 3D printing is particularly suitable for large-format objects with high wall thicknesses and high filling levels, since granulates can be fed continuously without interrupting the printing process. It should be noted that the material residence times in the screw are still comparatively long compared to classic extrusion and injection processes. This may have to be addressed by adding additional stabilizers to the plastic formulation of the granules used, or the screw volume sizes may have to be adapted to the desired throughput profile in order to minimize the residence times.

Granules are typically up to 90 % cheaper than an equivalent 3D printer filament, as granules are also by far the preferred feed option for most traditional plastic manufacturing processes such as injection molding and plastic extrusion. Many plastics are produced in granular form. The standard mass-produced thermoplastic granules used are in spherical or cylindrical form with a diameter of a few millimeters. In addition to lower raw material costs, a major advantage of pellet 3D printing is the ready availability of a wide range of already certified plastics and formulations in pellet form, including many recycled plastics.

Industrial granular 3D printing typically cannot achieve resolutions comparable to FLM. For most granule printers, the smallest achievable layer height is about 1 mm. Part complexity is limited, and overhangs are thus limited. Large die diameters in combination with large screw assemblies only allow material to be withdrawn to a limited extent during the production process between a point of sale and a starting point of extrusion. Therefore, a continuous extrusion path is particularly important here to prevent unsightly material residues due to leakage from the die.

Another advantage of screw-barrel-based extrusion is the processability of particularly soft materials such as TPE or even rubber compounds. In this way, it is possible to build highly elastic and very soft components with hardnesses of less than 80 Shore A, which can only be processed slowly or not at all in filament-based processes.

Finally, smaller granulate printers available on the market also allow the production of higher resolution objects with layer thicknesses of 0.1–0.2 mm and layer widths from approx. 0.2 mm due to the smaller screws and special deposition techniques. However, here, too, very close attention must always be paid to the temperature control and temperature load of the input materials.

In Figure 4.5 commercial granulate extruders and their suitability for different material classes are listed. As discussed before, in principle the reference of the material and the printer are mostly decoupled in granulate 3D printing.

Fused Pellet Fabrication	Printer	Extruder	Large size	Small size	XYZ	Delta	Robot	High temperature	Composite
Aectual	X	X	X				X		X
AIM3D	X	X		X	X			X	X
Arburg	X	X		X	X			X	
BLB Industries	X	X	X		X				
caracol-am	X	X	X				X		X
CEADgroup	X	X	X		X		X		X
Cincinnati Incorporated	X	X	X	X	X			X	X
discovery3dprinter	X	X	X	X	X				X
dyzedesign		X	X	X	X	X	X	X	X
colossusprinters	X		X		X		X		X
coinrobotics	X	X	X				X		X
cosineadditive	X	X	X		X				X
epeire3D	X	X		X	X			X	X
everplast	X	X		X	X				
filamentinnovations			X		X				X
hage3D	X	X	X	X	X*				X
hansweber	X	X		X	X		X		X
IEMAI3D	X	X		X	X			X	X
juggerbot 3d	X	X	X		X				X
Magna Recta ARKI	X	X		X	X				
millebot	X	X	X		X				X
periplast									
Pollen.AM	X	X		X		X		X	X
re3D	X	X		X	X				X
scmgroup	X	X	X						X
thermwood	X	X	X		X				X
titan3drobotics	X	X	X	X	X				X
Tumaker NXPro	X	X		X	X				
3dwasp	X	X	X	X		X			
massivedimension		X	X						
dohle-extruders		X							
direct3D		X		X					
UPE v4 pellet extruder		X		X					

* Movable plattform enabling 5 axis printing or endless printing on conveyor belt

Figure 4.5 Granule printers – designs and sizes

4.1.2 Powder-based Processes (SLS, HSS)

The materials available for powder-based processes are still very manageable compared to extrusion-based processes, but are increasing continuously as new sintering processes are developed. In order to be able to process a material in the sintering process, it is first necessary to convert it into a suitable uniform powder form. If a suitable powder is available, then it is necessary to check whether the powder can be processed sufficiently robustly in the existing powder sintering processes. Since both the powder and the associated additive powder sintering processes are significantly more cost-intensive than the extrusion-based processes, there is virtually no do-it-yourself (DIY)/hobby scene in this regard. In comparison, the number of open devices that can process various powder materials from different sources is also lower.

Powder printing is – even more than the extrusion-based processes – on the threshold of the transition from prototype to production process. However, the number of suppliers of suitable equipment and material combinations is significantly lower. Even in powder-based processes, the dwell time and temperature of the receiving layer in combination with the energy transferred into the last applied layer by laser (SLS) or IR radiation (HSS) are responsible for the adhesion between the layers. For each material, there are theoretical ideal process parameters dictated by the intrinsic material properties of the thermoplastic (fast or slow crystallization, high or low viscosity, high or low glass point).

These ideal parameters usually do not perfectly match the real process conditions (build room temperature distribution, energy absorption), since these themselves depend on the build-job or the number of products and the individual product geometries (layer thickness, wall thickness, high or low layer speed), the conversion to the print codes to be executed, as well as the printer capabilities (laser power/number of lasers/heating power/absorption coefficients of the matrix or inks, the type of powder coating process, build room temperature, traversing speed, etc.).

Ideally, both process and material are perfectly matched. Most genuine production solutions for the manufacture of functional components are therefore based on proprietary solutions that provide clear specifications for the material and the printing process, as well as for the permitted printing geometries.

In powder sintering processes, a distinction is made between two major process types (Figure 4.6):

1. The **laser sintering process (SLS)**: Here, one or more lasers selectively melt lines into a powder layer, which is repeatedly renewed by fresh material in the z-direction. The process is thus somewhat similar to the FLM (FFF, FDM) or SLA processes, whereby the line speed and the number of lines created simultaneously in a layer determine the build speed.

2. The so-called **high-speed sintering processes (HSS)**: In these processes, a selected part in each powder layer is selectively printed with a radiation absorption ampli-

fier (usually a pigmented ink). This allows selective sintering of the printed areas in a single process step under a planar radiation source, before the powder layer is replenished with fresh powder. The process is thus more similar to DLP printers for liquid systems, where entire layers can also be exposed in one step.

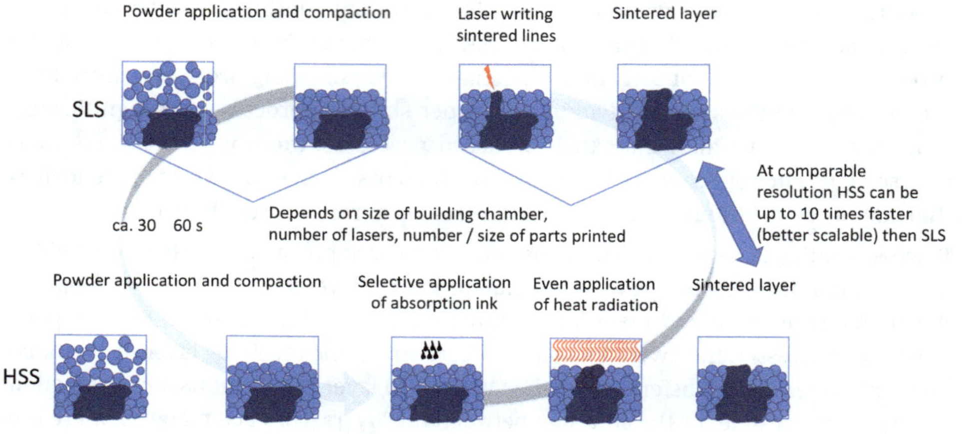

Figure 4.6 Comparison between SLS and HSS processes

Since the processes mentioned place different requirements on the plastic powders, not every process is compatible with every material. In the following discussion, the value chain of powder-based processes is presented, as well as an overview of printer suppliers, the printing technology, and their open and closed and/or certified materials.

4.1.2.1 Building Materials

Standard and engineering plastics in particular are fundamentally well suited as raw materials for sintering in additive processes due to their low melt temperature and low thermal conductivity. The low thermal conductivity makes it possible to localize the introduced sintering energy well and thus to achieve a high component resolution. However, the low thermal conductivity also makes it difficult to achieve uniform temperature control of the build space through the build space wall, which is why the build space size is comparatively limited in SLS processes.

For sintering processes in AM, particles with a diameter of 20 µm to about 100 µm are preferred. Below 20 µm, the risk of dust explosions increases disproportionately due to the low density of the polymers. Above 100 µm, the resolution of the components suffers visibly and tangibly. Particle size distribution can be controlled by sieve classification. This yields clean fractions with grain fractions that are typically 90 % equal to or smaller than the nominal diameter.

If amorphous plastics (such as PS, PMMA, or PC) are processed in SLS or HSS processes, then due to the high viscosity of the polymers, often only the surface of the powder particles is melted directly above the glass point, and components with good resolution but also slightly rough and porous surfaces are obtained. Functional components are only available to a limited extent from amorphous materials without additional post-processing – for example, by infiltration with reinforcing resins or post-sintering.

Crystallizing plastics are more suitable for a sintering process to produce functional components, as they experience a greater viscosity reduction above the melting point. The molten particles flow together, and smoother surfaces as well as denser components tend to be obtained. When cooling the melt, especially of highly crystalline plastics such as (co-)polyamides, polyesters and TPUs, if the temperature control is not ideal, the surface may collapse or stand up due to the crystallization shrinkage (pitting, curling, etc.). This can result in rough surfaces, or disruptions in the build process (lifting and shifting of layers), e.g. due to the build-up of sintered surfaces (curling). Due to the only slight pressure acting on the component surfaces in the building and sintering process, somewhat lower material densities tend to be achieved than with (injection-)molded plastic components. Similar to metal sintering processes, downstream post-sintering processes, if necessary under pressure in an inert medium, can increase component density, reduce porosity and significantly increase strength [Achten et al. 2020]. However, it is essential to take the resulting forced consolidation of the component into account in the design.

Thermoplastics can theoretically be melted and re-solidified any number of times. However, depending on the mode of operation of the additive sintering process, even the non-sintered support areas experience a certain amount of thermal damage/aging and therefore cannot always be fully recycled into the build process. This is primarily countered by flooding the construction space with inert gas (N_2, CO_2, argon) during the construction process, in order to largely suppress oxidative aging and discoloration under the influence of atmospheric oxygen. Nevertheless, the non-sintered powder material from the building process must often be mixed with fresh powder, so that ideally a stable material quality results from fresh and old powder. A 0.5:1 to 1:1 mixture of old powder to new powder is considered a guide value. The exact ratios of the powder mixture depend on the thermoplastics used, process parameters, and build times, and are essentially determined empirically.

Compared to HSS processes, SLS processes allow the processing of a wider range of materials, in particular the processing of high-melting materials. HSS processes are characterized by a temperature control that is particularly gentle on the material, since it is primarily the surface temperature that is adjusted here. Many of the powder materials processed in HSS processes, such as PA12 and TPU, therefore allow a particularly high return rate of the non-sintered powder into the process of 80 % up to 100 %.

The plastic powders are often still formulated with stabilizers, flow aids, and fillers before the powders are manufactured and/or before the powders are processed in additive processes. The formulation with fillers is limited to those that do not negatively affect the flow and sintering properties of the plastics. Otherwise common fiber-reinforced plastics are largely excluded from this.

The subsequent filling of powders with flow aids and fillers must also be carried out with care. Improved flow of the powders is often accompanied by significantly poorer product properties, since the fillers tend to hinder the coalescence of the melted thermoplastics and additionally represent starting points or defects for early crack growth under load. If the powders are to be filled to a higher level or with critical fillers, this is ideally done before the powders are produced. However, the choice of fillers in turn limits the choice of powder manufacturing process, and has a significant influence on particle geometry and particle size distribution.

The most important processes for producing powders for AM are briefly explained below.

Suspension polymerization

If polymers are produced in a suspension polymerization, then perfectly round particles of a certain size can be produced in a controlled manner by adjusting the number and size of the particles via suitable protective colloids and emulsifiers. This process is used almost exclusively for amorphous polymers with a high T_g (>50 °C), such as in the production of suspension PMMA and PS.

Precipitation process

Polymers can be dissolved in a solvent under the action of heat and then precipitated by selective cooling or the addition of a non-solvent. Depending on the choice of conditions, powders are formed which often have a nice spherical to potato-shaped particle form. In principle, this process is applicable to many amorphous and crystalline polymers. The best known application is the production of PA12 and PA 11 powders. A special form of precipitation process is precipitation polymerization, where polymers of a certain degree of polymerization become insoluble in their solvent during polymerization and precipitate. However, controlling the correct molecular weights and powder sizes via solvent composition is difficult. Typically, therefore, polymer powders obtainable from precipitation polymerizations are once again post-processed in a controlled manner into a solution precipitation process.

Spray-drying

Polymers in solution can be subjected to spray-drying. In this process, polymer solvent droplets are finely dispersed in a chamber by an "atomizer". Conditions prevail in the chamber (temperature/vacuum) at which the solvent evaporates rapidly. However, the prerequisite for such a process is that the resulting precipitated polymers reach the bottom of the spray dryer dry and non-sticky, to avoid caking of the particles. Overall, this process is widely used in the food industry with water-soluble polymers in combination with cooling towers, but is rarely used in the production of powders for AM.

A special process that is somewhat similar to spray drying is the mechanical atomization/droplet generation of low-viscosity polymer melts on a rapidly rotating roller or jet disperser. The fine melt droplets generated are then rapidly cooled to produce stable non-sticky particles, which can subsequently be collected.

Cryogenic grinding

In the cryogenic grinding process, polymers are cooled below their T_g and mechanically comminuted in suitable grinding units until a desired maximum particle size is achieved. Powder that is too coarse is filtered off in a sieve and returned to the process. Powder that is too fine is removed, for example, by air classifying. The result is mainly fractal powder surfaces, from which lower bulk densities and component densities as well as poorer flow properties follow. The process is basically applicable to a wide range of materials. Most elastic materials such as PEBA and TPU are processed into powders in this way.

Melt dispersion processes

In the melt dispersion process, a polymer melt is dispersed into an inert liquid (waxes, silicone oils, liquid polyethers) under high shear, for example in an extruder or a special disperser. The polymer melt is cooled, and the resulting solid particles are filtered off and freed from the dispersing liquid by washing/distillation. Spherical to potato-shaped powders are formed, similar to the precipitation process. The method is mainly used for the production of comparatively low-melting materials with a high drop in melt viscosity, such as TPU. The high drop in melt viscosity above the melt temperature is advantageous for dispersion and droplet formation. However, the temperature must not be selected too high to avoid decomposition processes.

The respective process conditions for the production of powders for AM must be specifically developed and optimized for each material and are mostly trade secrets of the manufacturers. In principle, suspension polymerization and precipitation processes allow the least formulation and product variations. On the other hand, the powder material PA12, which dominates AM with powders, is produced almost exclusively via a precipitation process. The most robust for formulation and product variations is the cryomilling process, but this also delivers the poorest powder qualities in a direct comparison [van den Eynde 2018].

4.1.2.2 Value Chains From Printers and (Certified) Materials

Thermoplastic powders for SLS and HSS processes are supplied by plastics manufacturers, formulators, powder manufacturers and printer manufacturers in the same way as for filament processes (Figure 4.7). In the case of materials from precipitation processes, the plastics manufacturers dominate, since complex chemical process technology is used here in comparison. These pass on the materials in part as “white label” powders to printer manufacturers, who then distribute them as certified materials with the correct parameter sets for their printers.

Cryogenically ground powders are largely manufactured by a few professional cryogenic mills and then distributed either directly as contract manufacturing by the material manufacturers, formulators, or also printer manufacturers. If special processes such as melt dispersion processes are used, or if the plastics are specially formulated beforehand, powders from independent sources may also be available, which are then in turn distributed either via printer manufacturers or directly as raw materials for open printer systems.

Powder provider	Process				Engineering plastic				High performance plastic		Filled		
	SLS	HSS	*Polyester**	*PE/PP*	*TPU/TPE**	*PC*	*POM*	*PA*	*PAEK**	*PPSU/PSU**	*Glass*	*Carbon*	*Other*
A. Schulman	X							X			X		
ALM (EOS)	X			X	X			X			X	X	X
AM-Polymers	X		X	X	X			X					
Arkema	X							X			X	X	X
BASF	X	X		X	X			X					X
Covestro	X	X	X		X								
CRP Technology	X			X	X			X			X	X	X
Diamond Plastics	X			X									
Eastman	X				X								
Evonik	X	X			X			X	X				
ExcelTec (Prodways)	X							X					
Hexcel	X								X			X	X
Huake3D	X			X	X			X					
Igus	X						X						
Launhardt	X	X			X								
Lehmann&Voss	X	X		X	X			X					
Lubrizol	X				X								
NeuMat Shanghai	X				X								
Oxford Performance Materials	X								X				
Sabic	X					X							
Solvay	X							X					
Tiger coatings	X												
Toray	X		X							X			

* Product family

Figure 4.7 Manufacturers of powders for SLS and HSS processes

However, especially for the SLS and HSS processes, sales via the printer manufacturers still dominate, since only precise process control in combination with high-quality material leads to good components (Figure 4.8). With the combination of open printers and free powders, the processor must first develop their own parameter sets at great expense, which can however be quite worthwhile for large contract manufacturers of 3D-printed products. While prices in the filament range for high-quality standard and construction plastics are €20–60/kg, the price range in the powder range is €50–150/kg. There may well be price differences of up to 50 % between open and certified powders, as the actual costs of powder production of around €2–10/kg, depending on the product and process, are not reflected by the price jump from raw material or formulation to powder.

System provider	Process		Printer	Print environment		Engineering plastic						High performance plastic		Filled		
	SLS	HSS		closed	Software	Polyester*	PE/PP	TPU/TPE*	PC	POM	PA	PAEK*	PPSU/PSU	Glass	Carbon	Other
3D Systems	X		X	X	X		X	X			X					
Aspect	X		X	X	X			X			X					
Beijing Long Yuan	X		X		X											
Dedibot	X		X		X											
EOS	X		X	X	X			X			X			X	X	X
Farsoon	X		X		X			X			X		X			
Formlabs	X		X	X	X						X					
HP		X	X	X	X		X	X			X					
LSS LaserSinterService	X		X		X		X	X			X	X		X		
Natural Robotics	X		X		X						X					
Nexa3d	X		X		X	X					X					
Prodways	X		X	X	X		X	X			X			X	X	X
Red Rock 3D	X		X		X											
Ricoh	X		X	X	X		X				X					
Sentrol	X		X		X						X	X				
Sharebot	X		X		X						X					
Shining 3D	X		X		X		X	X			X			X	X	
Sindoh	X		X		X		X	X			X					
Sinterit	X		X		X		X	X			X					
Sintratec	X		X		X			X			X					
Sondasys	X		X		X						X					
Tecnica	X		X		X						X					
TPM3D (Stratasys)	X		X	X	X			X			X					
Voxeljet		X	X	X	X			X			X					
Weirather	X		X		X		X	X			X					
Wematter	X		X	X	X		X				X			X	X	X
Wuhan Huake	X		X		X		X				X					
Xery	X		X		X						X					
XYZprinting	X		X	X	X			X			X					X
ZRapidTech	X		X	X	X			X			X			X		X

* Product family

Figure 4.8 Manufacturers of powder printers working according to the SLS and HSS processes

4.2 Products and Applications

According to a survey conducted in 2018 by the French AM service provider Sculpteo, the extrusion process (with 46 %) and the sintering process (with 38 %) are the most widely used AM principles worldwide (see Section 1.2.1). The following sections therefore present various applications from different sectors which use these additive processes.

4.2.1 Prototypes and Sample Components

Historically, AM processes were primarily used for the production of prototypes or samples and were superior to classic manual production in terms of time and costs. In the following sections, applications are presented that are not yet used commercially, and which demonstrate the multi-layered usability of additive extrusion.

4.2.1.1 Ventilation Grille

In principle, it is possible to create complex structures with FLM. However, the achievable fineness is very low compared to textile membranes. Since the ventilation grille presented in this section involves two-dimensional structures that are primarily generated in the xy-orientation for process-related reasons, the achievable resolution with FLM is limited by the nozzle diameter and the resulting extrusion. Furthermore, textile membranes are much more permeable to air than fine, additively generated structures, which also have high production times and thus high costs due to the complexity of the paths to be traced.

One of the strengths of FLM is that semi-finished products, for example, can be integrated in the building process. In most cases, this requires the design of a receptacle or connection in the part to be manufactured and the definition of a layer height at which the build process is stopped and the semi-finished product is inserted. After insertion of the semi-finished product, the building process is continued and the semi-finished product is thus integrated in the additively produced component. The design of the holder should be selected so that the inserted semi-finished product is flush with the defined layer height. If this is not the case, damage to the printing unit may occur in the further course of the building process. FLM systems with open installation spaces are advantageous for the integration of heat-sensitive semi-finished products (see Section 1.4.2.1). According to this principle, textile membranes can also be integrated into additively generated components to produce efficient ventilation grilles.

4.2.1.2 Large-format AM in the Design Process

The Villeroy & Boch company in Mettlach uses a large-volume FLM system for the development of bathroom ceramics. The system is from the Berlin-based manufacturer

BigRep. The digital 3D models built up in the design process must be converted into a scaled and polished sample for acceptance by decision-makers. Classically, these samples were built up by model makers using wood, foam and polyester. The analog process often resulted in samples that did not exactly match the designers' specifications. By using the BigRep ONE with a build volume of one cubic meter, the accuracy of the sample could be maximized and the manufacturing time for a bathtub sample is reduced to three weeks (Figure 4.9). The samples for the decision-making process must have the same surface finish of the targeted end product. Therefore, the surfaces of the additively manufactured sample must be smoothed over approximately three cycles by analogous application of putty and grinding. The samples are then spray-painted. The classic analog production of the true-to-scale bathtub samples originally took 6–7 weeks, with costs around five times as high.

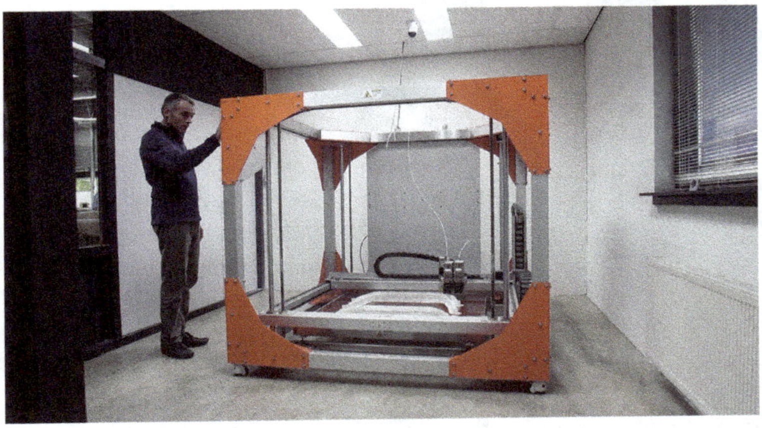

Figure 4.9 The BigRep ONE FLM system producing sample parts at Villeroy & Boch [BigRep 2022b]

4.2.2 Procedure with Dead-mold Casting

In the dead-mold casting process, the components to be produced are manufactured as a positive model in a material that can be dissolved primarily by the target material. For this purpose, the target material must be in liquid form. By dissolving the positive pattern and solidifying the target material, the actual component is produced in the casting process step. In addition to the lost-wax technique, casting with lost foam is also a common process for producing cast metal components. AM primarily uses positive wax samples, which are used in jewelry manufacturing (see Section 1.6). The wax is applied with high precision in drop form, with the respective layer being additionally leveled by a rotating milling blade. The layer thickness for these systems is 6–38 µm. In the following sections, examples are given which adapt the dead-mold casting for additive extrusion.

4.2.2.1 Soluble FDM Molds

Using FDM sacrificial tooling, the Stratasys company was able to develop a brake duct inlet intended for racing and produce it in hybrid AM. The component's function on the vehicle is to guide air from the front bumper to the brake caliper and to counteract the braking heat generated there. It was important to ensure that the air duct in particular had a smooth surface. FDM sacrificial tooling is particularly suitable for creating fiber-based composite applications in which the lamination process encloses the entire mold (see Section 1.6.1). It is thus advantageous to implement complex pipes or channel structures using this strategy (Figure 4.10).

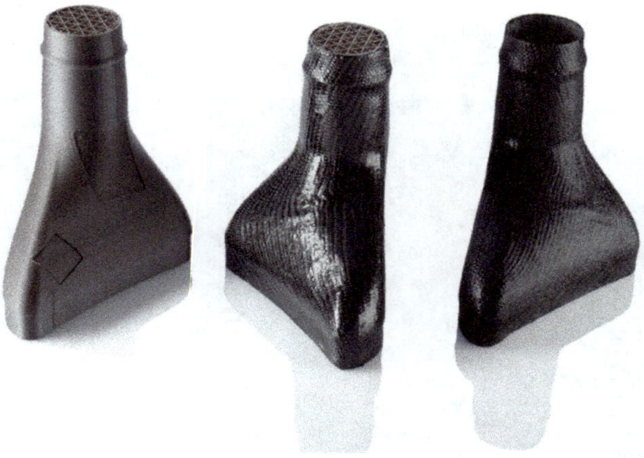

Figure 4.10 The three basic steps to the part in FDM sacrificial tooling (left to right): additively manufactured positive mold; laminated mold; and part after mold dissolution (© Stratasys Ltd [Stratasys 2021])

Table 4.20 FDM Support Materials for Dead-molds in Combination with Composite Applications

Support material	Maximum curing temperature (°C)	Glass transition temperature T_g (°C)
SR-30	80	110
SR-100	120	132
ULTEM Support	165	185
ST-130	121	132

Hybrid AM eliminated the need to create a two-part mold and join the halves of the part. The dead-mold for the component was additively manufactured on a Fortus 900 mc in 9.75 h. In this process, the mold is created using the ST-130 thermoplastic FDM support material. This thermoplastic material dissolves in a basic solution (pH >7). In addition to the ST-130 support material, three other support materials are suitable for the generation of dead-molds in combination with composite applications (Table 4.20). The lamination process is simplified by using a positive mold. Additional tools, which would be necessary with negative mold variants, are thus unnecessary. The composite component can be cured at a maximum of 98 °C and 90 psi when using the ST-130 support material. If a higher temperature is required, for example, multiple curing cycles can be used.

After creating the digital solid model of the dead-mold to be produced with closed openings and their extension, trim lines, and positioning marks, the digital preparation required for manufacturing can be carried out in the system control software in two ways:

1. Sparse style
2. Shell style.

With the sparse style strategy, the interior of the digital solid model is automatically converted into a three-dimensional structure. This strategy offers an optimum ratio between the manufacturing time and the load-bearing capacity of the mold. In contrast, the shell style strategy creates a wall in the digital solid model. This produces a thin-walled mold that is not as resilient. This strategy is preferably used in combination with an inflatable bladder or vacuum bag. The second type of digital preparation is particularly suitable for pipes or components with channels. If the mold to be produced cannot be manufactured in one piece, it can be digitally divided into several parts. These parts can then be manufactured individually and subsequently joined using hot air and the appropriate thermoplastic filament. If a support structure is required during the AM of the mold, this is created using a special support material. For the ST-130, this is the ST-130S, which must be mechanically removed from the mold. To facilitate this, the mold and the support structure should be heated to approximately 100 °C. If a higher surface quality should be achieved, the mold made of ST-130 can be milled before the lamination process. Furthermore, the mold should be sealed to prevent premature dissolution by the resin and its penetration into the mold. The Stratasys company recommends Zyvax Quickskin for sealing.

FDM sacrificial tooling can be used with the following three consolidation methods:

1. Vacuum autoclave
2. Heat-shrink tubing
3. Shrink band.

After curing of the composite, the additively generated mold is dissolved in a heated and circulated basic liquid, as in other soluble FDM support materials [Fischer et al. 2018].

4.2.2.2 Sustainable Disposable Coffee Cups

Starbucks, a company specializing in coffee products, reported that 2.6 billion disposable cups were used to sell coffee in 2006. The production of a paper-based disposable cup generates approximately 0.11 kilograms of CO_2, and only about 0.25 % of these cups are recycled after disposal. Inspired by boxed watermelons, which are given a rectangular shape for more efficient transportation, Jun Aizaki developed HyO-Cups. These disposable, biodegradable cups are created from fast-growing pumpkins combined with a translucent plastic mold made via FDM. Unlike the dead-mold casting, this plastic mold can be used multiple times. The pumpkin was chosen to have a strong outer skin and fibrous flesh. The pumpkins grow in the molding, additively created hollow mold. The resulting pumpkin is dried after the pulp is removed and can then be used. The HyO-Cups are easy to compost after use and are therefore also a circular product.

4.2.2.3 Hybrid AM Aircraft Seats

AM enables the production of complex geometries that are not or hardly representable with classic manufacturing technologies. This high degree of design freedom often makes it possible to combine it with generative design. Generative design as a digital design method uses specific software solutions or adapted programming of modeling kernels with sets of rules or algorithms. This systematic approach allows many different and complex design combinations to be generated automatically in a short time. Partially algorithms are used, which were compiled during the research of biological systems. Such algorithms are especially used in the research discipline 'artificial life' (AL). This research discipline was first named in 1986 by the American biologist Christopher Langton.

The combination of AM and generative design is also often used to create load-optimized lightweight structures, such as the aircraft seat frame presented in this section. The final aircraft seat frame should be produced in a metallic material. Complex geometries in combination with structures cannot be optimally produced via additive processes in the metal sector. Negative factors in this context can be the increased manufacturing time and the high post-processing effort. Other criteria that suggest hybrid AM as the optimum solution are the limited choice of materials and the component sizes that can be produced with additive metal-based processes.

Hybrid AM with investment casting can be a solution for such components. Here, wall thicknesses of up to 0.5 mm, high surface quality, and high dimensional accuracy are achievable. Investment casting is certified for applications in the automotive sector, medical technology, and aerospace. Investment casting is based on the principle of the dead-mold casting, whereby this mold consists largely of a wax sample that is provided with a shell by repeated dipping in a ceramic slurry. During the casting process, the wax model is burned out, creating a cavity that is then filled with molten metal. In addition to high-performance aluminum, magnesium, stainless alloys, and high-alloy nickel alloys are also used in investment casting. Depending on the complexity and volume of the component, a lead time of between two days and 36 weeks may be required.

Another similar solution is sand casting by means of additive generation of the sand molds. Sand casting is used in particular for larger components due to its lower cost structure. Here, the tolerances are lower and the surface rougher than in investment casting. Classic sand casting is widely used in mass-production of automotive parts and structural applications. The necessary sand mold consists, on the one hand, of fine sand and epoxy resin or similar binders. During the metal casting process, the epoxy resin is decomposed by the heat generated. The remaining fine sand can then be easily removed from the component.

Hybrid AM with metal casting technologies is primarily used for prototypes, small quantities and complex geometries. On the additive process side, stereolithography, SLS and inkjet-based additive systems with wax or binders are used. In commercial aviation, lighter aircraft are the key to lower fuel consumption and thus to reducing CO_2 emissions.

With the aim of achieving weight reduction in aircraft seat frames, the Autodesk company developed a lightweight aircraft seat frame at its technology center in San Francisco. The company's own Netfabb software was used to generate the load-optimized internal lattice structure required for this purpose. Classic aircraft seat frames are made of aluminum and are produced by casting, milling or assembling individual parts. The production of the digitally generated lightweight aircraft seat frame would not be possible via classical manufacturing or direct AM in metal. In cooperation with the investment casting company Aristo-Cast in Almont, the first step was to realize a section of the lattice structure using wax printing and magnesium material. During the evaluation of this sample, it was found that some struts of the lattice pattern had to be thickened in order to be cast optimally. The next step was to produce the investment casting mold of the entire aircraft seat frame on a VX 1000 from Voxeljet using Polymethyl Methacrylate (PMMA) powder and binder. This thermoplastic has the excellent burnout properties required for investment casting. The final cast aircraft seat frame was produced within four weeks and weighed about half as much as a conventionally manufactured frame, while the magnesium material also weighed 35 % less than the aluminum conventionally used. The hybrid additively produced aircraft seat frame could theoretically offer an airline with a fleet of approximately 100 commercial aircraft a potential saving of $200 million (Table 4.21). Over the operational span of the fleet, there would still theoretically be an emissions reduction of around 126,000 t of CO_2.

Table 4.21 Fuel Savings and Emissions Reduction From Hybrid Additive Aircraft Seat Frame [Autodesk, Inc. 2017]

Parameter	Airbus A 321 – 236 seats	Airbus A 380 – 615 seats
Weight saving Per aircraft	214 kg	557 Kilogram
Fuel economy Annually per aircraft	9.6 t	63 t
Reduction of CO_2 emissions Annually per aircraft	28.9 t	190.1 t

Table 4.21 Fuel Savings and Emissions Reduction From Hybrid Additive Aircraft Seat Frame [Autodesk, Inc. 2017] (*continued*)

Parameter	Airbus A 321 - 236 seats	Airbus A 380 - 615 seats
Financial savings Annually per fleet of 100 aircraft each	$1,569,365	$10,332,446
Financial lifetime saving Over 20 years per fleet of 100 aircraft each	$31,387,300	$206,648,920
Lifetime reduction of CO_2 Over 20 years per fleet of 100 aircraft each	57,800 t	126,000 t

4.2.3 Functional Components

AM is not a viable manufacturing solution for every component. Only if the part to be manufactured has one of the following four specifications should AM should be considered as a manufacturing process:

1. Integration of functions
2. Complex geometry
3. Personalization
4. Small batch series.

If the component has combinations of these specifications, the probability of meaningful use of AM increases. Furthermore, there is also the possibility that certain combinations of the specifications mean that the component can only be manufactured additively.

In principle, mechanical integration of functions in components can be implemented more successfully using processes based on the principle of sintering. However, processes based on the principle of extrusion can enable multilayered and complex functional integration, in particular through hybrid AM and material combinations.

4.2.3.1 Silent Underwater Propulsion

The technical problem to be solved by the development of the octopus siphon actuator (OSA) presented in this section was the search for an effective and silent underwater propulsion system. The actuator was intended to minimize negative impacts on the underwater environment and replace classic propeller propulsion with biologically inspired locomotion in and under water. The biological model of the actuator to be developed was the octopus. Through a biomimetic top-down process, the siphon used by octopods primarily for escape was technically adapted through FLM. The siphon in the octopus can be described as a compliant water pumping system. The OSA consists of a thermoplastic polyurethane bellows that performs contraction and relaxation by means of an integrated and internal cable structure. This cyclic movement draws water into the bellows via an intake opening and the contraction causes it to be ejected as a jet via an outlet nozzle. This effect makes it possible to achieve controllable locomotion. The actual movement of the bellows was realized in the demonstrator using miniaturized hydraulic pistons. These could be inserted into intended holders by turning the additively manufactured actuator inside out. The demonstrator presented at Hannover Messe 2013 was equipped with four OSAs whose output nozzles were aligned in the same direction. Directional changes can be made by specifically attached actuators or by rudders (Figure 4.11).

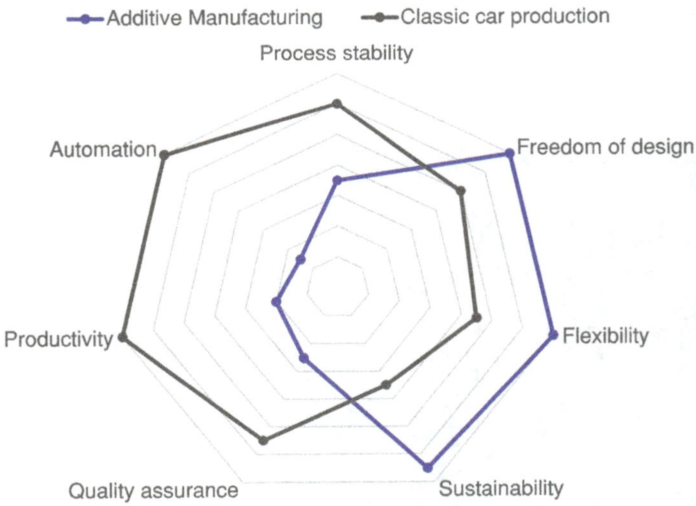

Figure 4.11 Schematic representation of a watercraft with three OSAs with different orientations (© Andreas Fischer)

The actuators of the demonstrator are also additively generated multi-material components. The outlet nozzles and the connectors to the watercraft were manufactured from PC (Makrolon 2808) using FLM. The bellows and the integrated cable structure of the actuator, in turn, were made of thermoplastic polyurethane (Desmopan DP 385) [Fischer et al. 2018].

4.2.3.2 Integration of Optical Fibers (Clear Cast Integration)

The starting point for 'clear cast integration', the integration of light-conducting materials in components produced by FDM, was the intensive use of LEDs in the automotive environment. In exterior vehicle lights, heat stress reduces the life of LEDs to

15,000–45,000 h. As a result, individual LEDs may fail in the assembly. However, German legislation does not provide for the replacement of individual defective LEDs. In principle, the relevant light module must be replaced once a certain number of defective LEDs has been reached. Another aspect that promoted clear cast integration was that LEDs require a larger overall housing for the necessary control electronics and cooling for the same luminous flux. The goal of optical fiber integration in additively produced components was a sustainable solution that could achieve similar degrees of freedom and a comparable appearance to LED modules with the use of a small number of light sources. In addition to the simplified replacement of defective light sources, the necessary overall package should also be smaller than a comparable one with LEDs. Optic fiber integration in components produced by FDM can be achieved either by semi-finished profiles like round cords or by filling a channel structure with cast material. If the optical fiber integration is implemented via semi-finished profiles, it is possible to integrate them directly in the AM process or to use them in a downstream step. However, the complexities of the channel structure that can be achieved by this strategy is very limited (Figure 4.12).

Figure 4.12 An opaque and compliant polyurethane-based round cord in an ABS specimen created via FDM for subsequent integration (© Andreas Fischer)

If the optical fibers are integrated via cast material, complex internal channel structures can be used. At certain levels of complexity, such structures can only be manufactured using AM, and they then offer a high degree of design freedom. In this variant, the optical fiber cast material is filled into the channel structure in a downstream step. A µPrint SE from Stratasys was used to manufacture the housing and the integrated complex channel structure of the prototype. Ivory-colored ABS was used for the housing and dissolvable support material. A reactive, two-component polyurethane-based casting material was used as optical fiber. This material can cure to a hard, soft or gel state at room temperature, depending on the mixing ratio. An opaque or transparent material state is also possible. The light transmittance of the UV-stable, flame-retardant and explosion-proof polyurethane material is around 90 %. The material has an IP67 protection rating and is therefore dust- and waterproof. In the prototype, the casting material was filled via an additively generated funnel at the light input opening, which could be removed mechanically after the process. The light exit side of the prototype was temporarily sealed with a film for the casting process to prevent leakage of the casting material. In the subsequent test of the prototype, a light-emitting diode lamp with 74,100 lux was used. However, only 311 lux could be measured at the light output side of the prototype. The reason for this, in addition to the small bending radii of the additively generated internal channels in some places, is the material used to generate the housing and the channels, which does not have the necessary reflective properties. An interesting side-effect of this choice of material was that the ABS housing resulted in light transmission, which made the internal channel structure from the outside visible when illuminated [Fischer et al. 2018].

4.2.4 Series Components

AM does not make sense for every application. If the applications exhibit certain typical indicators, an assessment should be made with regard to AM depending on the manifestation, frequency, and combination of these indicators (see Section 4.2.3). The probability for the use of AM is increased especially with combinations of the indicators, and with certain characteristics and or combinations of the indicators, it is possible that manufacturing via AM is the only option. If the focus is set on serial production, the indicators unit numbers and integration of functions in particular play a role. Integration of functions via AM can reduce assembly work in series production. If simple components are manufactured using FLM on low-cost systems and appropriate materials, unit costs can also be significantly reduced. Mechanical integration of functions can also be produced in principle by means of additive extrusion (Figure 4.13).

Figure 4.13 Two-color keyboard with integrated spring mechanism manufactured from ABS using FDM (© Andreas Fischer)

The resilience and functionality of such additively generated mechanical solutions strongly depend on the material used. In comparison, mechanical integration of functions are more often generated by SLS than by additive extrusion.

4.2.4.1 Functional Tool Housing

The starting point for the AM of a tool housing of the company WS Wieländer+Schill Engineering in Tuningen was a manual deburring system with a flexible shaft in combination with a two-hand safety device. The annual sales volume was estimated at 100 units for this system. As a result, AM of the housing was considered and tested. In the development of the deburring system and its housing, the targeted use of FDM drastically reduced the complexity of development and shortened the time to market. The AM prototype of the housing could be used for verification and validation at an early stage, and by using AM, the housing could be designed and manufactured in one piece. This approach allowed the necessary assembly to be reduced to sliding in the components and securing them with screws. Light-emitting diodes are installed in the front area of the deburring system, which optically signal the active state. In order to conduct this light signal through the surrounding housing, the front part of the housing was manufactured in white and the rest of the housing in blue (Figure 4.14).

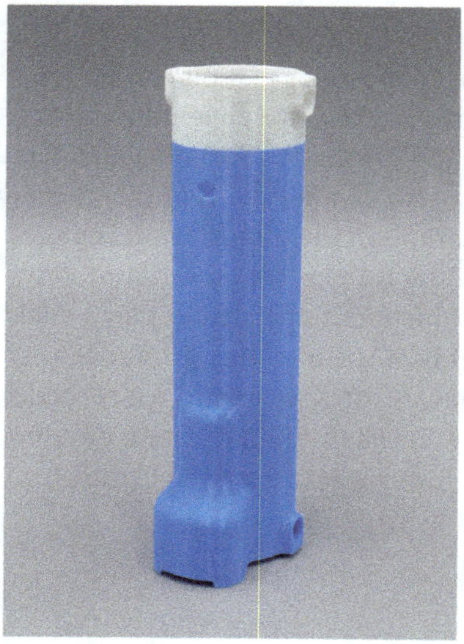

Figure 4.14
Two-color housing of a handle for a deburring system produced by means of FDM (© Andreas Fischer)

With FDM systems, a color change can basically be made in each layer. In this case, the systems are put into pause mode after completion of a layer and the model material is changed according to the color request. This change increases the manufacturing time and thus also the manufacturing costs for a multicolored component. In order to realize the appropriate color scheme for the housing, it had to be manufactured with the long side in the z-direction. The positive effects of this orientation of the housing during production were as follows:

- Reduced need for support structure
- Time reduction for the removal of the support structure
- Increased slip protection due to specific surface structure.

The positive effects of the specific production of the housing were countered by the following negative effects:

- Increase in manufacturing time
- Visible seam formation.

Seams are thermoplastic accumulations created by consistent starting points of extrusion for each layer. These seams can be easily removed mechanically, but further treatment of the entire surface of the component must then also be carried out to achieve a uniform surface appearance. The housings of the handle and the matching shaft adapter were produced using an HP Designjet Color 3D Printer with a z-resolution of 0.254 mm. ABSplus was used as the material for both components. The manufacturing costs of the handle housing were around €160. For the shaft adapter, the cost was €46. Using the FDM system, nine housings could be produced in 82 h or 12 shaft adapters in 27 h in a single production run. In the process, 28.38 cm^3 of ABS was used for the housing and 44 cm^3 of thermoplastic material for the support structure. The shaft adapter required 16.03 cm^3 of ABS and 3.82 cm^3 for the support structure [Fischer et al. 2018].

4.2.5 Furniture

While customized mass-production (mass-customization) has already established itself as the norm, the current trend is towards "hyperpersonalization". Increased data collection and better algorithms allow products to be tailored to the customer in ever greater detail. In thermoplastic processing, AM is playing an increasingly important role with regard to the production of personalized plastic components (see Section 4.2.4). In the furniture sector, all three personalization strategies of series production can be found:

1. *Diversity of variants* (series products with many variants)
2. Individualization (individualized products)
3. Custom manufacturing (special and one-off productions).

In particular, the custom manufacturing strategy is experiencing a rebirth – driven by hyperpersonalization. In this environment, there are enormous opportunities for AM. These are restricted by the current limitations with regard to the efficient production of large-volume components and the lack of furniture-specific materials.

4.2.5.1 Small Furniture and Lamps

The Hamburg-based company Engineering-Art produces small furniture and lamps using FLM. In addition to classically FLM systems, an ABB industrial robot is used as an axis system, which enables a 1 m^3 construction space (see Section 1.4.2.2). The thermoplastic material is fed to this system in pellet form. The pellet extruder of the industrial robot can process up to 2 kg of thermoplastic pellets per hour (see Section 1.4.1). The nozzle diameter of the robotic FLM system varies from 1 to 3 mm. By means of the RoboDK software, the specific additive G-code for the ABB industrial robot is converted into RAPID (Figure 4.15). Currently, the range of Engineering-Art consists of six lamps, which are additively manufactured directly with the possible customer requirements when ordered. The material used for the lamps is primarily ABS. The prices for the lamps are between €195 and €365.

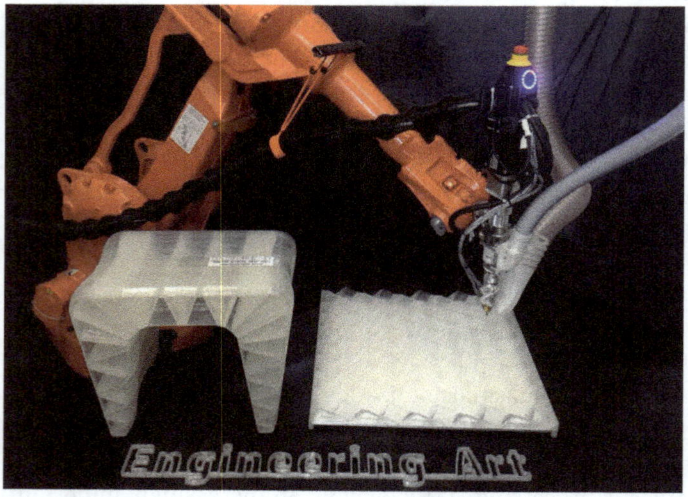

Figure 4.15 Industrial robot with plastic granulate extruder and an additively generated side table, developed by Ascan Aldag (© Ascan Aldag [Engineering-Art 2021])

4.2.5.2 Street Furniture

Founded in 2015 by Panos Sakkas and Foteini Setaki, "The New Raw" a research and design studio based in Rotterdam, creates urban furniture from recycled plastic using FLM and industrial robots. The "Pots Plus" seating collection, with four variants, was made from urban recycled plastic waste using robotic FLM. The seating also features an integrated planter. The geometries used for the seating furniture are based on ergonomic curvatures that are designed to allow a relaxed posture. The four variants are made of either recycled PP or recycled PE. The "Curved" variant accommodates up to two people while weighing 68 kg and saving 137 kg of CO_2. The "Triangle" version for one person saves 129 kg of CO_2 with a weight of 64 kg. The "Organic" variant is designed for up to three people, weighing 94 kg with 190 kg of CO_2 saved. The fourth variant, called "Straight", for a maximum of two people, saves 153 kg of CO_2 with a weight of 76 kg. The seating furniture in the "Pots Plus" collection has a thin-walled design and is therefore manufactured to save materials. The layers are created by thick extrusion beads, which are oriented perpendicular to the forces generated by the use of the seating furniture. This makes the seating furniture very resilient despite the low material input.

4.2.5.3 Increasing the Usability of IKEA Products for People with Disabilities

Under the leadership of its Israeli branch, the multinational furnishing group IKEA has developed supplements in cooperation with the non-profit organizations MILBAT and Israel Accessibility that increase the use of its own products by people with disabilities. The two organizations in particular specialize in the development of targeted accessibility solutions. For the "ThisAbles" project, a specific website has been created, where, in addition to the supplements, some furnishing products are listed that are particularly suitable in terms of accessibility. Furthermore, the supplements are divided into two groups, one of which is specific for children. The supplements can be downloaded from the website as a surface tessellation language (STL) file to be to be produced using FLM systems. The supplements are each assigned to specific IKEA products or series. The "New Developments" section lists a total of 15 additions that can be manufactured additively (Table 4.22); all use PLA with 10–15 % filling, and can achieve 0.2 mm resolution.

Table 4.22 Additions for Accessible IKEA Products

Name	Feature	Product	Material	Filling	Resolution
TAP-ON	Allows easy access to the faucet handle	BOSJON faucet	PLA	10–15 %	0.2 mm
PEN-HOLDER	Allows the adjustment of the mouthpiece while working on the computer	PÅHL series table	PLA	10–15 %	0.2 mm
BAG HANDLER	Simplifies the grip on the bag handles	IKEA bags	PLA	10–15 %	0.2 mm
SPOT ON SHELF	Eye-catching contour marking for shelves to assist people with visual impairment	KALLAX shelves	PLA	10–15 %	0.2 mm

Table 4.22 Additions for Accessible IKEA Products (*continued*)

Name	Feature	Product	Material	Filling	Resolution
FINGER BRUSH	A device that allows holding drawing and painting utensils, for people who have difficulty controlling their fingers	MALA brush	PLA	10–15 %	0.2 mm
SNAP CUP	A device that allows people with mobility difficulties to have a drinking vessel near them	MALM beds	PLA	10–15 %	0.2 mm
STUFF READER	A device that can be connected to a scanner pen by means of an adhesive tape that reads the contents of the shelf aloud to help people with visual impairments	KALLAX shelves	PLA	10–15 %	0.2 mm
CANE BY ME	A rack through which you can place the walking stick next to the bed so that it is available and convenient to reach	MALM beds	PLA	10–15 %	0.2 mm
CURTAIN GRIPPER	A rigid handle that facilitates the opening and closing of the bath curtain	Shower curtains	PLA	10–15 %	0.2 mm
FRIENDLY ZIPPER	Rings that connect to the zipper handle and facilitate its grip and operation	Zipper	PLA	10–15 %	0.2 mm
POPUP HANDLE	Handle for easy opening for people with disabilities on the wrist	Handles	PLA	10–15 %	0.2 mm

Name	Feature	Product	Material	Filling	Resolution
INSIDER	Three brackets connected to the mirror and allow to see the contents of high shelves from a low viewing angle	KALLAX shelves	PLA	10–15 %	0.2 mm
EASY HANDLE	Handle that allows the door to be opened with the forearm or the entire palm of the hand. The handle is available in two versions: For opening with the left or right hand depending on the disability of the user.	PAX cabinets	PLA	10–15 %	0.2 mm
MEGA SWITCH	Enlarged switch to turn the lamp on and off without the need for precise dexterity	RANARP lamp	PLA	10–15 %	0.2 mm
GLASS BUMPER	Bumper that protects glass doors from collisions	Glass cabinets	PLA	10–15 %	0.2 mm

In the "Developments for children" section, eight additively manufacturable extensions for IKEA products are listed. As in the case of the solutions for the products listed in Table 4.22, the material is PLA. The extensions are divided into five for hand motor support and three for mobility support. The extensions are intended for the PÅHL series, BUSUNGE series, TROFAST series, and IKEA children's chairs.

4.2.5.4 AM as a Circular Manufacturing Model

In terms of sustainable production, AM opens up several positive perspectives compared to classic injection molding. Basically, no tools are required for the production of components using AM, thus saving material and energy. Furthermore, real-time production or on-demand production is possible by means of AM. This means that products are only manufactured when they are actually needed, minimizing warehousing and storage. Locally positioned AM centers can also reduce goods transportation. Based on these sustainable positive effects, the accessories collection bFRIENDS was developed in cooperation between the London design studio Pearson Lloyd, the Austrian furniture manufacturer BENE and the AM start-up Batch.Works. Batch.Works is responsible for manufacturing the bFRIENDS collection, which uses FLM with 100 % recycled bioplastic from local discarded food packaging. The material used is recycled PLA from the company Reflow, which specializes in recycling bioplastics used in food packaging. In addition, a circular system has been established whereby the individual products used in the collection can be returned to Batch.Works to be recycled as a material for AM. The bFRIENDS collection consists of 21 accessories for the office, which are available in 10 colors (Figure 4.16).

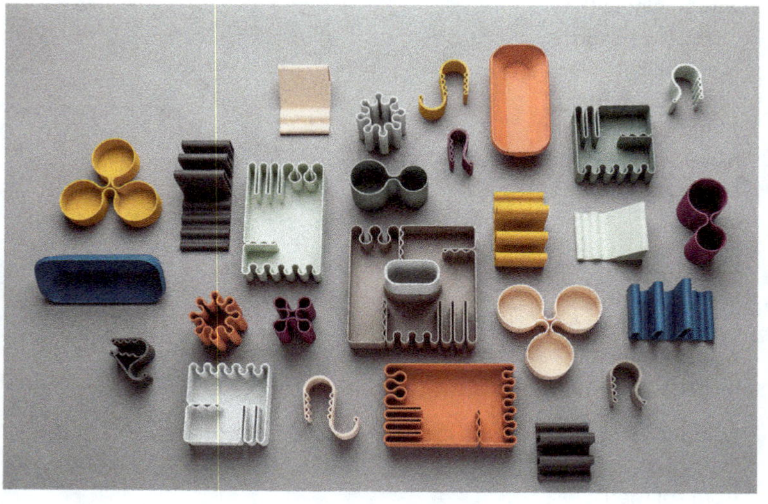

Figure 4.16 Additively manufactured bFRIENDS collection from the furniture manufacturer BENE (© BENE GmbH/Alex Sarginson)

The products are designed in such a way that no support structure is required for additive production (see Section 1.4.2.1). In some cases, this is made possible by a contour that runs straight up. This approach saves both energy and material for the support structure that is not needed. The development of the products was done within six months with direct integration of AM. The collection consists of storage systems, pen holders, trays, stands for mobile devices, hooks and clips to which personal items such as headphones can be attached. The bFRIENDS line of accessories from international workplace innovation leader BENE is one of the first additively manufactured with a circular manufacturing model. It is pointed out by BENE that PLA becomes very flexible above 50 °C and therefore dishwasher cleaning of the products is not possible. Furthermore, the usual grooved surface structure of FLM and process-related irregularities of the surface structure are noted.

4.2.6 Spare Parts and Accessories

The EU Ecodesign Directive, which came into force on 1 March 2021, regulates the improved availability of spare parts and focuses on the repair of electrical appliances instead of disposing them. In Germany, the directive has been integrated into the Energy-Related Products Act (EVPG). The directive focuses primarily on manufacturers of electrical appliances. However, this is to be extended in the future. According to the directive, spare parts must now be available for seven years after the last model has been put on the market. The spare parts must be available within 15 days, and corresponding repair instructions must be made available to customers online. In addition to energy efficiency, the aim of the directive is to improve reparability so that products can be used for a longer period. The EU directive applies to the product groups refrigerators, washing machines and dryers, dishwashers, electronic displays (including televisions), light sources and separate operating devices, external power supplies, electric motors, refrigerators with direct sales function, transformers, and welding equipment. The directive can also be seen as an impetus for more circular economy. In principle, the obligation to stock spare parts also leads to extensive warehousing and, in parallel, to further transport of goods to the repair service provider. The directive does not counteract tendencies in industry to scrap spare parts from certain points in time in order to prevent a gray market or price erosion of spare parts. When AM is considered in this context, the benefits of on-demand manufacturing and local production can lead to the desired circular economy (see Section 4.2.5.4).

4.2.6.1 Spare Parts for Segway X2

The Blumberg-based company SeWaTo offers guided Segway tours in the Black Forest, Lake Constance, and Switzerland with one of the largest off-road Segway X2 fleets in Germany. The Segways in the fleet showed high wear on the elastomer couplings, which transmit the power of the electric motor to the transmission, and are therefore a highly stressed component. In order to ensure smooth operation, these couplings had to be kept in stock at the company as spare parts.

Driven by the relatively high price of the original spare parts, the long delivery times, and the resulting inventory costs, local AM for the spare parts came into the company's focus as a solution. The original couplings consist of three injection-molded parts, with the mount and inner piece made of reinforced ABS and the actual force-transmitting component of the coupling made of flexible plastic. The flexible component is fitted onto the inner piece of the coupling. Since the inner piece hardly wears out and accordingly is not replaced, it was not considered for AM. The inner piece is attached directly to the electric motor and is inserted into the mount with the flexible component, which sits on the transmission box. Two of these elastomer couplings are required in each Segway X2.

In order to be able to use AM, the digital CAD data of each of the two spare parts was generated in several loops of measurement, testing, and AM (Figure 4.17). Here, the coupling mount was produced using FDM in ABS and the flexible component was produced using FLM in TPU. The mount of the coupling with a diameter of 50 mm and a length of 57 mm was suitable for FDM, because of the similar material of the original component and because of the geometry, which requires a support structure (see Section 1.3.2.2). The production of the mount of the coupling was done on an HP Designjet Color 3D Printer with an installation space of 200 × 200 × 150 mm using ABS P430L and the chemically dissolvable support material SR30L. A mount produced via FDM requires 41.75 cm^3 of ABS and 6.49 cm^3 of thermoplastic support material with a fabrication time of 4.06 h. Using the HP Designjet Color 3D Printer, 18 mounts could be produced simultaneously.

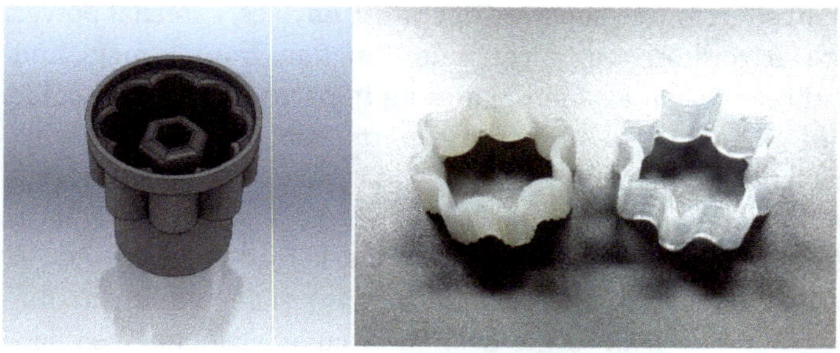

Figure 4.17 Left: CAD model of the Segway mount. Center: Original spare part from Segway. Right: Flexible FLM component made from DP 3690 AU (© Andreas Fischer)

In order to minimize the manufacturing costs for the flexible components, a Dremel 3D Idea Builder FLM system was used. The system has a simple printing unit and does not have a heated build platform (see Section 1.4.2.1). Neither of these factors had a negative effect on the fabrication of the flexible coupling components. Furthermore, the geometry of the flexible component did not require a support structure. However, due to the limited FLM system, only one component could be produced per build process at a time. The flexible components of the elastomer couplings with a diameter of 22 mm and a length of 18 mm were additively produced using Desmopan DP 3690 AU as a filament with a diameter of 1.75 mm. For the production of a flexible component, 4.57 cm^3 DP 3690 AU was required with a production time of 50 min.

The original mount was offered by Segway as a spare part for €69.00 and by third-party suppliers for €49.58. The price of the additively produced mount was reduced by €29.00 and €9.58 per mount, respectively. The original flexible spare part was offered by Segway for €34.90 and by third-party suppliers for €24.37. AM of this part reduced the price per part by €22.90 and €12.37, respectively. For a fleet operator with

20 Segways, a saving of around €878 per year can therefore be achieved for these two spare parts through AM. The additively produced spare parts were tested during test runs over 3000 km in the Black Forest. No excessive wear could be detected on the two components during the test period. The approach of additively manufacturing spare parts additionally also has a sustainable character due to local production, on demand production, saving of injection molds and the reduction of warehousing and goods transport [Fischer et al. 2018].

4.2.6.2 Miele Accessories

Under the brand 3D4U, German home appliance manufacturer Miele provides 3D data for accessories intended for AM [Miele 2022]. The target of this service are Miele's customers, who can additively manufacture the accessories for their Miele appliances using a FLM system. Miele's 3D data can be downloaded free of charge in standard tessellation language (STL) format via the digital platform *www.thingiverse.com*. This website was created in 2008 by Zach Smith and was intended as a parallel project to MakerBot Industries, a FLM system manufacturer. The platform primarily collects user-produced 3D digital data intended for AM, allowing people to use AM who are not capable of producing the 3D data. The platform includes a description and recommended basic parameters for FLM for each 3D4U file. Some of the accessory files can be personalized by customer motifs and parameters. PLA is the intended material for the accessories, as it has a wide distribution among private FLM users.

The accessories are divided into three groups: "Vacuum Cleaners", "Kitchen Gadgets", and "Do-It-Yourself". The "Vacuum Cleaner" group consists of five 3D files, whereas the "Kitchen Aids" and "Do-It-Yourself" groups consists of two 3D files each (Figure 4.18). The most popular accessory of the 3D4U line on *www.thingiverse.com*, with about 9800 likes and 110 documented conversions, is the coffee clip in the "Kitchen Helpers" group. This accessory is a resealable clip that additionally also allows dosing of the coffee beans. The clip is available as a 3D file in two variants, suitable either for bags of approximately 250 g or for bags of approximately 1000 g. In addition to coffee beans, the clip is also suitable for muesli or similar bulk goods. As this accessory is intended for food contact, it is advised to use nationally approved materials for the production of the accessory. In the EU, these are Regulations No. 1935/2004 and No. 10/2011, and for the North American market, Regulation 21CFR.

In second place, with around 5500 likes and 40 documented conversions, is the borehole cleaning aid from the "Do-It-Yourself" group. The base plate of the accessory was flow-optimized using computer simulation, so that the borehole cleaning aid is sucked by use onto wall. The accessory is attached to the handle tube of the Miele vacuum cleaner, and nine openings around the actual borehole effectively extract the dust.

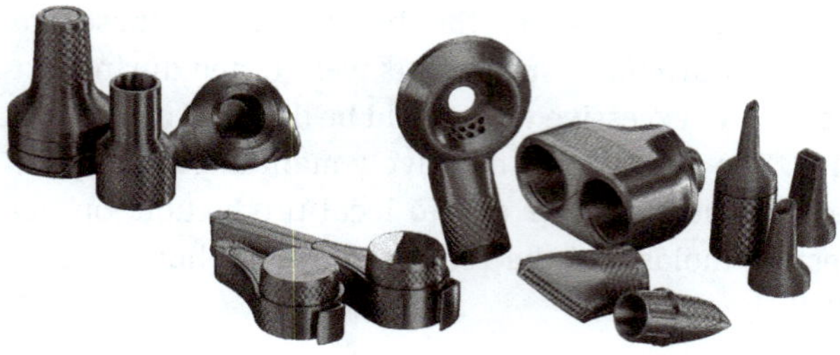

Figure 4.18
Additively manufactured accessories line from Miele (© Miele & Cie. KG [Miele 2021])

The "motif dispenser" accessory in the "Kitchen Gadgets" group can be personalized. The decor dispenser can be used, for example, to apply motifs to milk foam or pastries. The functionality here has been adapted from the "Comfort Clean" function of the Miele CX1. This is an automatic cleaning of the fine dust filter in which micro-vibrations, triggered by a mechanical pulse generator, loosen fine particles. Personalization takes place via the lid of the accessory. To do this, a digital template of the lid must be downloaded and then adapted by the customer using CAD software. For this purpose, this 3D file is available in the STEP format (Standard for the Exchange of Product Model Data).

Other accessories in the 3D4U line include the micro handle, bubble attachment, flexi nozzle, mono bracket, valuables separator, and twin adapter. Since June 2021, three 3D4U accessories (coffee clip, borehole cleaning aid, and valuables separator) have also been available in the Miele online store to enable customers without a FLM system to be reached with the accessories.

4.2.7 Transportation

One of the early adopters of AM was the aerospace sector – an environment where cost is a low priority and small volumes combined with high demands and technical focus are defining. One of the reasons for its use in the aerospace sector is AM's ability to produce near net shape in contrast to ablative manufacturing processes. Thus, costs can be reduced through material savings. In the aerospace sector, the focus is primarily on metallic materials and fiber reinforced plastic composite (FRP) in combination with AM (see Section 1.6.1). In the automotive sector, AM is increasingly developing its potential in manufacturing in addition to the vehicle development process. German chemical company BASF expects AM to become more established in the automotive sector, moving from simple additively-generated fixtures to manufacturing to small, additively-directly-manufactured volumes for niche vehicles. An evolution of AM toward mass production is also a possible scenario in the future. To achieve this, the manufacturing characteristics of process stability, automation, productivity, and quality assurance of AM would have to be significantly improved (Figure 4.19).

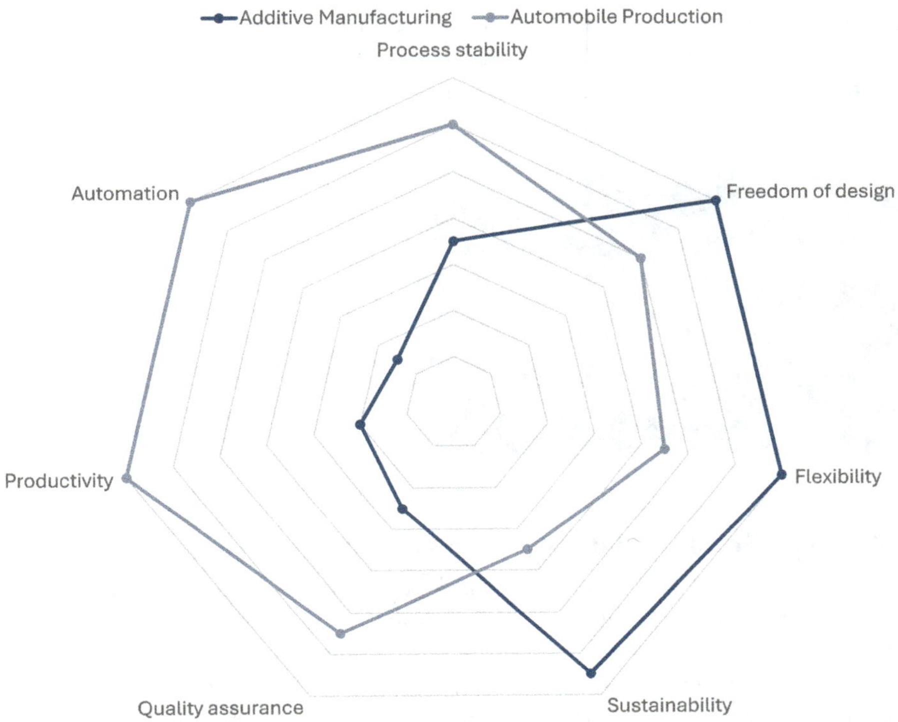

Figure 4.19 Comparison of the manufacturing properties of AM and classic automotive production (© Andreas Fischer, adapted from [Landua 2015])

4.2.7.1 Vehicle Body via FDM

The Urbee Hybrid concept car developed by the Canadian company KOR EcoLogic in 2013 can be described as the first vehicle with an AM body . The name Urbee stands for "Urban Electric with Ethanol". The vehicle, developed by Jim Kor, is based on a three-wheeled base and focuses on lightweight construction and efficient energy use. One focus of the body was to achieve a drag coefficient c_W of less than 0.15. By means of digital optimization of the body, a c_W of 0.149 was achieved (Figure 4.20). The body parts created via FDM were manufactured with a honeycomb-like hollow structure (see Section 1.3.1). By using this specific FDM hollow structure, it was possible to reduce the weight of the thermoplastic components. In parallel, manufacturing time and material usage could be saved by this production strategy.

The Urbee Hybrid uses an aluminum frame as a stability base, to which the additively manufactured plastic body parts are bolted. The size of the body parts was dependent on the available installation space of the FDM systems. The Stratasys Fortus 900 mc system, with an installation space of 914.4 × 609.6 × 914.4 mm, was used to produce the Urbee Hybrid body. Stratasys service provider RedEye On Demand handled the production of the necessary ABS body parts. Due to installation space limitations, each additively manufactured body part had to be assembled from approximately four components. A total of around 50 FDM components were required for the Urbee Hybrid, which were produced in around 2500 production hours.

Figure 4.20
Flow-resistance-optimized lightweight body of the Urbee concept car (© Stratasys)

4.2.7.2 Vehicle Frame and Body using Big-area AM

Founded in Phoenix in 2007, Local Motors broke new ground in vehicle development and customer integration. In addition to crowdsourcing (the outsourcing of internal company tasks to external voluntary users or customers), the vehicle plans were also published license-free (open source). The individual development steps required were worked out by competitions in the Local Motors member network, which comprised approximately 50,000 people in 2015. Using the crowdsourcing strategy, the members of the network were thus able to determine the technology and design of the vehicle. Another feature of Local Motors is its micro-factories, where customers assemble their vehicles themselves under supervision. This intensive integration of customers or members in the product development process creates a high emotional attachment to the product, which also generates positive sustainable effects such as long use.

For Local Motors' first vehicle, the Rally Fighter off-road vehicle, AM was used classically in the product development process, but not on the vehicle itself. The two-dimensional personalization of the Rally Fighter's body was achieved via car wrapping, which was also touted as a sustainable replacement for conventional varnishing. In 2014, Local Motors produced the Starti concept car, which features an additively manufactured frame and body. The Starti is based on the 2011 Renault Twizy electric vehicle, which reaches a top speed of 40 km/h with a maximum weight of 1400 kg. The frame as well as the body of the concept vehicle were additively produced from ABS in 44 h using big-area additive manufacturing (BAAM) from Cincinnati Incorporated (see Section 1.4.1).

4.2.7.3 Electric Motorcycle

The fully additively manufactured and functional electric motorcycle prototype NERA was developed and manufactured by the company BigRep in Berlin within 12 weeks and has a total weight of approximately 60 kg. The NERA prototype consists of a total of 15 components manufactured via FLM, whereby the basic mechanical functionalities as well as the appearance were defined in five sketch-based iterations. AM allows a high degree of design freedom (Figure 4.19) with regard to integration of functions and the product's appearance.

In the development process of the NERA motorcycle, the first step was to arrange the functionally relevant components such as the battery or electric motor. In the following step, the defined appearance was then applied to create the "fairing" and integrate functions in parallel. The fairing of the electric motorcycle features a reduced polygonal design (Figure 4.21), which minimizes areas characterized by the X–Y resolution of the FLM system. The non-additively manufactured components such as battery, electric motor, electronics, control systems, and wiring are embedded in the additively manufactured thermoplastic components.

Four types of thermoplastic filaments were used in the NERA prototype. A UV-resistant thermoplastic polyurethane with a Shore A hardness of 98 and a temperature resistance of up to 100 °C was used as a functional material for the tires, bumpers, seats, and handles. A high-temperature resistant bio-polymer, with softening resistance up to 115 °C, was used for rigid components, whereas the classic BigRep polylactide (PLA) was used for color accents. PET with glycol (PETG) was used for the light reflectors because of its translucent property.

Highlights of the AM electric motorcycle include:

- Flexible parametric bumper instead of suspension as shock absorber
- Personalized seat and chest support with attached flexible thermoplastic
- Airless tires with personalized tread made of flexible thermoplastic material
- Front tire with soft damping by means of an arch structure
- Rear tire with a strong hexagonal structure for load-bearing components.

Figure 4.21
Polygonal design appearance of the AM electric motorcycle NERA [BigRep 2022a]

4.2.7.4 Jet-powered Drone

Manassas-based Aurora Flight Sciences develops unmanned aerial vehicles (UAVs) for both the commercial and military markets. The company uses AM in development to produce components and tooling. Together with FDM system manufacturer Stratasys, a jet-powered drone with thrust vector control was to be developed and additively manufactured. The jet engine, able to provide 98 N of thrust, was to accelerate the additively manufactured drone to 241 km/h. By using AM, the production time of the unmanned aerial vehicle was reduced by half. A major contributing factor was the fact that no tooling had to be created. The wingspan of the drone is 2.9 m with a fuselage weight of 6.4 kg. The total weight of the drone with 3 L of fuel is approximately 14.5 kg. In total, the UAV consists of 34 components, 26 of which were additively manufactured. Using Stratasys Fortus systems, the wings (Figure 4.22) and the fuselage were manufactured. The material used was primarily acrylic ester-styrene-acrylonitrile (ASA) to achieve the necessary strength and stiffness at low density. The following main components were manufactured additively:

- Two wings (FDM)
- Fuselage (FDM)
- Payload module (FDM)
- Fuel tank (SLS)
- Thrust vector mechanism (FDM & DMLS).

The fuel tank was additively manufactured using SLS on the basis of polyamide. For temperature reasons, the exhaust duct cover had to be manufactured using ULTEM 1010. Due to the high exhaust gas temperatures of up to 700 °C, it was necessary to manufacture the thrust vector channel from a metallic material. The heat-resistant nickel–chromium alloy INCONEL 718, which was produced by direct metal laser sintering (DMLS), was used here. DMLS is used in particular for the additive production of porous components made of metal alloys. The drone's cruising speed ranges from 113–129 km/h with a maximum payload of 0.2 kg [Stratasys 2022].

Figure 4.22 Optimized inner wing design using complex, additively generated lightweight structures (© Stratasys)

4.2.7.5 Camper Concept

The VisionVenture motorhome concept presented in 2019 is based on a Mercedes-Benz Sprinter and was developed in cooperation between HYMER and BASF. Approximately 100 interior and exterior components were additively manufactured. Without the integration of AM in the design process of a customer-specific vehicle, the costs as well as the complexity would not have been justifiable. The short-term focus here was on the production of special parts for niche vehicles. In the long term, however, AM is also to be used for larger numbers of series-produced vehicles. AM was used on the VisionVenture to produce personalized wheel arch linings, body panels, interior lamps, furniture handles, wall hooks and mobile device holders. The two customer groups for the VisionVenture, vehicle rental companies and private individuals, often have the desire to personalize the vehicle. Lettering or logos, which are classically applied via coating or foiling, can discolor or even structurally compromise the material of the components. This can lead to replacement or repainting of the components. The wheel arch linings with a three-dimensional and customizable logo, produced with MJF, were additively generated using the material Ultrasint® TPU01 (Figure 4.23). The thermoplastic material has high shock absorption, flexibility, and elasticity, as well as thermal and UV stability. The Ultrasint® TPU01 components have highly detailed edges and contours and can be readily coated. The additively manufactured wheel arches were coated with Elastocoat C Polyurea spray coating to create a homogeneous surface and protect the component against scratches and stone chipping.

Figure 4.23 Additively manufactured wheel arch linings with three-dimensional personalization (© BASF SE)

The inner lamp of the VisionVenture was manufactured using SLS as well as the FLM process. The UL94-certified flame-retardant material Ultrafuse® PC/ABS FR was used for the housing. The lampshade was manufactured with the translucent material Ultrasint® TPU 88A and then coated with Ultracur3D® Coat F, which was specially developed for flexible 3D-printed components.

The following high-performance AM materials were used in the VisionVenture motorhome concept:

- Ultracur3D® Coat F
- Ultracur3D® EPD1006
- Ultrasint® TPU01
- Ultrafuse® 316L
- Ultrafuse® PC/ABS FR
- Ultrasint® TPU 88A.

4.2.8 Medical Applications

A key feature of AM is the very direct coupling with the three-dimensional digital representation of the component. Due to the layer-by-layer generation of the physical component, there are hardly any restrictions. Compared to conventional manufacturing, the material-guided manufacturing strategy does not require any product- or manufacturing-specific tools to create components. This in turn leads to components that can be personalized to a high degree in three dimensions (see Section 4.2.4). In the medical sector, a large proportion of products must be adapted to the respective patient. In conventional manufacturing, this is done by time-consuming post-processing of the products. In AM of medical products, the tightly coupled digital model can already be designed for load scenarios and biomechanical as well as patient-specific requirements. Post-processing may also be required for additively generated medical applications. These range from the removal of support structures and component refinement to the improvement of surface quality. Furthermore, it should be noted that the manufacturing-specific parameters such as the additive system, operating location, materials, and post-processing must be certified for the medical sector. By means of AM with thermoplastic materials, the following areas of medical applications are primarily focused on:

- Orthotics
- Prostheses
- Medical devices (for example, drilling and sawing templates).

4.2.8.1 Valves for Breathing Masks

Starting from a hospital in Sassuolo, Italy, a consortium of different companies was formed in 2020 to support healthcare professionals in the treatment of COVID-19 patients. At the time, there was an acute shortage of safety equipment for physicians and nurses. The goal of the initiative by Mares, Ferrari, Solid Energy, and Nuovamacut was to convert Mares snorkel masks into respirators (Figure 4.24). These were to be designed to facilitate breathing for patients as well as physicians and nurses. The advantage of this approach is that the masks perfectly cover facial areas through which infection can occur, such as the mouth, nose, and eyes.

In a first step, the snorkel masks were digitized and reverse engineering was performed. This enabled an accurate digital design of the connector to the existing components. The first prototypes were produced within 24 h using FDM in ABS on Fortus 370 and 450 systems. These were then tested at the hospital in Sassuolo. The developed and additively manufactured connector ties two commercially available FFP3 filters to the snorkel mask. To optimize the AM of the connectors, they were designed so that hardly any support structure was needed. This increased small batch production, as overall manufacturing time was saved (see Section 1.3.2.2). The Honeywell Twin A2P3 FFP3 filters used are readily available commercially and have an effectiveness of one month. In addition, both the snorkel mask and the connector can be fully disinfected. The digital data of the connector was additionally made available online to support local health services during the pandemic.

Ferrari handled the additive production of the connectors in its prototype department at the Maranello plant. The company Nuovamacut was responsible for logistics and research. The design and assembly of the respirator was carried out by the company Solid Energy. During the project, over 1000 respirators were produced and distributed to various hospitals through the Italian Civil Defense.

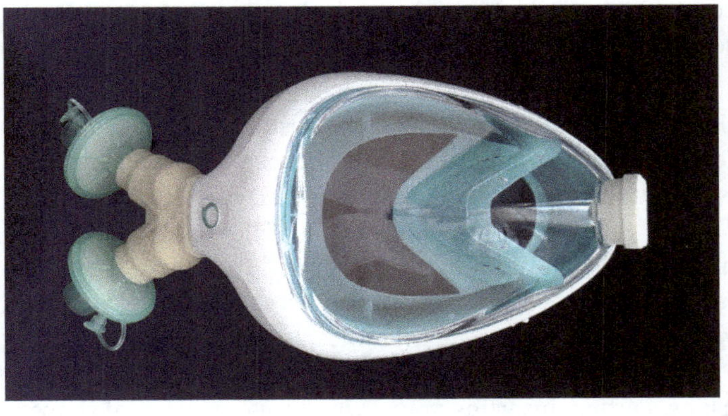

Figure 4.24 Conversion of a snorkel mask to a respirator with FFP3 filters and additively manufactured connectors [Solid Energy 2022]

4.2.8.2 Implants Made of High-performance Plastic

The Munich-based company Kumovis offers an open AM system specifically for medical technology, called Kumovis R1. This FLM system can process eight high-performance polymers, such as PEEK, PPSU, and PEI. The Kumovis R1 can heat its installation space homogeneously to a maximum of 250 °C via an air flow, thus achieving high component strength and dimensional stability. In addition, local temperature management can be used to individually influence the adhesion of the layers. By means of an integrated filter system and laminar air flow, the additive system offers a cleanroom environment in the installation space and thus prevents contamination with contaminant. Furthermore, the system can be cleaned according to medical standards. Contamination of the used thermoplastic filament in the printer is also prevented. To demonstrate the range of applications of the Kumovis R1, a patient-specific cranial implant was additively manufactured from the high-performance plastic PEKK [Fischer 2020].

4.2.8.3 Person-specific Prostheses and Orthoses

Since its founding in 2016, the Munich-based company Mecuris has been developing and offering patient-specific AM prostheses and orthoses via an online platform. The prostheses and orthoses were primarily manufactured using SLS, making them cheaper and faster to produce. Both product groups are personalized based on digitized body segments of the patients and adapted to their needs. In addition to the desired functionality, the patient can also influence the design, color and structures of their customized prosthesis or orthosis.

The company also launched the first AM prosthetic feet for children and adults. These can be customized in size and design via the current Mecuris solution platform (MSP) to be individually created in terms of size and design. The company's portfolio also offers various prosthetic covers for transtibial amputees, which are resistant to water or salt water. These can also be additionally personalized by means of coloring and structures.

From 2021, Mecuris will focus on the digitalization of the prosthesis and orthosis sector and establish the Mecuris solution platform for this purpose. This platform provides three digital tools online for orthoses or prostheses. "Mecuris3D Correction" allows digitization and digital correction of the joint angle of the patient scan. The correction function can be used for positioning the lower leg, the whole leg or the hand. "Mecuris3D Modeling" is intended for modeling and creating positioning and functional orthoses for the lower and upper extremities as well as the torso and head, and includes the following options: perpendicular alignment, up and down milling, smoothing, marking, cutting, inserting your own three-dimensional shapes, connecting, and measuring. "Mecuris3D Creator" is designed for digital configuration and design of the final orthosis blank. The following options are available in the digital tool: determining a cutting line and wall thickness of the orthosis, setting slotted and round holes for a locking system, defining an area for perforation holes, and designing the model with structures or motifs. Data exchange takes place in the standard tessellation language (STL) format. Mecuris additionally offers additive service providers for manufacturing. However, the data can also be used by means of milling, which is widely used in the field.

4.2.9 Sports Equipment and Fashion

AM is also influencing the fashion industry and in particular the footwear sector. The high degree of personalization possible with AM is certainly an overriding factor from the point of view of the end consumer. However, companies such as Adidas or New Balance also see potential for series production. Shoe designers, on the other hand, can implement novel and unique shoe concepts thanks to the high design freedom of AM (Figure 4.19). For running and sports shoes, requirements such as correct fit, stability, and durability must be taken into account. These requirements are in turn strongly dependent on the used material, material combination, and additive process. FLM, for example, offers a high material selection and the possibility of material combination. Compared with SLS, however, FLM components have lower stability for the same material. Since material combination is excluded in SLS for process engineering reasons, cushioning systems must be solved here via complex structures. Compressed, non-fused powder in these structures is in turn difficult to remove completely without residue, which is the challenge in this context.

Adidas uses the digital light synthesis (DLS) process for the production of midsoles. This is a UV polymerization process that can produce high-resolution components at high speed. After processing, the materials used can be chemically modified by the application of heat to obtain technical and mechanical properties for industrial applications. The midsole of the Futurecraft 4D consists of a grid geometry that provides different levels of cushioning. The remaining parts of the athletic shoe are made of regular materials and are conventionally manufactured. Other models featuring Adidas' AM approach include the Y-3 Runner 4D and Alphaedge 4D.

Other footwear manufacturers that have incorporated AM into footwear include the following:

- New Balance (990 Sport Sneaker)
- Reebok (Liquid Floatride Run)
- Nike Football (Vapor Carbon 2014 Elite Cleat).

Covestro Germany AG offers recyclable TPU for AM, which is available in both powder and filament form. The material is characterized by high rebound and abrasion resistance, which makes the material suitable for creating both the upper material and the soles of sports shoes. If the complete sports shoe is made from one material, it can also be completely recycled in one step, including the polyurethane adhesive required during production (Figure 4.25).

Figure 4.25
Design study of an sports shoe consisting of recyclable TPU and a shoe sole manufactured using SLS with 83 Shore A TPU (© Covestro Deutschland AG)

4.2.9.1 Sustainable Surfboard

The WYVE brand has introduced HEXA, a partially additively manufactured surfboard, which is a sustainable and customizable product. Today's surfboards, which are mass produced, have the same impact on the environment as, for example, laptops. Approximately 6 kg of carcinogenic and toxic waste is generated in the production of an approximately 3 kg surfboard. Customized surfboards are produced via so-called shapers. In the first step of personalizing the HEXA surfboard, the outer contour is selected based on the shapers' knowledge, body measurements and surfing habits. Then, the inner honeycomb structure is digitally optimized by the parameters of weight, density, distribution, and bending. In the next step, the honeycomb structure core of the surfboard is manufactured using FLM. Currently, PLA is used for this, but the extent to which recycled PET from shredded plastic bottles, which are collected in the ocean, can be used is also being investigated. In the final production step, the additively manufactured core, which does not require a support structure during production, is cast with glass fibers and a 50 % bio-based resin. Finally, a hot coating is applied. The HEXA surfboard is translucent, and the honeycomb structure remains visible.

4.2.9.2 Force-absorbing Bicycle Helmet

The Canadian company Kupol developed an additively manufactured, force-absorbing core for bicycle helmets. The Kollide Safety System consists of three layers, which are designed to move in order to optimally absorb force impacts. The middle layer consists of kinetic shock absorbers, which are made of flexible material using FLM. The second layer is made of a non-resilient material and is based on the same twisted hexagonal structure as the kinetic shock absorbers. The third layer is the actual helmet shell. Movable dampers inside the bicycle helmet are also designed to mitigate the rotational movement of the head in the event of an impact and increase the fit. The helmet shell is designed with penetrations to ensure good air permeability. Kupol is also a participant in the Montreal-based consortium Kollide, which developed the football helmet of the same name, also based on the Kollide Safety System (Figure 4.26).

Figure 4.26
A Kollide football helmet with transparent helmet shell and additively generated force-absorbing structure (© Cyrille Cambien)

4.2.9.3 Hygiene Warning System

Developed in 2020 by NASA at the Jet Propulsion Laboratory of the California Institute of Technology, the PULSE trailer is designed to help reduce the spread of COVID-19. To do so, the system focuses on reducing hand–face contact through warning vibrations. Contaminated surfaces that are touched with the hands can lead to disease if those hands are subsequently used to touch the eyes, nose, or mouth. Unconscious and habituated behaviors that lead to such hand–face contacts are difficult to suppress or control. In developing the PULSE pendant, care was taken to use simple and inexpensive components to allow easy reproducibility even by private individuals. The component list as well as the data for the housing in STL format are available as open source. The additively manufactured housing includes an infrared proximity sensor with a range of 31 cm for motion detection and a motor for vibration. The closer the movement is detected to the sensor, the stronger the triggered vibration. Furthermore, a slide switch, a 3 V battery with holder, and an oscillator circuit have been installed. The approximately 4 cm-long PULSE pendant is worn 15–30 cm below the chin. The digital data of the case can also be used as a basis for customized and redesigned pendants. For the private production of the housing, NASA recommends FLM in combination with PLA.

4.2.9.4 Polyamide Spectacle Frames

AM of spectacle frames is currently mainly carried out using SLS in PA. The reason for this is the relatively high resolution and resilience of the produced components. Classically, the additively produced spectacle frames are processed via downstream finishing steps such as smoothing, deburring, coloring, and re-drilling of holes. The Berlin-based company MYKITA can be seen as a pioneer of additively manufactured spectacle frames. In 2011, the company launched the Mylon collection, the first serially and additively manufactured spectacle frames. The frames are made of PA using SLS. Since the introduction of the MYLON collection, sales have been in the double-digit range. In 2013, this amounted to around 20,000 frames.

In contrast, the company FrameLapp focuses on ergonomically adapted spectacle frames in cooperation with opticians. The customer parameters recorded by the opticians are sent to FrameLapp for digital fitting and AM. The recorded and controlling parameters include lens height and width, web width, and temple length. The matching lenses are ordered by the optician in parallel with the production process. Approximately 100 spectacle frames can be produced on each additive system in 24 h. Sales of FrameLapp spectacle frames are around 1000 units per year.

The company Monoqool relies on lightweight construction and integration of functions for its additively produced spectacle frames. The AM spectacle frames weigh around 10 g, making them up to 30 % lighter than those produced in the conventional way. Furthermore, the construction of the spectacle frame hinges were solved without screws using an additive approach.

The company FramePunk in Berlin offers additively manufactured spectacle frames including prescription lenses for about €249. Each shape of the spectacle frames in the collection is available in five sizes and 17 colors, with a delivery period of 5–14 days. For a surcharge of €100 the spectacle frame is adjusted to the customer, through the parameters of nose angle, web width, frame width, and temple length. Furthermore, it is possible to integrate three-dimensional names or logos.

Hoya Vision Care's Yuniku spectacle frames are maximally customizable through a combination of face digitization, algorithms, and AM. When the spectacle frames and lenses are designed separately, the options for customization are limited. In the case of Yuniku glasses, the first step is for the optician to perform facial digitization. Based on this data, the algorithm calculates the optimal position of the lenses in relation to the eyes. A spectacle frame is then selected from the collection. This is then digitally adjusted to the position of the lenses and, based on further information, to the customer. The digital result can then be viewed using an augmented-reality mirror on the customer. If the evaluation is positive, the model is manufactured using SLS in PA at additive service provider Materialise. One of the biggest challenges in manufacturing spectacle frames using SLS is color design. The additive spectacle frames are mainly manufactured in white PA powder and also feature a porous surface. This can be counteracted by combining coloring and smoothing. The dyeing process is also intended to prevent discoloration. Material combinations as well as multicolor or patterns are not possible due to the limitations of the additive process and have to be solved subsequently by assembling.

4.2.9.5 Personalized Handbags

Turin-based XYZ Bag combines AM and traditional handbag making. A PA frame is combined with straps and a lockable leather body. In this process, the PA frame is manufactured using MJF, and is then painted by hand. The basic body of the frame is a strongly rounded and upwardly open cuboid. The basic body can be personalized with manifold structures, three-dimensional texts, reliefs and logos. It is also possible to purchase pre-designed handbags. There are currently six pre-designed handbags, which can also be additionally personalized and range in price from €270–360. The XYZ Bag handbags are manufactured only after the order is placed, and are therefore also partially sustainable.

4.2.10 Manufacturing Systems Engineering

AM enables greater flexibility and shorter manufacturing times for components, particularly in special-purpose machinery manufacture. In addition, cost reductions can often be achieved for complex components. In addition, AM enables material-saving, with lighter and novel design solutions due to the ability to generate complex geometries. WS Wieländer+Schill Engineering in Tuningen, Germany, for example, uses SLS in combination with PA to produce robot-based dampened deburring tools. The company now offers four dampened deburring tools and variations as a catalog product. These tools can be adapted to existing industrial robots as well as spindles. Thanks to the specific design and the associated elimination of classic connecting elements, a weight saving of up to 65 % can be achieved. Furthermore, the systems use additively generated and pneumatically operated bellows for the damping mechanism (Figure 4.27).

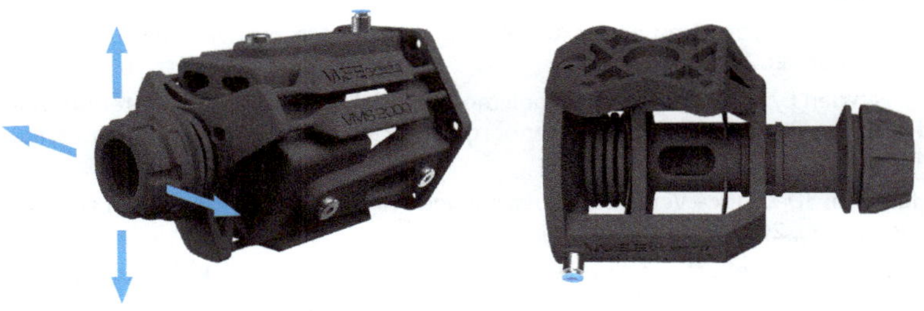

Figure 4.27 Additively manufactured robot-based dampened deburring tool VMS 2000 with four directions of movement and AMS 2000 with axial movement and visible bellows [Wieländer+Schill 2021]

References

3D Druck.com: DIY: Incorporate textiles into FDM 3D printing to make ventilation grilles. 26 August 2019. *https://3druck.com/diy/diy-textilien-in-fdm-3d-druck-einbinden-zur-herstellung-von-lueftungsgittern-4784851* (retrieved 27 September 2021)

3D-Systems: 3D printers for dentistry. 3D-Systems, Inc., *https://de.3dsystems.com/3d-printers/dental* (retrieved 27 February 2022)

Achten et al. 2020: D. Achten, Dirk, T. Büsgen, R. Wagner, J. Künzel, M. Kessler & B. Mettmann, Method of manufacturing a treated 3D-printed article. Patent specification EP3727801A1. Patent proprietor: Covestro Deutschland AG. European Patent Office, 28 October 2020

Asche 2020: S. Asche: XXL gearbox from the 3-D printer. VDI News, 9 March 2020. *https://www.vdi-nachrichten.com/technik/additive-fertigung/xxl-getriebe-aus-dem-3-d-drucker* (retrieved 27 February 2022)

Autodesk 2017: Making the Unmakeable. Combining Additive Manufacturing with Proven Metal Casting Processes. Autodesk, Inc., 2017. *https://aristo-cast.com/Downloads/Aristo-cast-making-the-unmakeable-additive-casting-ebook-en.pdf* (retrieved 8 February 2022)

BASF 2022: The #vanlife of the Future - Realized by BASF for HYMER. Concept Van HYMER VisionVenture. BASF 3D Printing Solutions GmbH, *https://forward-am.com/Use Cases-and-whitepapers/the-vanlife-of-the-future-realized-by-basf-for-hymer* (retrieved 28 February 2022)

BASF 2019: The camper of the future. Press release, BASF SE, 10 October 2019. *https://www.basf.com/global/de/media/news-releases/2019/10/p-19-357.html* (retrieved 28 February 2022).

BASF SE: Additive manufacturing is driving a new automotive frontier. *https://automotive-transportation.basf.com/global/en/automotive/stories/additive-manufacturing-is-driving-a-new-automotive-frontier.html* (retrieved 7 January 07, 2022).

BENE 2021: bFRIENDS – Design with ecological added value. BENE GmbH *https://bene.com/de/bfriends-de* (retrieved 20 December 2021)

BigRep 2022a: NERA: The First Fully 3D Printed E-Bike in The World. BigRep GmbH *https://bigrep.com/nera-e-motorbike* (retrieved 28 February 2022)

BigRep 2022b: Villeroy & Boch 3D Printing Industrial Design Prototypes. E-Book. BigRep GmbH, *https://bigrep.com/ebooks/two-centuries-of-design-meets-3d-printing* (retrieved 12 January 2022)

Coxon et al. 2018: S. Coxon, R. Napper & M. Richardson: Urban Mobility Design. Elsevier, Amsterdam 2018

Dosch 2020: A.-L. Dosch: 3D-printed consumer products. Shoes from the 3D printer – a future model? Industry of Things, 11 May 2020. *https://www.industry-of-things.de/schuhe-aus-dem-3d-drucker-ein-zukunftsmodell-a-1068438* (retrieved 11 March 2022)

Ebner 2016: New form of eyewear manufacturing. 3D printing: Ophthalmic optics in upheaval? Eyebizz, Ebner Media Group GmbH & Co. KG, 14 Decembe 2016. *https://www.eyebizz.de/brillen/3d-druck-augenoptik-im-umbruch* (retrieved 13 March 2022)

Engineering-Art 2021: Industrial robot with pellet extruder. Engineering-Art, *http://engineering-art.de/roboter* (retrieved 23 November 2021)

van den Eynde 2018: M. van den Eynde, Expanding the polymer material palette for laser sintering. The importance of powder flowability. Doctoral thesis. Arenberg Doctoral School, Faculty of Engineering Science, October 2018

Feldmann 2016: C. Feldmann, 3D-Druck – Verfahrensauswahl und Wirtschaftlichkeit. Decision support for companies. Springer Verlag, 2016

Ferrari 2022: Ferrari continues its efforts to fight the Covid-19 pandemic. Ferrari website, *https://corporate.ferrari.com/en/ferrari-continues-its-efforts-fight-covid-19-pandemic* (retrieved 8 March 2022)

Fischer 2020: A. Fischer: 3D-Druck & Co. – Potenziale der additiven Fertigung. In: M. Lang & M. Müller, From Augmented Reality to AI. The most important IT topics you need to know for your business. pp. 203–229. Carl Hanser Verlag, 2020

Fischer et al. 2018: A. Fischer, S. Gebaue & E. Khavkin, 3D Printing in the Enterprise. Decision models, best practices and application examples. Using the example of fused layer modeling (FLM). Carl Hanser Verlag, 2018

HOYA 2022: Yuniku. You are unique. You are Yuniku. HOYA Lens Deutschland GmbH. *https://www.hoyavision.com/de/yuniku* (retrieved 13 March 2022)

IHK 2021: Circular economy. The new Ecodesign Directive came into force on March 1, 2021. IHK Düsseldorf, *https://www.duesseldorf.ihk.de/innovation-umwelt-energie/aktuelles/die-neue-oekodesign-richtlinie-ist-am-1-maerz-2021-in-kraft-getreten-4971030* (retrieved 30 December 2021)

Kilimann 2015: S. Kilimann, Local Motors. Community, develop a car. ZEIT Online, 3 August 2015. *https://www.zeit.de/mobilitaet/2015-08/local-motors-auto-regionale-produktion-berlin* (retrieved 22 February 2022)

Kollide 2022: Kollide – A 3D Revolution is Coming. *https://www.kollide.ca* (retrieved 14 March 2022)

Kumovis 2022: Kumovis R1. First FLM 3D printer specifically for medical technology. Kumovis GmbH, *https://kumovis.com/de/3d-printer* (retrieved 08.03.2022).

Kupol 2022: Kupol. Toward the future of protective gear. Kupol, Inc., *https://www.kupol.ca* (retrieved 14 March 2022)

Landua 2015: S. Landua, 3D-Druck – Technologie und Potenziale. Contribution by Volkswagen AG. Fachkonferenz "3D-Druck – Additive Fertigung in der Automobilindustrie", Augsburg, 20–21 October 2015

Matschinski 2020: A. Matschinski, Virtual Symposium on AFP and AM. TU Munich, Chair of Carbon Composites (LCC), September 2020

Mecuris 2022: Creating orthoses digitally. Independent – intuitive – cost-effective #YouCreate. Mecuris GmbH *https://www.mecuris.com/msp-orthesen-digital-erstellen* (retrieved 10 March 2022)

Miele 2021: 3D4U – Miele offers exclusive accessories from the 3D printer in its online store for the first time. Miele & Cie. KG: Press release No. 048/2021, 23 June 2021. *https://www.miele.de/de/m/3d4u-miele-bietet-exklusives-zubehoer-aus-dem-3d-drucker-erstmals-im-online-shop-an-5611.htm* (retrieved 26 January 2022)

Miele 2022: 3D4U – Know-how meets 3D printing. Miele & Cie. KG: *https://www.miele.de/haushalt/3d4u-3d-druck-anleitungen-10585.html* (retrieved 03.01.2022)

Moore 2013: R. Moore, How the Razor Blade Model Works in 3D Printing. The Motley Fool, 10 July 2013. *https://www.fool.com/investing/general/2013/07/10/how-the-razor-blade-model-works-in-3-d-printing.aspx?source=epthstlnk0000001* (retrieved 27 February 2022)

NASA 2020: PULSE – A Pendant to Warn You When You Touch Your Face. 2020. NASA/California Institute of Technology, *https://medeng.jpl.nasa.gov/covid-19/ventilator/UMS/pulse/Pulse_Brochure_JPL%20COVID-19.pdf* (retrieved 3 December 2022).

Pearson Lloyd 2021: BENE bFRIENDS. Pearson Lloyd Ltd, *https://pearsonlloyd.com/project/bene_bfriends* (retrieved 20 December 2021)

Schram et al. 2019: B. Schramm et al.: Medical applications of additive manufacturing. In: H. A. Richard et al.: Additive Manufacturing of Components and Structures, pp. 21-40. Springer Vieweg, 2019

Shaik et al. 2021: Y. P. Shaik, J. Schuster & A. Shaik, A scientific review on various pellet extruders used in 3D Printing FDM Processes. Open Access Library Journal, 8, 1-19. 2021. DOI: 10.4236/oalib.1107698

Solid Energy 2022: Our help in fighting COVID19 snorkeling masks converted to protect doctors as well. Solid Energy, *https://www.solidenergy.it/covid-19/covid-19-emergency* (retrieved 3 August 2022)

Stratasys 2022: Shaping The Future of Flight. Case Study. Stratasys Ltd, 2022. *https://www.stratasys.com/de/resources/search/case-studies/aurora* (retrieved 7 March 2022).

ThisAbles 2021: Accessibility for IKEA products. *https://thisables.com/de* (retrieved 30 November 2021)

The Gourd Project 2021: The Gourd Cup. *https://www.thegourdproject.com/about-1* (retrieved 27 September 2021)

The New Raw 2021: Crafting plastic waste with robots. *https://thenewraw.org* (retrieved 28 December 2021)

Umweltbundesamt 2019: Spare parts from the 3D printer. Umweltbundesamt, 28 March 2019. *https://www.umweltbundesamt.de/themen/ersatzteile-aus-dem-3d-drucker* (retrieved 27 February 2022)

Müller 2020: A. Müller, Special mechanical engineering: How 3D printing is transforming manufacturing. AUTOCAD Magazine, 19 March 2020. *https://www.autocad-magazin.de/sondermaschinenbau-wie-der-3d-druck-die-fertigung-transformiert* (retrieved 14 March 2022)

Wieländer+Schill 2021: Innovation and technology. Tools with which you grow beyond yourself. WS Wieländer+Schill Engineering GmbH & Co. KG, 2021

WYVE 2021: High Performance Surfboard As Unique As You Are. The Hexa Surfboard By Wyve. Press Kit 2021. *https://uploads-ssl.webflow.com/5ee3d848b71cb4f59074ccba/60a3f72ea6a4127d80fd355e_2021_%20WYVE%20PRESS%20KIT_ENGLISH_LD.pdf* (retrieved 3 November 2022)

XYZCUBE 2022: XYZBAG. XYZCUBE Sas di Nicola Annalisa & C. *https://www.xyzbag.com/technology3d* (retrieved 14 March 2022)

Zukunftsinstitut 2021: Glossary Individualization: Trend terms and definitions. Zukunftsinstitut GmbH, *https://www.zukunftsinstitut.de/artikel/megatrend-glossar/individualisierung-glossar* (retrieved 22 November 2021)

Index

Symbols

A

B

C

D

E

G

H

I

Q

R

T

U

V

W

Y

www.ingramcontent.com/pod-product-compliance
Lightning Source LLC
LaVergne TN
LVHW081929200926
848484LV00009B/2397
* 9 7 8 1 5 6 9 9 0 2 2 9 5 *